DROUGHT AND THE HUMAN STORY

Dedicated to the memory of
Ronald Leslie Heathcote
1934–2010

A meticulous scholar and a true gentleman

Les's favourite definition of drought, as given to him by farmers while
undertaking research in semi-arid USA and Australia:

'That's when two cottonwood trees fight over a dog!' (Nebraska)
'That's when two mallee trees fight over a dog!' (Eastern Australia)

Les Heathcote on Field Work in the Karakum Desert, Turkmenistan, 1976

Drought and the Human Story
Braving the Bull of Heaven

R.L. HEATHCOTE
Flinders University, Australia

ASHGATE

Published by
Ashgate Publishing Limited
Wey Court East
Union Road
Farnham
Surrey, GU9 7PT
England

Ashgate Publishing Company
110 Cherry Street
Suite 3-1
Burlington, VT 05401-3818
USA

www.ashgate.com

British Library Cataloguing in Publication Data
Heathcote, R. L.
  Drought and the human story : braving the Bull of Heaven.
  1. Droughts. 2. Droughts--History. 3. Drought relief.
  4. Drought forecasting.
  I. Title
  363.3'4929-dc23

Library of Congress Cataloging-in-Publication Data
Heathcote, R. L.
  Drought and the human story : braving the bull of heaven / by R.L. Heathcote.
    p. cm.
  Includes bibliographical references and index.
  ISBN 978-1-4094-0501-6 (hbk) -- ISBN 978-1-4094-0502-3 (ebook)
  1. Droughts--History. 2. Droughts--Social aspects--History. 3. Human ecology--History.
  4. World history. I. Title.
  QC929.24.H43 2012
  363.34'92909--dc23

2012018720

ISBN: 9781409405016 (hbk)
ISBN: 9781409405023 (ebk – PDF)
ISBN: 9781409472650 (ebk – ePUB)

Printed and bound in Great Britain by the
MPG Books Group, UK.

Contents

List of Figures

List of Illustrations

List of Tables

Foreword

This book is the capstone of a lifetime's research and scholarship by one of Australia's most outstanding geographers. Ronald Leslie Heathcote, known universally as Les, was raised in Derbyshire in the UK and graduated with a BA in Geography from University College London in 1955. Two years of national service in the army followed, including a spell of duty in Cyprus where exploration of his surroundings awakened a lifetime passion for the study of arid environments. A Fulbright Scholarship enabled him to develop this interest at the University of Nebraska where he was awarded an MA in Geography in 1959. He then enrolled for a PhD in Geography at the Australian National University, undertaking a comparative study of land appraisal and settlement in outback Queensland and New South Wales. The degree was awarded in 1963 and the 'book of the thesis', *Back of Bourke*, became an Australian geographical classic. By this time Les had taken up his first lecturing position in the Department of Geography at University College London. In 1966 he left London to take up a post in Geography at Flinders University, Adelaide, in the first year of the university's operation. Les spent the rest of his academic career at Flinders, retiring as a Reader in Geography in 1997 but continuing active research and writing as an Adjunct Associate Professor in the School of Geography, Population and Environmental Management until his death in 2010.

During his years at Flinders, Les established himself as a scholar of international significance in his chosen fields of arid land management and the human perception of drought and other natural hazards. The titles of some of his many publications reflect the direction and maturing of his work: *The Arid Lands* (with C.R. Twidale, 1969); *Australia* (1975); *Natural Hazards in Australia* (edited with B.G. Thom, 1979); *Perception of Desertification* (edited, 1980); *The Arid Lands: Their Use and Abuse* (1983); *Land Water and People: Geographical Essays in Australian Resource Management* (edited with J.A. Mabbutt, 1988). And when the International Geographical Congress was held in Australia in the bicentennial year of 1988, Les was the natural choice to mastermind the book published to mark the occasion, entitled *The Australian Experience: Essays in Australian Land Settlement and Resource Management*.

His curriculum vitae also lists 45 chapters in books and 43 journal articles, many in international geographical journals. Typically, it doesn't even bother to list his conference papers, as Les was never much concerned about padding-out his resume. It also makes little mention of the many influential reports he wrote for government bodies, all of which reflected his strong conviction that land management policy needed to be based on sound academic research.

Les's work has always radiated meticulousness and authority. He believed in evidence and he relished the hard work of collecting that evidence – particularly in the field. He also believed in not pronouncing on a topic until you really understood it. So when Les wrote or spoke, people sat up and took notice. He was – in the best possible sense – an old fashioned scholar. His publications were renowned for their schematic diagrams, designed by his own hand, summarizing the complexities of change in an array of interrelated variables over time and space as deciphered by Les's unrivalled understanding of the processes involved. At Flinders University they were affectionately known as 'Lesograms'.

Les Heathcote's stature as a scholar is reflected in the honours bestowed on him by his peers in Australia and internationally, most notably the Honors Award of the Association of American Geographers for 'sustained perceptive study of arid and semi-arid lands, including their environmental perception, especially in Australia and the United States', in 1989, and the Griffith Taylor Medal, the highest award of the Institute of Australian Geographers, for 'distinguished service to professional geography in Australia', in 1997. He was also awarded the John Lewis Gold Medal of the Royal Geographical Society of South Australia for 'geographical research and literature', in 1998, and the J.P. Thomson Medal of the Royal Geographical Society of Queensland, for 'long and dedicated contribution to the discipline of geography especially in the study of arid and semi-arid lands', in 2000. He was elected a Fellow of the Academy of the Social Sciences in Australia in 1981, President of the Geographical Society of Australasia (South Australian Branch) in 1972–1973, and President of the Institute of Australian Geographers in 1984–1985.

Les Heathcote's own academic apprenticeship brought him into contact with legendary geographers such as H.C. Darby at the University College London, who towered over British historical geography in the first half of the twentieth century, and Griffith Taylor and Oskar Spate in Australia, who did so much to create and shape geography as an academic discipline in this country. In his own work, culminating in this, his last and most comprehensive project, Les has passed on that legacy of scholarship to new generations.

Clive Forster
Adelaide
20 April 2010

Preface and Acknowledgements

Les said, in a previous book, that the scourge of drought had often been greeted by society with 'indignant surprise'. This, grief at his sudden death in February 2010, and an inability to express myself nearly as well as he did, are my excuses for a preface which in no way does justice to this book, which brings together much of what he studied and learned in a long academic life.

Les and I were students together at University College London from 1952–1955. Since meeting Les again after 46 years, our six years together from 2003–2010 have been much involved with the preparation of his book. Our variety of experiences was endless, whether sitting in Exeter University library looking through old journals, photographing grubbed-up orange plantations along the River Murray or rainforest vegetation changes in Borneo due to climate change, or studying old maps and checking up on place-names with drought connections in Devon. He also spent many happy days in the UK Meteorological Office library, providentially only ten minutes by car from our house.

The book summarizes his deep and abiding interest in arid landscapes and their effects on humans, both historically and in the present day. Hugh Prince, a tutor and later colleague at UCL, very recently summarized Les's unique contribution to geographical research, sparked, as he reminded me, by Les bicycling through the dry zone in Cyprus when he was in British Army Intelligence during a lull in the fighting during the Emergency.

Les believed his book was the first attempt to study the human conceptualization of drought in a world setting, and he hoped that it had a meaningful message for us all. It examines the role of drought in human history and how, and why, human attitudes to it have varied so much.

He points out that its mitigation is more difficult today, with increasing demands for water from the world's increasing population and per capita consumption, and the possible threat of 'water wars' between states. Global warming, he thought, would make these challenges greater.

Yet his message is a positive one. He suggests that much can, and should, be done. However, much closer co-operation is needed between physical and social scientists, in which the former's technological expertise and global knowledge is allied to the latter's understanding of the sociological contexts by which modern societies operate, and in which the general public must, essentially, be involved in finding answers.

He believed that 'The Bull of Heaven' was a beast that could, and must, be overcome.

I think the book is a fitting memorial to a life of painstaking devotion to geographical research, and I am delighted that his daughters, Caroline and Elizabeth, and Caroline's husband Sasa, have prepared the manuscript for publication, in cooperation with Alaric Maude, who has skilfully and meticulously masterminded the whole process, working with other colleagues at Flinders University, where Les spent most of his academic career. To all of them my grateful thanks.

Had he lived, Les would have prepared a list of people whose help and guidance he wished to acknowledge. I cannot do this for him, so on his behalf my thanks to all of you for your part in this book. You will know who you are.

Sheila Heathcote
Adelaide
24 March 2010

Chapter 1
Drought in World History

Drought has a long history. The first heroic legend, the Epic of Gilgamesh, which dates from the second millennium before Christ, tells how King Gilgamesh of Uruk in Mesopotamia fought with, and defeated, drought in the form of the Bull of Heaven. Written records of drought in China go back at least to 206 B.C., and in Australia there is abundant evidence of droughts both before and after the First Fleet arrived in 1788 [bringing the first permanent European settlers to Australia]. (Heathcote 1969: 175)

Many years ago, I used these sentences to introduce a study of the perception of drought in Australia, and ten years later I returned to the topic to chart any changes (Heathcote 1988). In this book I want to bring up to date, and broaden, my understanding of how drought is currently regarded around the world and how and to what extent its challenges to human welfare are being met. To do this I want to examine the problem of defining droughts; to consider some past patterns of droughts in space and time; evaluate their impacts; identify who is affected; review management strategies, both private and official; and finally to try to forecast the future for droughts. The perspective will be global in so far as my experience and researches permit, and I hope that this book may make a contribution towards helping global societies in both the developed and developing world face up to and cope with the undoubted future challenges that droughts will provide.

Increasing Concerns about the Role of Drought in Human Welfare

> Unless we take action, water will be the next global crisis, emerging from the shadow of climate change and today's food and fuel crises. (Thunell 2008: 3)

Thunell, as Executive Vice-President of the INFC (the private sector arm of the World Bank Group), was raising only an example of the more recent concerns for the challenge of drought. Yet humanity's experiences of droughts certainly appear to date from earliest records – as witness the Epic of Gilgamesh noted above, and the Biblical seven fat kine consumed by the seven lean kine of Pharoah's Dream. The closing decades of the twentieth century brought droughts to the media headlines once more. First were the extensive droughts and associated desertification and famines of the 1970s in West Africa, particularly the Sahel, which brought graphic television coverage of parched lands, dying crops and livestock, and emaciated people into the living rooms of the western nations. Second were the prolonged and continuing debates in the 1980s on the question of climate change, supposedly resulting from the accumulation of greenhouse gases. A third stimulus was the United Nations declaration of the decade of the 1990s as the International Decade of Natural Disaster Reduction, with drought as one of the disasters. Finally, the twenty-first century opened with extensive scientific concerns to try to document the extent and speed of global climate warming and the resultant implications for climate change, with specifically a fear for the increasing incidence and extents of droughts.

But before embarking upon a detailed study of drought, however, what can we find immediately available in the literature and history books? A search of any encyclopaedia produces abundant evidence of the phenomenon of drought and its associated adverse impacts upon societies for as long as there has been recorded history. From the Microsoft Encarta 2004, for example, a list of countries where drought has been and is experienced includes most of the world's nations. From the 'constant drought' of Sicily, to the 'constant threat of prolonged drought alternating with devastating floods' of Mozambique, through the 'ongoing drought that has lasted for decades' in Mali, to the seasonal vulnerability of India due to the failure of the southwest monsoonal rains, to South Africa where 'drought is relatively common'. An internet search for 'drought' produces several thousand sites and emphasizes that there is obvious evidence of the widespread occurrence and impacts of droughts around the world, while a crude check of references to drought in the Encyclopaedia Britannica Yearbooks shows that Canada, Ethiopia, Honduras, Somalia and Tajikstan each recorded three droughts between 2000 and 2009, and Djibouti, Eritrea, Lesotho, Mauritania, Romania and the United States recorded two. Another 20 countries recorded one drought in these years.

Factual reports of natural disaster occurrences world-wide, prior to the new century, had shown droughts to be increasing in frequency over time. Significant international drought disasters totalled at 62 in the 1960s, rose to 139 in the 1970s and to 237 in the 1980s (Blaikie, Cannon, Davis and Wisner 1994: 32). Even allowing for improving reporting over time and the possibility of national governments exaggerating the occurrences and impacts in hopes of international disaster relief, the virtual doubling of events each decade is troubling. Such disasters have caused deaths and considerable economic hardship at national levels (Table 1.1). Over the period 1963 to 1992 there were 21 drought events where at least 100 persons died, and drought ranked 7th in number of events (with floods topping the list with 202 events). But in terms of substantial economic damage (i.e. losses of at least 1 per cent of annual gross national product) the 53 such droughts were the third most numerous disaster events, while for disasters where at least 1 per cent of the national population was affected, the 167 such drought events topped the list. So, while drought as a disaster was not the worst killer, it affected more people than any other disaster and was third in importance for substantial economic damage.

Table 1.1   Drought in the Global Disaster Context 1963–1992

Characteristics of disasters	High death toll (at least 100 dead)	Large numbers affected (at least 1% of national population)	Significant damage (at least 1% of annual gross national product)
Worst disaster (no. of events)	Floods (202)	Droughts (167)	Floods (76)
No. of drought events	21	167	53
Drought rank in all disasters	7th	1st	3rd

Source: Based on data from Hewitt 1997: 60–61.

More recent data have confirmed the role of droughts in natural disasters:

> Recent statistics compiled by the International Decade for Natural Disaster Reduction … indicated that drought accounted for 22 per cent of the damage from disasters, 33 per cent of the number of persons affected by disasters, and 3 per cent of the number of deaths attributed to natural disasters. (Wilhite and Vanyarkho 2000: 246)

The same authors went on to note that these figures only included the international aid efforts, and the substantial financial costs of national drought relief efforts were not included. Examples of such costs were given as $8 billion (US) spent by the USA in drought relief in 1974–1977 and a further $6 billion in 1988–1989, $940 million (Australian) spent on official drought relief in Australia in 1970–1984; and R450 million (South African rand) spent in South Africa in 1984–1985 (Wilhite and Vanyarkho 2000).

In Australia, drought has been a long-time companion. Aboriginal legends referred to its impacts and since 1788 those impacts, both positive and negative, have concerned Australian society. In 1813 it helped to push pastoralists from their drought blasted ranges on the Cumberland Plain west of Sydney to cross the Blue Mountains in search of greener pastures (Perry 1963: 29). In 1865 the South Australian Surveyor General Goyder was ordered into the field to delimit that colony's current drought as part of proposed official drought relief measures. His work initiated a division of the colony into climatically safe versus risky areas which continued to have relevance for at least the next century (Heathcote 1981). In 1888 – the Centennial Year of European settlement – drought was devastating southeastern Australia (Davison et al. 1987), and the regional disaster of the 1890s droughts brought the first successful attempt at regional government in New South Wales, with the creation of the Western Division in 1901. This new unit was intended to offer more liberal land settlement laws to cope with the high incidence and impact of droughts. In the 1930s drought added to the miseries of the Depression years; in 1944 it helped to dust the New Zealand snowfields with Australian top soil, and in February 1983 it helped to bring Victorian Mallee soils to Melbourne in a dust storm, as it had done before in 1902 (Illustration 1.1). In 1988 it merited a six-page article in the Bicentennial Commonwealth Year Book (ABS 1988), and from the 1990s through the first decade of the twenty-first century it has been documented again in eastern Australia, with effects culminating in what has been claimed as the worst drought on record in 2007–2008 (see Chapter 7).

A glance at the newspapers can provide evidence of recent and increasing concerns for water supplies and the impending droughts. In the United Kingdom the front page of *The Daily Mail* of 29 June 2005 described an 'Emergency five-point plan as hosepipe ban spreads' as 'the country suffers its worst drought since 1976'. Companies providing the water were criticized for allowing massive water leaks from their pipes and complaints were aired that 'water supply should be for the common good; not for profit, yet up to a quarter of the average householder's water bill goes straight to shareholders'. *The Independent on Sunday* described on 3 July 2005 'How British golfers and tourists are turning Spain and Portugal into dust' through the demand for golf courses, luxury holiday homes and associated swimming pools, while in the USA the tide of urban development advancing into the Nevada desert was reported as 'Crops vs craps row pits farmers against Las Vegas' (*The Independent*, 30 July 2005). In this case the city of Las Vegas was trying to divert agricultural water from northern Nevada into its municipal reservoirs and groundwater reserves. By 3 December *The Independent* was claiming on its front

Illustration 1.1 Dust from Eroding Farmlands approaching Melbourne, Victoria, on 8 June 1983

page that 'Today protesters unite in 30 nations – this is what lies ahead if nothing is done'. The necessary action concerned the need to face the anticipated impacts of climatic change, namely 'Killer Storms', 'Rampant Disease', 'Rising Sea Levels', 'Devastated Wildlife', and 'Water Shortages'. The latter was linked with droughts becoming more common as shrinking glaciers, which fed the major river systems, provided reduced melt waters to feed the rivers.

From the newspapers the message seems to be that droughts can and do cause national distress, and their onset can be aggravated by rapidly increasing human demands for water, ranging from general urban expansion to specific new tourist developments. To this increasing demand is added the inefficient distribution from existing supplies, increasing competition between users, together with fears for future supplies affected by the effects of climate change and global warming.

So what are droughts and how do we measure them and their impacts? To do this let us step back and look at the chequered history of the rise and fall of human societies on earth. Did drought in fact have a role in this history?

Droughts and the Fortunes of Human Societies?

Human interest in the history of humans on earth has given rise to many attempts to understand the rise and fall of past societies, each in its heyday locally apparently omnipotent, but in time reduced to ruins and but a fleeting mention in the annals of global history. For Arnold Toynbee, his 12-volume *A Study of History* (1934) was a demonstration in his words of the role of the linkage between the challenges (many of them environmental) facing past societies and responses made by the societies in attempts to meet those challenges. W. Kirk tried a similar overview (Kirk 1951) and offered graphs of the varied responses and subsequent tracks of the societal fortunes, with the successful adapters progressing and in some cases improving their lot, whereas the failures found themselves unable to adapt or modify their resource use systems sufficiently to avoid the deterioration of living standards and eventual collapse of their society (Figure 1.1).

A recent illustration of the process can be found in Diamond's study *Collapse: How Societies Choose to Fail or Succeed*:

> First of all, a group may fail to anticipate a problem before the problem actually arrives. Second, when the problem does arrive, the group may fail to perceive it. Three, after they perceive it, they may fail even to try to solve it. Finally, they may try to solve it, but may not succeed. (Diamond 2005: 421)

A more recent example of such general studies is the 'Integrated History and Future of People on Earth' (IHOPE), which aims to review the last 10,000 years of human occupation of the globe, and focus at a higher temporal and spatial resolution on the last 2000 and the last 100 years. The aim is:

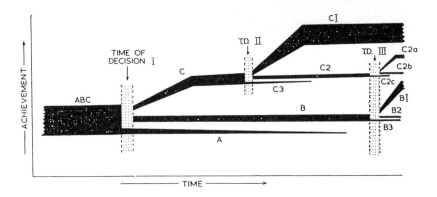

Figure 1.1   Kirk's Graph of the Evolution of Human Societies
Over time a society comes to a Time of Decision [TD] where challenges (many of which are environmental) require modification or adaptation of the existing achievements/lifestyle of the society. Three possible responses are postulated: 'A' where the response does not meet the challenge and the society eventually disintegrates; 'B' where the response is sufficient to at least maintain the society until another major challenge confronts it; and 'C' where the response contains innovations which provide an improved level of wellbeing or achievement. At each successive challenge the society faces the three possible challenges with the fate of the society indicated.
Source: Kirk 1951.

> An integrated history of the Earth – of climate, atmospheric chemistry and composition, material and water cycles, ecosystem distribution, species extinctions, land use systems, human settlement patterns, technological changes, patterns of disease, patterns of language and institutions (including political and religious institutions), wars and alliances and other variables. (Costanza et al. 2005: 19)

All this was to be done with the goal of understanding 'why some societies succeed and other fail in the face of environmental challenges'. Three key research questions were to be posed: first, 'how can we use integrated knowledge of past and present human perceptions of, and behaviour in, the environment, for understanding the future?'; second, 'what are the complex reasons for the emergence, sustainability or collapse of coupled human-environment systems?'; and third 'how do we evaluate alternative explanatory frameworks (including complex systems models) against observations of highly variable quality and coverage?' (Costanza et al. 2005).

While not directly linked to this international study, one of the aims of this book is to attempt to provide some understanding of how peoples' attitudes to drought may have changed through time and how those changes may appear to be relevant to the strategies to cope with forthcoming droughts. While much of the past may be open to conflicting opinions and explanations, there seems to be little doubt that droughts will continue to confront humanity into the future, and in intensities possibly not previously witnessed. First we might ask are there specific instances of societies collapsing in the face of drought threats?

Drought and the Fall of Civilizations?

What of the past? Are droughts recent phenomena confronting societies or are they merely continuing challenges to the human occupation of the earth? What is the evidence from history, and beyond history, archaeology. Some examples from the distant past are relevant.

Mycenae – An Early Victim of Droughts?

The Mycenaean Empire's power over Greece and the eastern Mediterranean peaked in the fourteenth and fifteenth centuries BC but by 1200 BC it had collapsed. In the opening chapters of their book on *Climates of Hunger: Mankind and the World's Changing Weather* (1977) Bryson and Murray considered the role of drought in the collapse of the Mycenae civilization. As they explained: 'Quite abruptly, before 1200 BC, Mycenaean power began to decline. In 1230 BC the main palaces and granaries of Mycenae itself were attacked and burned' (Bryson and Murray 1977: 4).

At the same time, c.1200 BC, the Hittite empire was also in decline and suffering drought. Bryson and Murray agreed with Rhys Carpenter (who had in 1966 made the first assessment) that the collapse was: 'not an enemy incursion, but a dispersal – a diaspora – of the Mycenaean inhabitants of the Peloponnese caused by drought, including a resort to violence by a drought-stricken population' (Bryson and Murray 1977: 5). The associated regional variation in the rainfall to create such a regional drought they found to be possible from their review of historical climate records, specifically a regional drought over the eastern Mediterranean in 1954–1955. They concluded: 'We found no cases that were not consistent. In short, a drought would have caused the decline of Mycenae. We have proof that the proposed Mycenaean drought pattern can exist. We also have evidence that the pattern did dominate about 1200 BC' (Bryson and Murray 1977: 16).

In 1982, at the onset of a sequence of drought years which was to remind the contemporary world of drought impacts and to stir the climatologists into a reassessment of drought's antecedents and global linkages (as we shall see in Chapters 3 and 4), a further study of the decline of the Late Bronze Age civilization extended the societal impacts of drought from merely Greece to the whole of the eastern Mediterranean and beyond:

> We may speculate that in the third quarter of the thirteenth century [BC] groups
> from the Greek mainland … and western Anatolia … fled from drought conditions

with which they could not cope. Some invested themselves in Cyprus from where they attacked Egypt c. 1232 B.C. Others fled westward to Italy, Sicily, and Sardinia … Conditions perhaps worsened towards the end of the century igniting a second, more devastating wave which rent the entire political and economic fabric of the Late Bronze Age civilizations … Climate conditions may have remained unfavourable until the late ninth century, when the recovery of civilization was almost simultaneous at all the old centres. (Weiss 1982: 183–4)

Michael Wood, investigating the collapse of Eastern Mediterranean cultures as background to his study of the Trojan War, suggested that drought was a strong contender among the many proposed explanations for the end of the Bronze Age civilizations (Wood 1985). A more recent researcher agreed, suggesting that a one year drought similar to that in the 1950s would have placed Mycenean agriculture in a 'precarious position … [while] A short, intense cycle of three to four such years would have been disastrous' (Fagan 2004: 184). More recently, the impact of such droughts has been claimed to have possibly contributed to the collapse of the Minoan civilization on Crete, which seems to support Weiss's broader view (Le Comte 2009: 228). Drought, in combination with whatever weaknesses already existed in the societies, appears to have been the final trigger to widespread societal collapse.

The Mayan Civilization as a Drought Victim?

In central America, from c.50 BC until c.900 AD human societies developed in the Yucatan Peninsula what is termed the Mayan civilization. There is some evidence of an earlier collapse of the society in the south around 250 AD (Dahlin 1983), but the culture reached its apogee of power from 300 to 900 AD. It was a sophisticated, urbanized culture with impressive public buildings based upon a complex agriculture of slash and burn cultivation in the subtropical jungle. Although the society possessed neither knowledge of the wheel nor of metal tools, massive pyramid temples were constructed in its cities, celestial observations led to a solar calendar of 365 days, and elaborate hieroglyphic scripts on temples and stele carved a niche in world history.

Water supplies, however, were critical to agriculture. Ninety per cent of the annual rains fell between June and September and they were usually reliable. Because the northern peninsula is basically a limestone plateau there were few streams at the surface, but in the limestone bedrock were subterranean river systems which were tapped by excavated tanks and reservoirs and access via natural sinkhole lakes ('*cenotes*'). These have been shown for one city (Tikol) to have had sufficient storage to support a population of 10,000 for at least 18 months. The agricultural importance and religious significance of these water sources is illustrated by the accumulated offerings of precious objects and the skeletons of sacrificial victims found in their sediments.

The civilization flourished from 300 to 950 AD and by c.750 AD supported a population of c.13 million. However, between 750 AD and 950 AD it crumbled, 'cities imploded, their populations perished or dispersed into small villages scattered across heavily cultivated landscape' (Fagan 2004: 232). Various explanations have been offered:

> From internal warfare to foreign intrusion, from widespread outbreaks of disease to a dangerous dependence on mono-cropping, from environmental degradation to climate change … all may be relevant but recent evidence points to drought. (Peterson and Haug 2005: 323)

> Disease, in all its possible ramifications; impoverishment of the cultivated lands due to the removal of essential nutrients by over-cropping [because the Mayans employed slash and burn agriculture]; soil loss from runoff; contamination of aquifers by salt water intrusion from rising sea levels; and over-population may also have been a contributing factor … [The collapse was] unfortunately associated with an unusual climatic period when several drought episodes occurred within a limited timeframe … importantly, severe droughts … remain an ongoing feature of Yucatan climatology. (Hunt and Elliott 2005: 393–4 and 406)

In effect the civilization had developed in a 'seasonal desert and depended upon a consistent [summer] rainfall cycle to support agricultural production'. Yet, fluctuations in those regional rainfall patterns over the Yucatan Peninsula, identified by evidence from deep sea and lake deposits, showed that a series of droughts, each lasting three to nine years, could be identified as occurring about 760, 810, 860 and 910 AD. In other words, these events coincided with the decline of the culture. Not surprisingly, therefore, these droughts collectively were thought to be a major factor in the collapse of the civilization, with the southern areas going first as they lacked the underground water sources of the north (Haug et al. 2003: 1731).

The Riddle of the Mesa Verde Peoples

Over roughly the same period as the Mayans were establishing their culture, in the southwest of what was to become the USA a Pueblo Indian culture was slowly developing and beginning to occupy what the Spanish Conquistadors labeled the Mesa Verde – green tablelands. The colour came from the piñon and juniper forests, which covered the Four Corners Country as it is now called (where Colorado, Utah, Arizona and New Mexico meet) and which tend to hide the semi-arid regional climate. The tablelands were crossed by deep canyons which were to hide the evidence of the Pueblo culture and its apparent collapse until the late nineteenth century, when a government survey in the 1870s and roaming cowboys in 1888 stumbled across abandoned field walls and structures on the tablelands and well preserved ruins tucked into the natural caves in the walls of the canyons (Illustration 1.2).

Illustration 1.2　Long House, Metherill Mesa, Mesa Verde National Park

The Long House was the largest of the cliff dwellings and shows the magnificent adaptation of dwellings to the large natural cave, within the security of the canyon wall. Water was obtained from streams in the canyon floor.

Archaeological evidence suggested that hunters and gatherers moved into the tablelands from the south c.100 AD and by 900 AD had begun to cultivate the plateau surfaces for corn, squash and beans as well as domesticating the native turkey. Over 3800 prehistoric sites have provided evidence that at the peak of the occupation over 19,000 people were supported by the area. Houses and subterranean religious chambers (*Kivas*) dotted the tablelands, but from 1200 AD there appears to have been a shift back to the natural caves in the canyons which had offered shelter to the original immigrants. Just why this move took place is not clear, and just why those 'new' locations were themselves abandoned from c.1300 AD remains unclear also. The archaeologists have suggested various reasons; the impacts of major droughts on the crops and wildlife resources were among the earliest explanations. In fact the dating of past human settlement sites by dendrochronology owed much to the Pueblo Cultures found in the Four Corners Country, since the ruins contained multiple timber beams whose growth rings offered insights into fluctuating seasonal rainfalls. A megadrought from 1276 to 1299 was documented from those timber beams and this coincided with other evidence of food shortages, but there were other possible explanations.

The most recent study of one of the Puelbo sites (Sand Canyon) noted that:

[D]roughts occurred at various times during the Pueblo habitation of the Four Corners region Some could even be called megadroughts. These might not have been more severe than droughts of the modern period, such as the Dust Bowl years, but they lasted much longer, sometimes for several decades, and their effects were broadly felt throughout the western United States. (Kohler et al. 2008)

In addition, however, there was evidence suggesting multiple factors in settlement failure and flight from the Mesa Verde:

There is no single, simple cause for this depopulation ... [It] was a cascade of events that include climate-induced immigration from peripheral regions resulting in overpopulation ... in turn generating resource depletion ... [and a] decline in maize productivity that affected both carbohydrate and protein intakes. These changes provoked conflict, which in turn induced more scarcity ... [and out migration, which left the remaining settlements] vulnerable to aggression. In the end, violence and famine provided potent motives for departure. (Kohler et al. 2008: 153)

With some of the settlements showing occupation for only 30 years, for whatever the reasons, the Mesa Verde by the end of the thirteenth century had obviously become a difficult region in which to survive and the remnants of the culture appear to have fled south to the better watered valley of the Rio Grande.

The Life and Death of the Garamantes of the Central Sahara

The central Sahara has had a fluctuating climate and series of ecosystems over the millennia, as evidenced by a combination of geological, geomorphological and paleo-climatic research. From 200,000 to 70,000 years ago (BP), a humid climate saw lakes and a savanna vegetation occupied by Paleolithic hunter-gatherers, who were forced out by increasing aridity from 70,000 to 12,000 BP. A humid climate phase reappeared from 12,000 to 5,000 BP and the hunter gatherers reoccupied the mountains and lake sites, but from 5,000 onwards the seasons began to dry and the lakes to shrink. Yet at this time, specifically by 2000 BP and lasting until 500 AD, recent archaeological work has provided evidence of a civilization of nomads and warriors, the Garamantes, known to the Romans, occupying the desiccating savannas. Their livelihood was based upon oases, where irrigation supported wheat, barley, millet and sorghum with the water brought from nearby mountains using *foggaras* (*qanats* – or excavated underground water conduits). Urban centres existed, occupying what is now known as the Fazzan of Libya and the Sand Sea of Ubari. When the surface waters failed about 500 AD, the Garamantes were finally defeated (White and Mattingly 2006). Interestingly, the very deep groundwater lying beneath the territory of the Garamantes but unknown to them has, since the

1980s, been supplying Tripoli, but the levels in the aquifers are falling rapidly. History is about to be repeated?

These historical cases are but a small sample of the increasing body of evidence of the historical role of drought in the fluctuations in the fortunes of civilizations now long gone. The cases raise interesting questions; the role of drought is certainly open to debate in all of them, but to sharpen the investigation of the contemporary role of drought we need to go back to basics and examine the various definitions of drought itself. This is because, as we shall see, it has been recognized in various forms in the past and present, and promises to be similarly difficult to pin down in the future.

Chapter 2
Defining Drought

How in the hell can the old folks tell
It ain't gonna rain no more?

(1920s song)

It is clearly impossible to measure accurately the precise nature, distribution and effects of natural disasters, for inconsistency, obscurity and vagueness continually cloud the view … [Data exists however and when sifted and correlated] give a picture which at least approximates reality.

(Freeberne 1962: 67–8)

How can we define drought? Is it just a shortage of water? If so, we need to remind ourselves that the first philosopher to consider that water might be the basic and vital element in nature was Thales of Miletus (b. 624 BC). According to Boorstin, Thales got his notion 'perhaps from seeing that the nutriment of all things is moist and kept alive by it … and from the fact that the seeds of all things have a moist nature, and that water is the origin of the nature of most things' (Boorstin 1998: 22). As the basic 'nutriment', then, how have the experts defined the lack of it?

There is a concern that the term drought may be mistakenly applied to significant seasonal variations in rainfall … Considerable research and analysis will be required before it is possible to develop a set of objective criteria for defining drought. (Drought Policy Review Task Force 1989: 14) [This from an Australian official task force set up to try to separate the occurrence of genuine drought impacts from the claims of self-defined victims.]

Drought is a persistent and abnormal moisture deficiency having adverse impacts on vegetation, animals, or people. (US NDPC 2000: 7) [This from a US national overview of the scope of drought impacts.].

A drought is a departure from the average or normal conditions in which shortage of water adversely impacts ecosystem functioning and the resident populations of people. The terms drought and aridity are sometimes used interchangeably and, therefore, incorrectly. Aridity refers to the average conditions of limited rainfall and water supplies, not to the departures there-from which define a drought. (Ffolliot et al. 2002: 52) [This from climatologists trying to sort out the unexpected from the expected versions of dryness.]

Drought is a creeping slow-onset natural hazard that is a normal part of climate for virtually all regions of the world; it results in serious economic, social and environmental impacts. Its onset and end are often difficult to determine, as is its severity. Drought affects more people than any other natural hazard. Lessons from developed and developing countries demonstrate that drought results in significant impacts, regardless of level of development, although the character of those impacts will differ profoundly ... The hazard or natural event is best characterized by the frequency of meteorological drought at different levels of intensity and duration. (Wilhite 2000b: 1–2) [This from an experienced social scientist as part of the introduction to the two volumes on drought which he edited.]

So, while there would seem to be some variations in the detail of the definitions, there seems to be reasonable agreement that drought as a shortage of water has to be unexpected and has to cause significant hardship to all life forms.

The Initial Confusions

Recognizing the importance of water, however, is a far cry from identifying how little of it can be vital to the survival of an animal, a plant, a human being or a human society. Yet such a problem is inherent in any attempts to define and put a value on that environmental deficiency which is traditionally labelled 'drought'.

Let us assume then that drought is usually defined in terms of an abnormal but temporary moisture deficiency leading to a significant ecological stress and/ or human distress. Most definitions, however, are relative, that is the supply shortage of moisture is assessed in relation to the demand for that moisture, so that absolute drought thresholds will vary not only according to individual water needs (whether plants, animals, humans, or societies and nations), but also within nations, depending upon the specific moisture needs of the rural properties or urban water managers. Thus for Britain an 'absolute drought' is defined as occurring after 15 days each of which had less than 0.01 inch [0.25mm] of precipitation, and a 'partial drought' is a period of '29 consecutive days the mean rainfall of which does not exceed 0.01 inch [0.25mm] per day (Herbst et al. 1966: 264). However, for Bali it is six days without rain (Wilhite et al. 1987: 15)! Extending the problem of definition to the plant world, a drought for a rainforest tree might be 'two weeks without rain whereas a drought for a desert shrub might be two years' (Kirkpatrick 1994: 60).

Consequently there may be as many different definitions for drought as there are uses for moisture, so that a universal quantitative definition of drought is of doubtful value and neither meaningful nor possible, for drought means different things to different people. A British scientist, called in to investigate drought impacts on the pastoral industry in the semi-arid interior of eastern Australia in the 1930s, found that:

Droughts, like deserts, are rather hard to define, and in this part of the world especially so. Theoretically a drought can be measured by comparison with a normal year; but there is no such thing as a normal year or normal rainfall here: an average rainfall perhaps, but normality is a matter of opinion … Moreover, in practice, one is forced to conclude that the measure of a drought depends on the number of stock carried on a property. (Ratcliffe 1963: 194)

In effect, he was hinting that excessive numbers of livestock carried might reduce the landscape to the appearance of a drought through the lack of forage, a point which will reappear in this narrative.

Some 20 years later, another Australian researcher compared the historical media reports of drought occurrence with the cumulative rainfall deficits from official records (Foley 1957). His technique was refined by research from the Commonwealth of Australia's Bureau of Meteorology which compared the media reports with annual rainfalls in the lowest decile (i.e. rainfalls in the lowest 10 per cent of records) and again a useful coincidence was noted:

We were surprised to find that the very crude rainfall index of drought given by the first decile range of rainfall total for the calendar year gives such a good correspondence with the [reported] occurrence of major droughts, remembering that in some areas drought may occur or continue with rainfall in the second, third or higher decile ranges and that drought occurrence seldom coincides with the calendar year. (Gibbs and Maher 1967: 16)

Subsequently the Bureau of Meteorology tried to tackle the problem by the introduction of what was called a 'Drought Watch' program. In introducing the program the Director, W.J. Gibbs, one of the authors of the previous study, acknowledged the difficulties:

There is no universally agreed definition of drought but a broad definition which may be acceptable to most people is 'severe water shortage'. This definition requires a further definition of 'shortage' which in turn implies specification of the amount of water needed. The amount of water needed depends on the nature and extent of animal and plant communities using the water, so that the concept of drought cannot be divorced from the use to which water is put. Conditions which a vegetable farmer would regard as drought would not worry the pastoralist. The availability of water depends largely on rainfall, although losses such as evaporation or wasteful use of water and gains such as storage in the soil, in artesian basins or reservoirs, must be taken into account … [T]he only objective method of specifying drought is to nominate minimum water needs for a particular purpose. (BOM 1965)

A recent survey in the United States continued this theme:

Drought to a climatologist or meteorologist is generally associated with lower than normal precipitation followed by a decrease in available soil moisture. To the groundwater hydrologist it is declining water tables and aquifers. For the citizen, municipal manager, and land operator (users) it is a negative impact on his/her ability to: manage and assess the risk of wildfires, operate an economic agricultural operation, generate power, provide recreation, manage wildlife resources, sustain a viable resource, maintain adequate water supplies, or satisfy a life style. (US NDPC 1999: 14)

Similarly, the Australian Bureau of Meteorology has more recently reiterated the belief that the perceptions of drought differ according to implied water uses: 'Meteorologists monitor the extent and severity of drought in terms of rainfall deficiencies. Agriculturalists rate the impact on primary industries, hydrologists compare groundwater levels and sociologists define it on social expectations and perceptions' (BOM 2001).

In this context then: 'There is little to be gained by reproducing a long list of conflicting definitions that merely illustrate the diverse interests of those who investigate drought. Most definitions anyway can be resolved into a generic statement that drought is caused by an imbalance between water supply and demand' (Agnew 2000: 7). In fact such a list, of over 150 definitions, was critically reviewed by Wilhite and Glantz in 1985. Confronted by the invitation to provide an analysis of drought impacts in Australia in the 1960s, I took an easy way out and defined drought on the same basis as an anthropologist, who had faced the politically difficult problem of defining Australian Aborigines. He used the social definition as 'any person who identifies himself or is identified as an aborigine'. On the same lines, I defined drought as occurring in Australia whenever it was said to occur, in other words a 'definition by acclamation' or, more explicitly perhaps, a 'shouting index' (Heathcote 1967: 27). In fact, as noted above, another scientist had used a similar technique to identify these 'droughts by acclamation' by using newspaper reports as well as meteorological records for the historic occurrences (Foley 1957).

In other words there are many 'stake holders' with a legitimate interest and concern for the recognition and definition of drought, a theme which will reoccur throughout this book. Nonetheless the potential confusions of the generic definition need to be addressed; to cope with this ambiguity, scientists have tended to refer to four types of drought.

The Four Types of Drought

Meteorological Drought

Meteorological drought is usually identified as precipitation recorded as below a statistical level of significance over a specific period. The significance is primarily statistical in the sense of mathematical deviations from 'normalized' data or averages of probabilities of occurrence. Such a drought would show 'significant decline from climatologically expected and seasonally normal precipitation over a wide area' (Carr 1966: 6). The problem remains, however, as to how long a record is needed to provide data to define a normal value against which to judge the abnormal events. In 1935 the World Meteorological Organization (WMO) decided that a 30-year period from 1901–1930 should stand as the norm for calculations of divergences. But subsequent 30-year periods have not provided the same averages. This indicates that, based upon these arbitrary periods the weather, and by implication the climate, is not stable but changing over time. A recent comment illustrated the concerns: 'We are certain, however, that the thirty-year time frame over which we define climate 'normals' is not sufficient to characterize climate-related risks, especially in the case of relatively rare, extreme events such as droughts' (Graumlich and Ingram 2000: 235). Attempts therefore to document such changes and variations over time face problems, especially in attempting to demonstrate climate change and its relevance to droughts.

In fact, successive thirty year periods can be shown to provide conflicting evidence. In the case of the Sahel of West Africa, where as we have seen droughts became global concerns in the 1970s, a researcher calculated meteorological droughts for the period 1931–1990, using the norm for the 1931–1960 period and then recalculated the droughts using the norm from the 1961–1990 period of records. The resulting definition of drought or non-drought years over the total 1931–1990 period differed. 'Because the 1940s and 1950s were wetter than normal in the Sahel, using a base period of 1931 to 1960 to calculate the SPI [Standardized Precipitation Index, which he used to define drought occurrence] produces a higher average for the annual rainfall and hence a greater incidence of drought' [as deviations from it] (Agnew 2000: 10). Using the 1931–1960 norm resulted in 19 drought-free years for the whole period 1931–1990, whereas using the 1961–1990 norm resulted in over twice the number of drought-free years (41) being identified for the whole period. So, which was correct? The researcher did concede however that the types of drought recognized were functions of occurrence over time:

> The SPI drought thresholds recommended here therefore correspond to 5 per cent, 10 per cent and 20 per cent probabilities. Hence drought is only expected 2 years in 10 and extreme drought only one year in 20. This, it is believed, is a more reliable drought frequency ... and it corresponds to the employment of the term *abnormal occurrence*, as used in other branches of environmental science. (Agnew 2000: 9)

It would appear then that even applying quantitative methods to defining meteorological drought can be problematic.

There are further problems in any definitions, basically to do with the thresholds required before drought is identified. First, what level of moisture deficit is required for drought to be recognized; second, is there a minimal period of time for the deficit to continue before drought is recognized; and third, over what area must the deficit continue for the required period before drought is recognized and reported? The WMO has suggested that drought could be identified where a region had '60% or less of annual precipitation, for over two consecutive years for over 50% of the spatial extent of the region' (WMO 1986: 2). They argued that societies could usually cope with one year of such a drought but would be adversely affected if it extended into a second year.

In Australia, as we have seen above, an early study suggested that annual rainfall in the lowest 10 per cent on record (the first decile) occurring over at least 10 per cent of the continent constituted a 'significant' drought (Gibbs and Maher 1967). More recently, the Australian Bureau of Meteorology has accepted the basic concept and in its periodic *Drought Review* publication (on-line since 2007) examines precipitation over three month periods for specific areas and reports drought (serious to severe rainfall deficiencies) as occurring when that precipitation is among the lowest 10 per cent on record.

In the United States one of the most influential drought definitions has been the Palmer Drought Index. Palmer, a meteorologist, was researching drought impacts upon farming in the American Mid-West in the 1930s and admitted subsequently that his index became particularly significant in June 1934, at the time when the US Congress was voting public funds for the purchase of drought-stricken starving livestock. Using his data for central Iowa, he identified a variety of intensities of drought from mild to extreme:

Mild drought = 'some of the native vegetation almost ceases to grow.'

Moderate drought = 'the least drought-resistant members of the native plant community begin to die and the more xerophytic [drought-resistant] varieties start to take their place.'

Severe drought = 'only the xerophytic varieties of native vegetation continue to grow. And vegetal cover decreases.'

Extreme drought = 'Drought-resistant varieties gradually give way to open cover. More and more bare soil is exposed.' (Palmer 1965: 45)

He then went on to quantify these surrogate conditions of the native vegetation into values reflecting the link between the effectiveness of the precipitation and the moisture holding capacity of the soil as a crop moisture index (Table 2.1). As a result he was able to provide a formula which relates 'accumulated weighted difference between actual precipitation and the precipitation requirement of evapotranspiration'(i.e. establishes when precipitation is no longer effective enough to support plant growth) (Wilhite and Glantz 1985). With minor modifications the index became accepted internationally as the Palmer Drought Severity Index and has been used to show that it is possible to document the chronological changes in the extent of drought at global scales (Dai et al. 2004). These become important statistics in the debates about the impacts of global warming as we shall see later.

Table 2.1   Palmer's Drought Index as Developed for the USA in the 1930s

Drought Index	Description
0	Normal moisture conditions (for the location analysed)
−0.5 to −0.9	Incipient drought
−1.0 to −1.9	Mild drought
−2.0 to −2.9	Moderate drought
−3.0 to 3.9	Severe drought
−4.0 or above	Extreme drought

Source: Palmer 1965.

Defining drought continues to pose problems for scientists, particularly trying to distinguish between seasonal dryness and droughts. In the early 1980s criticism was made of the use of the term 'seasonal drought' by one author for the 'rhythmic cycle of dryness in the Tropical Savanna and Mediterranean climatic regimes'. The critic argued that the 'seasonal water budget deficit ... [of the regions] is an "expected" occurrence, not an abnormality' and suggested that more appropriate terms would be 'seasonal aridity', 'seasonal dryness' or 'seasonal moisture deficits' (Steila 1981: 373). The targeted author defended himself by arguing that: '"Drought" can and does mean "state of dryness due to lack of rain", seasonal or otherwise, just as properly as it signifies Steila's [definitions]' (McBryde 1982: 347). He went on to suggest that 'respected geographers, botanists and climatologists have long referred to Köppen's B climatic regions [where the climate restricts vegetation growth] as where plants perish by "death from drought". Xerophytes of the desert and dry-margin plants are undeniably drought-resistant, and are so defined in commonly used American dictionaries' (McBryde 1982: 347).

Steila replied by picking up on McBryde's omission of the recognition of the abnormality of drought and stressed the need for quantitative measures of the moisture deficiency and recognition of the role of evaporation and evapotranspiration as reducing the efficiency of any levels of precipitation. He praised the meteorological models for measuring the 'state of dryness' as formulated on a set of common principles:

1. drought is neither a continually nor seasonally occurring phenomenon;
2. drought-status is a function of environmental moisture status, i.e. the sum of the environment's available water relative to the environment's water demand (PE) [Precipitation Effectiveness];
3. environmental moisture status components, viz., PE, precipitation, and soil water, are measurable;

4. the drought concept characterizes intervals (times) when negative deviations from expected EMS [Environmental Moisture Status] occur; and

5. a range of negative deviations in EMS that approximate the norm is not considered representative of drought (Steila 1983: 194).

He did have to admit, however, that category 5 was the most difficult to defend.

Hydrological Drought

Hydrological drought is usually recognized when surface and sub-surface moisture volumes are significantly depleted, streams and rivers dry up, and reservoirs levels fall dramatically along with declines in groundwater levels. However, the threshold of significance varies with location and demand for moisture and the definitions as a result tend to be user biased and arbitrary. In a recent global review, Wilhite commented that 'Hydrological droughts are usually out of phase or lag the occurrence of meteorological and agricultural droughts'. He went on to provide the example of the Missouri River Basin in the USA, where 4–5 years of 'normal precipitation' would be needed 'over the basin to bring reservoirs [depleted by the previous severe meteorological drought years of 1987–1992] back to normal levels' (Wilhite 2000a: 11).

Illustration 2.1   The Castlereagh River, New South Wales, Australia
Photographed in September 1960 at the onset of a two-year drought which was to cover the largest area (39.8%) of the continent in 1961.

Agricultural Drought

Agricultural drought is usually recognized when precipitation or hydrological conditions result in significant reduction in crop yields and/or distress or death of livestock from lack of water or feed. Again the occurrence is related to specific crop and livestock moisture needs and is usually defined after a period of reduced moisture availability has extended beyond normal expectations. A typical definition would be that an agricultural drought would be recognized 'when soil moisture and rainfall are inadequate during the growing season to support healthy crop growth to maturity and to prevent extreme crop stress and wilt' (Carr 1966). Various statistical formulae have been developed for specific agricultural activities. Perhaps the best known is the Palmer Drought Severity Index noted above.

Depending upon the measurement period, however, some anomalies might result. These could occur merely because of the bad timing of rainfalls with regard to the growing season even in an above average year of rainfall. Thus, in terms of total moisture shortages, the agricultural drought could occur before any meteorological drought was recognized, and a meteorological drought might be well established before any significant hydrological drought became apparent.

In one sense the definition of an agricultural drought is the most difficult of all, since so much depends upon the moisture demands of the crops or livestock raised, the management strategies of the farmers or graziers, as well as the nature of the intended market. For example, reporting upon the occurrence of drought in the grazing areas of Botswana, a researcher commented:

> Drought probability is also determined by the productivity of the environment, by government investments that make grazing more accessible, or by changes in livestock numbers and, hence, in feed requirements. All these factors change with the passage of time. Moreover, since the occurrence of a drought itself affects subsequent environmental productivity and livestock numbers, the probability of occurrence of the next drought will be substantially determined by how much time has elapsed since the previous one. (Sandford 1979: 37)

More recently, research in the USA has focused on distinguishing between the causes which create 'true drought emergencies and those that are normal cyclical conditions' and, taking its cue from work in Australia, recognized that the causes or 'triggers' for drought could be seen as both supply-type triggers, reflecting moisture deficiencies caused by acts of nature such as lack of rain or excessive temperatures, as well as demand-type triggers reflecting drought impacts, such as available water supply (US NDPC 2000: 5). Even the supply-type triggers were defined differently between the east and west halves of the country. In the west they were identified as precipitation less than 60 per cent of the normal (for the season or 'present water year'), while in the east they were identified as precipitation less than 85 per cent of normal over the previous six months. This suggested perhaps a smaller 'buffer' against drought in the east? A Palmer Drought Severity Index of

−2.0 or more was also recognized and 'more rigid triggers such as 5th percentile drought [precipitation for the year in the lowest 5 per cent of records] might be appropriate reflecting truly unusual circumstances' (Ibid.: 5.) The demand-type triggers included available water supply less than 60 per cent of normal or soil moisture below minimal crop requirements. This latter criterion would reflect previous farm management demands on soil moisture.

Socio-Economic Drought

Finally, and leading on from the above, a fourth socio-economic drought has been recognized. Here is the situation where agricultural drought has resulted from human resource management. This has placed impossible demands upon the normally available moisture supply, resulting in severe stress upon the community. Here drought is the result of the mismatch between society's moisture demands and the normally accessible natural supply. This might reflect human ignorance of the normal conditions, or deliberately risky management decisions hoping to capitalize upon better than normal seasons, or hopes for the benefit of sympathetic official disaster relief if the speculative venture fails, or desperate management choices driven by lack of possible alternatives (Glantz 1994). In effect this type of drought is the result of unwise human activities – i.e. asking for more moisture from the environment than would normally be available, and as a result society makes itself more vulnerable.

One example must suffice to show the difficulties involved in sorting out the management factors from the meteorological conditions. In South Africa the official drought management policy recognizes three types of drought:

1. Disaster drought is a set of rainfall-soil-crop conditions for which poor rainfall results in crop yields that will not be exceeded 7 per cent of the time (1:14 years).
2. Severe drought is a set of rainfall-soil-crop conditions for which poor rainfall results in crop yields expected to occur 7–25 per cent of the time. [i.e. 1:4 to 1:14 years].
3. Mild drought is a set of rainfall-soil-crop conditions resulting in yields expected to occur 25–50 per cent of the time. [i.e. 1:2 to 1:4 years].

No drought occurs when rainfall-soil-crop conditions result in yields exceeding the median yield.

Sorting out the criteria necessary to identify the rainfall, soil conditions and crop yields is a sizeable challenge. As the official drought policy admits 'management for drought risk entails implementation of drought policy; monitoring drought severity; technology transfer; researching econometric models; income and tax levelling; and following trends and developments in world trade agreements' (de Jager, Howard and Fouche 2000: 269). In such a context, sorting out the real causes of an apparent drought impact could be tricky, and not least because there could be many interested parties.

The Complex and Often Conflicting Views of the Stakeholders in Droughts

> Drought is thus integrally linked with society, although it is often convenient for analytical reasons to consider it an independent variable. (Spitz 1980: 126)

Competing demands can stretch water resources beyond their capacity. And what's good for the environment or perhaps a fish – full reservoirs help migrating fish over dam tops and out of the way of the turbines – is not simultaneously optimal for a microchip manufacturer, a potato farmer, or a householder. (Ward 2002: 152).

With such a variety of drought types and impacts, it is not surprising that many different groups (stakeholders) view drought differently. For the urban lay person drought is remote, only relevant when water restrictions are imposed upon domestic uses, unless of course they own or have a share in a rural property and can claim drought losses on the property against their income from other ventures for tax purposes. For the farmer or pastoralist it is an economic stress threatening the viability of the enterprise and their future livelihood or even survival. For the city water engineers it is also an economic and political stress with the potential to require added capital expenditure for increased storages, alongside criticisms of the apparent lack of past concern for future supplies. For politicians in opposition it may be a convenient stick with which to beat the government for supposed past mismanagement, or a means to lobby for financial support for one's constituency, while for the government it may be a threat to its survival, a challenge to its welfare program or even an excuse to obtain international disaster relief (Glantz 1976, Heathcote 1986b). For the scientists it may be a phenomenon to be studied in isolation or for its role in global ecosystems, especially nowadays as evidence of trends in climate variability or change, or as part of a global program of scientific investigation of the possibilities of drought mitigation or prediction. As such, occurrences may be welcomed as offering the chance of research by the scientists, but be far from welcome to people whose living depends upon the errant rainfalls. Indeed, for the philosophers and writers drought may be seen as one of the major natural challenges facing societies, and in the face of such challenges the outcomes of personal or communal mitigation strategies might reflect the role of fate, or human ability to rise to the stresses, or even illustrate the dysfunctional relationships between humans and their environment.

We shall examine some of these issues later in Chapter 9, in the context of drought as recognized in art, science and literature, but before that we need to look at the means by which droughts are identified and documented.

Chapter 3
Documenting Drought

The question of whether or not an extended dry spell has, in fact, become a drought causes considerable debate among meteorologists, farmers and public officials.

(Warrick et al. 1975: xiii)

The Characteristics of Droughts in Time and Space

Droughts exist in time and over space, but are there minimal dimensions which need to be reached before they are recognized and registered? Bearing in mind the different definitions and four categories of drought, we might expect the duration of drought to have distinctive characteristics, a beginning and an end, and the spatial dimensions should be identifiable, whether as a core or edge of the area affected.

In the general context of climate, four scales of phenomena are recognized (Table 3.1) of which only the largest two categories, Meso and Macro climate, need concern us, since these are the scales at which the drought phenomena and their impacts on human activities are generally recognized.

Table 3.1   Classification of Climate Phenomena by Spatial Scale

Designation	Micro climate	Local climate	Meso climate	Macro climate
Horizontal distribution	$10^{-2}–10^2$m	$10^2–10^4$m	$10^3–2\text{x}10^5$m	$2\text{x}10^5–5\text{x}10^{7\text{m}}$
Vertical distribution	$10^{-2}–10^1$m	$10^{-1}–10^3$m	$10^0–6\text{x}10^3$m	$10^0–10^5$m
Example	Greenhouse	Thermal belt on a slope	Mountain Basin climate	Climate zone
Lifetime of corresponding meteorological phenomena	$10^{-1}–10^1$sec	$10^1–10^4$sec	$10^4–10^5$sec	$10^5–10^6$sec

Source: Winterhalder 1980: 155.

By comparison with most other natural disasters, in terms of the area affected and the duration of the effects drought has been identified as a 'pervasive' rather than an 'intensive' phenomenon. By comparison for example with tornadoes, it is slower in onset, affects a much larger area but with less energy input per unit area, and lasts a much longer time (White 1974). Such terms of course are relative. Yet, on the geological time scale, the onset of climatic changes resulting in increased drought frequency has been shown to be quite rapid. Deep sea sediments off the northwest coast of Africa between 20°N and 18°W contain the pollen blown from the adjacent land vegetation over many thousands of years. A recent study by a paleoclimatologist:

> [F]ound that every 1,500 years or so, within periods as short as 80 years, ocean temperatures off Africa plummeted and seasonal rains essentially ceased over the continent. Africa became colder and drier and remained that way for centuries … [indicating that] larger-scale, longer-term climate change such as the one that turned northern Africa from a landscape dotted with crocodile-filled lakes 9,000 years ago into the vast Sahara today, took not thousands of years but less than a century. (DeMenocal 1997: 2)

The changes were claimed to document the 'shifts of the Saharan [desert]–Sahelian [savannah] border from 14°N (maximally expanded Sahara) to 23°N (maximally squeezed Sahara)' (Hooghiemstra 2004: 7).

Even more rapid change of regional atmospheric temperature, some 8°C in 30 years, has been recorded from Greenland ice cores. Similar rapid temperature changes must have been partly responsible for rapid drought onsets, such as the 'severe drought over the Great Plains of the United States in summer of 1980, which saw a heat wave and drought develop in approximately a week' (Namias 1983: 38).

If the onset may be fairly rapid, how long do the changed conditions need to last before a drought is recognized? Earlier we have seen that drought might be defined in terms of as little as a week of abnormal conditions; yet, for Australia as we have seen, three consecutive months of precipitation in the lowest 10 per cent of records are needed for an official meteorological drought to be recognized.

Once evidenced, how long can droughts last? Apart from those that initiated the thousands of years of arid climatic conditions forming the deserts which produced the massive sandstone deposits of the Permian and Triassic geological periods (some 260,000–180,000 years ago), what is the evidence from human history? Dendrochronology has shown that for the southwest of the USA and Nebraska over the period 1220–1957 AD some 738 years of climatic conditions could be reconstructed, during which many droughts were evident. The average duration of drought was over 12 years, but 12 droughts lasted over ten years, three lasted over 20 years and one lasted for 38 years! The average period without droughts was 24 years. (Schultz and Stout 1977).

Can droughts disappear over night? In effect, given a sufficiently large downpour, a logical answer for a meteorological drought might be yes, but not for an agricultural drought where the plants affected may in fact have died and it may take months for a newly planted crop to come to harvest.

Droughts can occur anywhere in the world. Britain, for example, perhaps not the first country to come to mind when droughts are mentioned, in fact suffered a major drought over 1975–1976, which was well documented by scientists (Doornkamp et al. 1980) and by victims (Cox 1978). The impacts provided interesting examples of the varied socio-economic impacts of a drought affecting a western society, as we shall see in Chapter 5. Globally, however, there does seem to be a tendency for the semi-arid regions of the world, particularly those with a marked dry season, and especially in the sub-tropical high pressure belt north and south of the equator, to be particularly vulnerable to drought. Historically this has been so and has continued through the twentieth century when it has often been associated with famines (Garcia et al. 1981, Glantz 1976, Heathcote 1986a, Mainguet 1999), but whether drought has been the sole causal factor for the famines is debatable, as we shall see later.

If droughts can occur anywhere, how frequent are they and how long can they last? Within the last 150 years three extended periods have been identified as drought prone and coinciding with persistent El Niño events of three or more years duration. First was the 1890s–c.1905 with droughts on the Great Plains of USA, in north China, Indonesia, northeast Brazil, southeast England, central and eastern Russia, southern and eastern Africa, India and Australia. Second was the 1920s–30s with droughts again on the Great Plains (the original Dust Bowl era), in north China, northeast Brazil, southern Africa and Australia. Finally, third was the late 1960s–70s with droughts in West Africa, North America and Australia (Allan 2000). These persistent El Niño events (for three years or more) seem to have occurred four or five times per century, and this seems to fit the models of climatic fluctuations reasonably well (Allan and D'Arrigo 1999: 115). So a major meteorological drought period with global implications might be expected on a frequency of 1 in 20 to 25 years. Shorter drought events of course will be more frequent, as we shall see.

Recognition vs Registration?

So far we have considered drought as a phenomenon which could be identified on the basis of some quantitative measurement and its existence documented in space and time. Yet the global history of droughts may not be as complete as we might expect. There may be gaps and there may be some 'ring-ins'. There seem to be two problematic processes involved: first the process by which drought is recognized, and second the process by which it is registered and, by implication, accepted, for the society's historical record.

In one sense we have already looked at the process of drought recognition but before going further we need to consider whether recognition alone is sufficient for drought to be documented, or whether some elements of registration – some associated phenomena or context – are necessary for the drought to be defined as significant?

Obviously droughts in geological time went largely unrecorded except in those cases where continuous meteorological 'droughts' became climates of aridity and affected regional natural erosional processes which stamped aridity upon the resultant landscape. There is no doubt that extensive droughts did and can create arid environments by transforming the abnormal conditions into normalcy. Indeed, part of the global warming scenarios is for regional drought frequencies to increase and result in the threat of increasingly arid environments or, at least, the desertification of previously humid environments, as we shall see later.

The Problem of Registering Drought

But what kinds of associated phenomena seem to trigger the acceptance of drought as significant and the registration of the drought as an historical fact? Researchers working in the history of west Africa have suggested that: 'The perception of drought is inextricably inter-woven in the social and political fabric of societies. Hence, considerations other than climate frequently influence whether an event is recorded or not. For example, a drought which coincides with an important socio-political event such as the death or coronation of a new king is more likely to be remembered and therefore recorded and preserved in some way' (Tarhule and Woo 1997: 603). In this case it is the association of drought with some other event which ensures its recognition and record.

One phenomenon often associated with drought is famine and for studies of historic droughts in India the association is strong. At least five basic historical sources have been tapped for evidence:

1. Oral traditions of 'legendary famines', e.g. Durga Deves famine in the Deccan which began in 1396.
2. Indian traditional literature 'including the Vedas, the Jataka stories, and the Arthashastra'.
3. Inscriptions on temples and palaces, which were used to document droughts and famines in Tanjore at Alangudi in 1054 and at Tirrukkadayur in 1160, and Tirkkalar in North Arcot in 1390–1391.
4. Writings of Muslim historians for the Deccan area in the fifteenth century, e.g. droughts in 1423–1424 and 1472.
5. European writings from the sixteenth century onwards. (Murton 1984; 72–3).
6. Yet, as we have shown already, definitions of drought are many and varied, reflecting different perceptions of drought itself. The evidence from West Africa suggested similar conclusions: 'drought concepts vary between cultures and evolve in response to changing environmental and social conditions. This implies that similar evidence, or records of historical droughts in different cultures could potentially refer to quite different phenomena' (Tarhule and Woo 1997: 603).

One of the first major droughts to make the global television screens, with pictures of sun-drenched bare earth, dead livestock and starving people, was that which affected the Sahel of West Africa from the early 1970s. Yet news of that disaster took time to reach the western world and the media. Concerned researchers at a meeting hurriedly called at seven weeks' notice in 1973 commented that 'The news [of the disaster] might just as well have been brought to Europe by trans-Saharan caravan' (Dalby and Harrison Church 1973: 4). They guessed that the delay was the combination of the location in remote areas of Africa and the fact that the victims were only subsistence farmers and pastoralists!

Reports of droughts may even carry moralistic overtones. One researcher has suggested a genealogy of moralistic interpretations of climate and its effects upon societies going back to at least Aristotle, and including philosophers such as David Hume and Immanuel Kant. Using evidence from Western travelogues and climatologists he postulated that the west persistently 'devised mechanisms to ensure that cultural relativism rarely implied cultural equivalence and that the temperate world would exercise dominion over tropical realms. Among these [mechanisms], moral climatology, I believe, had a key role to play' (Livingstone 2002: 174).

Morality might influence the process of reporting itself. Study of the reports from Christian missionaries on the changing seasonal conditions in the areas of their activities in the nineteenth century has suggested such biases in that reporting. The reports of the agents of the Missionary Society coming out of central southern Africa over the period 1815–1900, particularly for what is now Botswana, and boosted by reports from Dr David Livingstone himself, suggested extensive desiccation was in process. However, the reports had a moral tone which blamed the natives for mismanagement of resources, when in fact other scientific evidence suggested that the missionaries were reporting on an environment at the end of periods of intensive but short term droughts (Enfield and Nash 2002: 44).

For attitudes in China from the seventeenth to nineteenth centuries some researchers have suggested a similar moral meteorology, where there was assumed to be a link between the weather and the morals of the society: 'Rainfall and sunshine were particularly susceptible to the behaviour of people. Immoral conduct would probably lead to poor rains and even drought, whereas moral behaviour was likely to assure good rains and bountiful crops' (O'Connor 1995: 12). Within imperial Chinese society, as regards levels of responsibility, the behaviour of the Emperor was most significant, then that of the bureaucrats followed by that of the ordinary people. In 1732 the Emperor Yongzeng was quoted as commenting that: 'Floods, droughts and famines in a particular locality could have many reasons: bad government at the centre, errors by local officials, "vexatious commands by the authorities," or "degenerate customs among the local population"' (Ibid.).

Certainly, where the effects of a drought were considered sufficiently serious the victims resorted to prayer and religious activities which might imply also penance for implied past wrong-doing. For evidence of historical droughts, researchers have in fact used such activities to document the occurrence of droughts. One recent example drew upon a marble tablet in a Sicilian church which documented past religious processions or 'rogations'. By way of explanation, the Catholic Church allows for processions: 'either for invoking protection against dangers, such as war, famine, and plague or for invoking protection against natural hazards, such as earthquakes, volcanic eruptions, and droughts. Usually the procession is the last act of a series of liturgical initiatives' (Piervitali and Colacino 2001: 227).

Those initiatives apparently began with the people most immediately affected organizing their own 'private rogations', but 'when the drought is prolonged, the image of the [community's patron] Saint [the 'intercessor'] is exposed in the Church and public rogation is carried out usually as a three-day rogation (*triduum*) or a nine-day period of devotion … Finally, when the dry weather and rainfall shortage persist, a procession [of the community and civil authorities through the community] is proposed' (Ibid.: 228). In one sense the nature of the rogation was a sign of the intensity of the drought threat, and the researchers explained that following a scale developed for similar Spanish rogations, 'the procession with the intercessor corresponded to a high intensity [drought] event' (Ibid.: 229). Interestingly, the researchers noted in 2001 that the most recent procession for the threat of drought occurred in central Sicily on 29 October 1999.

Documenting Droughts in Space and Time: the Global Picture

The Search for Dates

> Since geological time is not salami, slicing it up has no particular virtue. But if it is to be sliced there is no need to botch the job, and chronometric dating provides the guidelines. (Vita-Finzi 1973: 47)

Within human history what has been the evidence of past droughts? The search for that evidence has been stimulated by the increasing impacts of droughts upon world societies and the current concern over possible climate change (with its relevance for future droughts), and has cast a wide net. From historical records and documents have come the fluctuating levels of lakes and river flows, diaries of travellers and stay-at-homes, official reports and company records. From research in the field has come archaeological data, biological data such as dendrochronology and pollen analysis, geomorphological data such as sedimentology and varve [seasonal sediments] analysis from lakes and most recently from the sea itself (Table 3.2). Such an array of evidence, however varied, has the potential not only to pin drought events to a specific time, but also to provide some clues as to why they occurred when they did.

An example of the range and inter-linkage of the available data comes from a study of droughts in Central Africa over the period 1300 to 1800 AD. The study used:

Table 3.2   Techniques for Dating Past Droughts

Technique	Measured parameters	Age range	Matrix	Results
Historical archives, archaeological and phenological data	Records and field evidence, including plant flowering times	From present to thousands years BP	Current environment and human artifacts	Reasonably precise dates obtained? Seasonal trends from plant data.
Dendro-chronology	Plant growth-rings	From present to 1000 years BP	Trees	Reasonably precise dates obtained?
Sediments	Glacial melt laminae (varves) Ice cores	From present to 10,000 years BP	Lake and playa sediments Glaciers and Ice Sheets	Reasonably precise dates obtained?
Weathering	Sedimentary characteris-tics	1–10,000 years BP	Soils and stratified deposits	Relative ages possible
Radiocarbon	C^{14}	0–c.45,000 years BP	Organic materials	Relative and numerical ages estimate
Thermo luminescence (TL)	TL emissions	100–1,000,000 years BP	Quartz and feldspar grains	Numerical age estimate
Flourescence	Intensity	> 400BP	Corals	Numerical age estimate
Oxygen isotope stages (OIS)	OIS	OIS 1 = last 10,000 years; OIS 2 = 25,000–10,000BP; OIS 3 = c.60,000–25,000BP	Deep sea sediments and ice caps	Reasonably precise data

Source: Modified after Isdale et al. 1998, Thackeray 2005, Williams et al. 1993, Appendix.

1. The low or summer River Nile levels which correlate with precipitation levels in the interlacustrine region of East Africa. These records go back, although somewhat imperfectly, to 622 AD.
2. An analysis of tree rings of an ancient yellowwood (*Podocarpus Palactus*) in Natal, which began its growth about 1300 AD and which indicated that

it fulfilled the basic dendro-chronological principle of sensitivity, [i.e.] exceptional growth, indicating wet years, and very little growth, suggesting drought.

3. Documentary reports of droughts in Indian Ocean localities and in Central Africa, variable, again imperfectly, since the late fifteenth century.

4. Oral references, especially from Mutapa historical traditions, dated by combining the genealogical method and written sources (Webster 1979: 148).

Interestingly, because the oral evidence of the Mutapa tribal elders considered only a drought of three years duration would cause the tribe to move its location, the researcher adopted a three year minimum for his estimate of the significant drought events.

Under the auspices of the US National Oceanic and Atmospheric Administration (NOAA), research to provide global estimates of climatic conditions since 1500 AD, using some of these techniques, has been made available (Bradley and Jones 1992) and global maps including drought areas are available from the 1950s onwards (Figure 3.1).

A recent study has applied the Palmer Drought Severity Index (PDSI) to historical meteorological data on a global scale, and has claimed that because the Index shows a good correlation with both moisture in the top meter of soil in the summer and stream flow characteristics the extent of past droughts can be indicated. The findings provide an interesting background to the global warming debates:

> The global very dry areas, defined as PDSI < −3.0, have more than doubled since the 1970s, with a large jump in the early 1980s due to an ENSO-induced precipitation decrease and a subsequent expansion primarily due to surface warming, while global very wet areas (PDSI > +3.0) declined slightly during the 1980s. (Dai, Trenberth and Qian 2004: 1117)

The study claimed that 'these results provide observational evidence for the increasing risk of droughts as anthropogenic global warming progresses and produces both increased temperatures and increased drying' (Ibid.).

The Search for Contemporary Droughts

So much for history, but what of the contemporary reporting of droughts? Prior to the development of satellite observations of earth any attempts at identification of drought affected areas had to rely upon national weather data which provided compatible elements, such as rainfall and air temperature, from which estimates of the potential evaporation could be calculated. One such attempt appeared in 1972 in the *Agroclimatic Atlas of the World*, which claimed to show the 'Vulnerability of major agricultural areas to drought based on a difference between annual

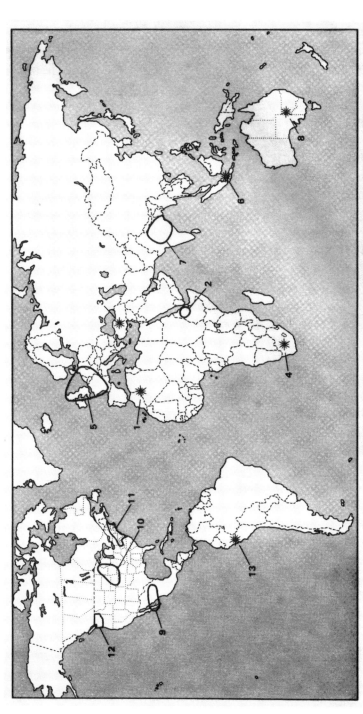

Figure 3.1   Climate Impacts Map: 1957 Droughts

Numerals on the map refer to brief descriptions of drought impacts. For example '8. Australia – Mar thru Jan 58. Reported in Oct that drought could cut wheat production 50%; heat wave late Dec thru early Jan in Sydney area; New South Wales esp. affected; rains broke drought in late Jan 58'.
Source: Climate Impact Assessment, US Department of Commerce, National Oceanic and Atmospheric Administration.

precipitation and potential evaporation' with areas of 'frequent intensive drought' differentiated from areas of 'infrequent non-intensive drought' (Goldsberg 1972). Although at a global scale, the map (reproduced in Kogan 2000 and here as Figure 3.2) showed that effectively all the major agricultural areas of the world were vulnerable to drought.

With the advent of satellite imagery, however, the possibility of providing real-time images of the earth's surface and specifically those areas appearing to be drought affected, without needing detailed ground information, became a reality. The basic method, much refined and improving continually, has been to compare the reflectivity of the earth's rural surfaces, whether vegetated or not, as an indicator of any drought stress upon the vegetation or surface hydrology. Using the low incident solar reflectivity of vegetation in the visible bands of the solar spectrum (VIS), and comparing with the high reflectance in the near infrared bands of the solar spectrum (NIR), the amount of moisture in the vegetation can be estimated and the presence or absence of agricultural and possibly hydrological drought stress calculated (Kogan 2000). Such a basic description hides complex processing of the data against ground truth (actual physical inspection of the sites, where available) to provide average conditions for relative assessments, but the results are sufficiently useful to be the basis of reports of current and forecasts of developing drought situations around the world.

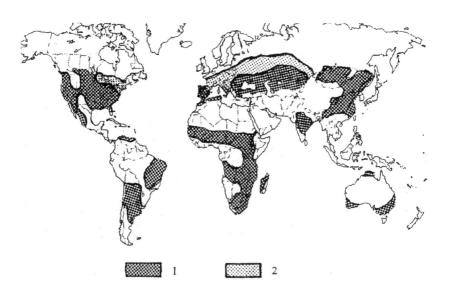

Figure 3.2   Vulnerability to Agricultural Drought c.1972
Based upon the difference between annual precipitation and potential evaporation. *Key*: 1-dark areas 'frequent intensive drought'. 2-lighter areas 'infrequent nonintensive drought' *Source*: Kogan 2000.

The US National Oceanic and Atmospheric Administration (NOAA) extended its research to use such data together with media reports to provide global coverage of annual drought-affected areas from the 1950s. Similar maps of annual rainfalls from 1950 to 1970 showing meteorological drought affected areas as 'much below average' (second decile) and 'very much below average' (first decile) were provided in Lee 1979, and are illustrated in Figure 3.3. Lee commented that 'the figures do show that the application of decile analysis on a global scale has value as a method of objectively comparing one period with another; as such it provides a reasonable basis for assuming that a global drought watch system has practical value and is feasible given the technology of the computer age and modern communications' (Lee 1979: 184).

The technical capabilities of satellite imagery are constantly being elaborated and a recent review suggested that the Advanced Very High Resolution Radiometer (AVHRR) imagery of NOAA satellites 'can be used to monitor drought conditions because of their sensitivity to changes in leaf chlorophyll, moisture content, and thermal conditions'. Since the 1980s the imagery has helped to identify droughts in the USA in 1988, 1989, 1996 and 2000; in the former Soviet Union in 1991, 1996 and 1998; in Argentina in 1988–1989, and 1989–1990; and in China in 1992–1993 (Kogan 2005: 79). Discussion of these possibilities can be found in Chapters 5 to 8 of Boken et al. 2005.

The most recent global coverage is conveniently provided by the University College London's *Global Drought Monitor* which provides a monthly global map with areas of minor to exceptional drought indicated and the total population in the current map area under exceptional drought enumerated (http://drought.mssl. ucl.ac.uk/drought.html).

Droughts in Space and Time: The North American Experience

In terms of the period of available records and their quality, the United States has perhaps the best documented record of past droughts. Many of those records are based on tree-ring analysis in the western USA. They reflect the importance of the arid southwest terrain where long-lived humid mountain forests were adjacent to lowland arid and semi-arid plains and plateaus, with varied scrub and grassland vegetation, which rapidly reflected changing rainfall regimes. To these were added the records of the fluctuating levels of the main drainage system of the southwest, the Colorado River. Thus a recent study could claim that multi-decadal droughts had been experienced in the Sierra Nevada and western USA over the period 900 to 1400 AD, with 1020–1070 AD and 1250–1360 AD as distinctive drought periods, and that from 6000 BC to 1 AD there had been droughts in 5970, 5881, 5591, 4058, 3948 and 1257 BC (Graumlich and Ingram 2000, Woodhouse and Overpeck 1998). Of particular interest was the drought from 1276 to 1313 AD, which was sometimes referred to as the 'Great Drought', and which coincided with the abandonment of the Anasazi Pueblo Indian settlements in the Mesa Verde area.

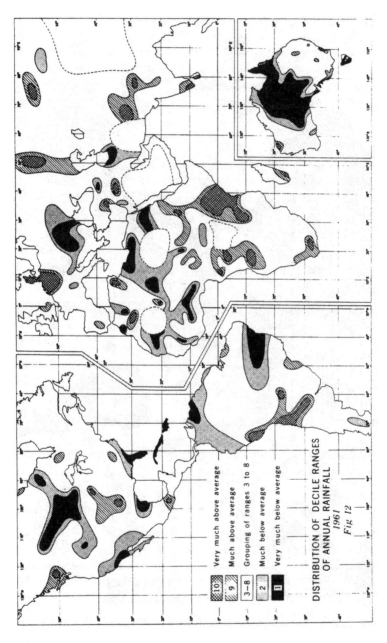

Figure 3.3 Distribution of Global Decile Ranges of Annual Rainfall 1961

One example from a series of 21 maps of global decile ranges for annual rainfalls from 1950 to 1970, based upon calendar year data supplied by the National Center for Atmospheric Research in the USA, and contained in the Appendix to Lee (1979).

Over the central Great Plains, the longest continuous coverage of drought occurrences was from 1200 to 1960 (Table 3.3). Here were listed 21 periods with five or more years of drought, while the maximum duration was 38 years from 1276 (the 'Great Drought' noted above). Significantly, with the onset of nineteenth century European settlement, decadal droughts from 1822 raised fears of a 'Great American Desert' (Brown 1948), that from 1884 initiated the first official drought relief efforts (see Chapter 7) and that from 1931 were to usher in the years of the 'Dust Bowl' (Worster 1979). In turn, each were to make their mark upon the human history of the region, both when the explorers were seeking new lands to occupy and then when the new settlers brought extensive rain-dependent agriculture on to the naturally sub-humid/semi-arid grasslands (Woodhouse and Overpeck 1998: 2698 and 2703).

Table 3.3   Droughts on the Great Plains from 1200 to 1960

Initial date	Duration (years)	Initial date	Duration (years)
1220	12	1688	20
1260	13	1728	5
1276	38	1761	13
1383	6	1798	6
1438	18	1822	11
1493	6	1858	9
1512	18	1884	12
1539	26	1906	8
1587	19	1931	10
1626	5	1952	6
1668	6		

Source: Wilhite 2003: 747, 3.4. Data are from tree-ring analysis for western Nebraska, showing 21 drought periods of five or more years duration.

Despite the efforts of several individual researchers, not until 1999 was a national drought monitoring service available in the USA. However, in 1998 a National Drought Policy Act created a National Drought Policy Commission and in 1999 the US Drought Monitor was launched (http://droughtmonitor.unl.edu/). Based upon a combination of precipitation, soil moisture, stream flow data and vegetation condition it provides a bi-weekly assessment map of areas of drought, classified from 'Abnormally dry' (heading into or recovering from drought) to 'Exceptional drought' (Table 3.4). Interestingly, the maps attempt some recognition of the types of drought impacts, identifying short-term impacts, such as on agriculture and grasslands, and long-term impacts, such as on hydrology and ecology.

Table 3.4   US Drought Monitor Classification System

Drought	D0	D1	D2	D3	D4
Type	Abnormally dry	Moderate drought	Severe drought	Extreme drought	Exceptional drought
Palmer drought index	−1.0 to −1.9	−2.0 to −2.9	−3.0 to −3.9	−4.0 to −4.9	−5.0 or less
CPC soil moisture model percentiles	21–30	11–20	6–10	3–5	0–2
USGS weekly streamflow percentiles	21–30	11–20	6–20	3–5	0–2
Percent of normal precipitation	<75% for 3 months	<70% for 3 months	<65% for 6 months	<60% for 6 months	<65% for 12 months
Standardized precipitation index	−0.5 to −0.7	−0.8 to −1.2	−1.3 to −1.5	−1.6 to −1.9	−2.0 or less
Satellite Vegetation index	36–45	26–35	16–25	6–15	1–5

Source: (Wilhite et al. 2005, Table 9.2).

From that time, however, an increasing number of scientific research projects have broadened and deepened the knowledge of past as well as current drought indicators. Most of the research has been either sponsored or coordinated by the National Oceanic and Atmospheric Administration in association with the National Geophysical Data Center and the National Climatic data Center. Examples are:

> http://www.ngdc.noaa.gov/paleo/drought.html
> http://www.ngdc.noaa.gov/paleo/drought/drght_graumlich.html
> http://www.ngdc.noaa.gov/paleo/usclient2.html
> http://www.ngdc.noaa.gov/paleo/pdsiyear.html
> http://www.ngdc.noaa.gov/paleo/pdsi.html

Included in these studies are plans to extend the record of drought for the conterminous USA back to 1700, and to develop more comprehensive data for the range of drought characteristics as a help in drought forecasting.

Droughts in Space and Time: The Australian Experience

As we have seen in Chapter 2, research into droughts in Australia suggested in 1967 that meteorological droughts (annual rainfalls in the lowest 10 per cent on record), when occurring over at least 10 per cent of the continent, have in the past coincided with damaging agricultural droughts (Gibbs and Maher 1967, ABS 1988).

The landmark study by Gibbs and Maher (1967) applied the above definition to the rainfall records from 1885 to 1965, and showed that while drought had never covered more than 50 per cent of the continent, droughts affecting up to 20 per cent of the continent had occurred in 60 per cent of the years studied, and only 22 per cent of the years were free of drought. Australia then is familiar with drought. Taking the study further, and using the Commonwealth Bureau of Meteorology's ongoing annual rainfall maps which supplement the earlier maps in Gibbs and Maher, it possible to survey the continental patterns from 1900 onwards (Figure 3.4). At the continental scale, meteorological drought occurred somewhere on the continent in 80 per cent of the 115 years from 1885 to 1999. For 31 per cent

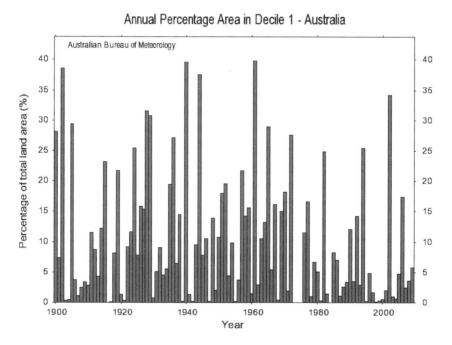

Figure 3.4   Droughts in Australia 1900–2009
Bars indicate area of Australia with annual rainfall in the first decile, i.e. lowest 10 per cent on record.
Source: Compiled from Australian Bureau of Meteorology data.

of that time over 10 per cent of the continent was affected and some significant societal impact was likely. The maximum sequence of years (four) with over 10 per cent of the continent drought affected was 1926–1929, a period of considerable stress for new farming settlements in southeastern and southwestern Australia. This also proved to be the kiss of death for the newly created political unit of Central Australia (Table 3.5 and Heathcote 1987). Yet the area of the continent affected alone did not necessarily increase the human impact, for the largest drought affected area occurred in 1961 with 46 per cent of the continent in the first decile, but because this was mainly in the arid and semi-arid parts of the continent, agricultural production and domestic water supplies were relatively unaffected.

Conversely, at the continental scale, the maximum sequence of years relatively free from drought (i.e. with less than 10 per cent of the continent in drought and years which I shall call 'benign periods'), has occurred recently – from 1983 to 1993 (11 years), the longest such period since the start of records in 1885.

At the regional scale of the states, however, the picture is different (Table 3.5). The whole of Victoria was drought affected in 1982, 90 per cent of the Northern Territory in 1961, 88 per cent of Tasmania in 1914, 81 per cent of New South Wales in 1940, 70 per cent of Queensland in 1902 and 70 per cent of South Australia in 1902 and 1961, while 66 per cent of Western Australia was affected in 1936. This varying spatial pattern reflects the latitudinal contrast between the tropical and temperate weather systems – since it has been rare for the summer rains of northern Australia to fail in sequence with the winter rains of the south. This probably accounts for the lowest maximum area drought affected being in Western Australia, the only state spanning the widest temperate/tropical latitudes.

The spatial variation of the patterns is not merely latitudinal, however, for there is evidence of longitudinal contrast between droughts in western and eastern Australia. Of the 115 years of records (1885–1999), there were no droughts in either area in 26 of the years; of the remaining 89 years when drought was affecting one or the other area, the eastern states (NSW and Qld) had droughts in 31 years (34.8 per cent of the drought years) when Western Australia was unaffected and, conversely, Western Australia had droughts in 15 years (16.9 per cent) when the eastern states were unaffected. Further, if we include years when the eastern states had drought, and Western Australia had drought but on less than 10 per cent of its area, the total rises to 54 or 60.6 per cent of the drought years. In other words, drought rarely affects the eastern and western halves of the continent at the same time.

A striking example of this longitudinal contrast was 1982, when 58 per cent of New South Wales was in drought but Western Australia was completely drought free and had a bumper harvest. Interestingly, fluctuations in Western Australia's wheat harvests do not coincide over time with those of the eastern states and bear out the longitudinal differences noted above. The correlation coefficient for the historical record comparing Western Australian wheat yields with the eastern states is -0.05 compared with New South Wales and 0.35 with Queensland, whereas between the eastern states correlations of wheat harvests are 0.55 to 0.88 (IAC 1986).

Table 3.5   Drought Sequences in Australia 1900–2009

Area	Maximum area in drought in any one year		Maximum sequence of years with drought on 10% or more of area		Maximum sequence of years with drought on less than 10% of area	
	%	Year	N	Period	N	Period
Australia	39.8	1961	4	1926–1929	7	1995–2001
			3	1950–1952	6	1983–1989
				1963–1965	5	1930–1934
					4	1978–1981
Western Australia	69.8	1924	4	1956–1959	9	1915–1923
			3	1950–1952	7	1995–2001
						2003–2009
Northern Territory	94.2	1961	3	1927–1929	14	1971–1984
				1934–1936	12	1997–2008
				1963–1965	11	1916–1926
Queensland	61.1	1900	4	1926–1929	11	1971–1981
					10	1983–1992
					7	1995–2001
New South Wales	83.5	1940	2	1901–1902	11	1983–1993
				1914–1915	10	1947–1956
				1918–1919	9	1968–1976
				1920–1923		
				1937–1938		
				1943–1944		
				1979–1980		
Victoria	96.5	1967	2	1907–1908	22	1945–1966
	95.5	2006		1943–1944	11	1983–1993
	94.4	1982			10	1915–1924
						1928–1937
South Australia	77.4	1929	5	1925–1929	11	1983–1993
					7	1995–2001
Tasmania	88.2	1914	2	1933–1934	11	1995–2005
				1940–1941	9	1973–1981
				1982–1983	8	1951–1958

Drought areas = areas with annual rainfall in first decile.
Source: Compiled from Australian Bureau of Meteorology data.

This longitudinal 'oscillation' of drought impacts has relevance for overall drought impacts as we shall see.

What of the so-called 'benign' periods, when a sequence of years passed with drought never affecting more than 10 per cent of the state area? For the period up to the end of the twentieth century, Victoria tops the list with 20 years from 1945 to 1964, the Northern Territory had 19 between 1971 and 1989, and Queensland had 17 years between 1983 and 1999. In the latter case this covered a period in the early 1990s when extensive and catastrophic agricultural drought impacts were being claimed for the state's agricultural and pastoral activities. Does this illustrate a failure of the data or is the contrast erroneous or what? Some suggested explanations will come later.

Droughts in Space and Time: The African Experience

Immediately prior to the arrival of humans on the scene, evidence from the major lakes in East Africa (Lakes Malawi, Tanganyika and Bosuntui) suggested 'a continental drought of extreme severity around 75,000 years ago. The apparent coincidence of this event with DNA-based dating of the origin of modern humans, raises the fascinating possibility of a strong climatic influence on human evolution and dispersion' (Johnson, Talbot and Odada 2005: 8). Further, over the last 1,100 years the Lake Naivasha region of Kenya had frequently suffered droughts that were far longer and more severe than those of the past century (Ibid.). More recent droughts have been recognized for Kenya, with 16 years of drought noted between 1945 and 1992, in other words a third of the time (Glantz 1994: 80).

For North Africa, Tunisia 'had at least 26 major drought-related famines during the period from 100 AD to the late nineteenth century' (Glantz 1994: 124). In Morocco agricultural drought has been identified every 1 in 3–4 years with the worst effects occurring if the drought extended into a second year (Swearingen and Bencherifa 2003: 778). In Ethiopia records of droughts every seven years from 253 BC to 1 AD, and every 9 years up to 1500 AD, have been claimed. Thereafter more reliable records suggested a frequency of 1 in 7 years until mid-twentieth century, but with the frequency increasing to 1 every 2 years by the turn of the century. However an apparent increase in drought frequency may have various causes.

> Northwest Africa's increasing drought hazard is primarily the result of changes in society. The region's climate has remained essentially the same during the past century. However, population levels and land-use patterns have radically changed. Farmers have progressively intensified their land use – both horizontally and vertically [i.e. occupying more land on the plains and in the mountains]. This has gradually eroded traditional buffers protecting society from drought. (Mersha and Boken 2005: 133)

For the continent as a whole, 1963 to 1967 saw 5 major droughts affecting 1 per cent of the continental population, while 1988 to 1992 saw 15 such events also affecting 1 per cent of the population (Diaz 2005).

Dry spells (effectively droughts) have been described as endemic to southern Africa (Table 3.6), usually identified as when rainfall is 75 per cent or less of normal precipitation, and classed as severe if the drought extended over two consecutive seasons (Vogel 2003: 839). 'In southern Africa, the worst drought of the century occurred during 1991–1992 and affected 100 million people' (Anyamba et al. 2005: 71).

Table 3.6   Dry Spells in Africa in the Twentieth Century

Period	South Africa (a)	Zimbabwe (b)	Ethiopia (c)
1900s and 1910s	1905–1906 to 1915–1916	1911–1912	
1920s and 1930s	1925–1926 to 1932–1933	1923–1924	
1940s and 1950s	1944–1945 to 1952–1953	1946–1947	1957–1958
1960s and 1970s	1962–1963 to 1970–1971	1972–1973	1964–1965 1972–1973
1980s	1980–1981 to 1990	1981–1983 1986–1987	1983–1984
1990s	1994–1995	1991–1992	

Sources: (a) Vogel 2003; (b) Unganai and Bandason 2005; (c) Diaz 2005.

Droughts in Space and Time: The Chinese Experience

As we have already noted, the record of drought occurrences in China is one of the longest in the world and, despite some reservations over the veracity of the reports, there can be no doubt that drought has been a major hazard since human occupation. Scientific records over the last 100 years have confirmed the significance of drought events and recent commentators have reaffirmed this. Reviewing the period 1949–1995, 12 'major drought disasters' were recognized, each covering at least one of the three main river valleys. Half of those droughts occurred between 1986 and 1994 and reinforced the comment that 'drought and flood are the major climate disasters concerning both the Chinese government and the general public' (Qian and Zhu 2001: 426). More recent researchers have provided further evidence (Table 3.7). They also provide specific data on drought impacts:

> The drought damage in 1943 is a natural disaster with the maximum deaths in historical record, the starved people reached 300×10^4 caused by the drought damage. The damage of the drought in 1986 is the severest autumn drought in Guangdong since the founding of the People's Republic of China in 1949, the

area damaged by the drought reached 112 x 10⁴ ha in Guangdong at that time. Influenced by the severe drought in 1991, most of the small-sized reservoirs dried up, even the water level of Xiufenjiasng reservoir, the largest in the Pearl River catchment, dropped below the dead water level. (Huang and Ziang 2004: 18)

Table 3.7   Droughts in China over Last 100 Years

Province	Years of drought
Fuzhou and Fujian	c.1930
Guangzhou and Guangdong	c.1913; c.1943; c.1955; c.1963; c.1977; c.1985; c.1991
Kunming and Yunnan	c.1906; 1931, 1933, 1937; 1960, 1963, 1969; 1987; 1992, 1997

Source: Adapted from Huang and Zhang 2004: 15.

Drought frequency appears to be increasing, particularly in northern China, partly as the result of increasing human pressure on resources: 'drought trend and human activity in northern China have induced more severe environmental problems. Environmental change and human activity through the hydrological cycle has resulted in the runoff variation in Northwest China, no-flow in the lower reach of the Yellow River and the salinization of soil in Northwest China' (Qian and Zhu 2001: 443). In other words, these look suspiciously like socio-economic droughts and any claims for changing climates would need to check the meteorological records against any 'shouting index' of drought occurrences.

So we might conclude that in the general process of identifying and documenting drought, there seems to be a demonstrable link between human pressure upon environmental resources and the reports of drought occurrences. This apparent link is of relevance not only to our next consideration of the causes of drought but also to our later analysis of the attempts of societies to mitigate its impacts.

Chapter 4
The Causes of Drought

We have already recognized four types of drought phenomena, so we need to separate out four causes, four sets of criteria which explain the differentiation. In effect, as we shall see, the four categories derive from the various effects of the first – meteorological drought, which depends for its mechanisms upon the variations in solar radiation and the 'natural' movements over the surface of the earth.

Why does drought occur? Like most aspects of the global climate, the answer seems to be complex and the relevant variables inter-connected. Basically the answer seems to be a function of the scale of the investigation. If we look beyond the globe itself, there are suggestions that solar radiation received (as the result of solar flares/sun spots) might change, first from changes in the energy transmission properties of space itself, second from changes in the distance between the earth and the sun, and third from the changing axis of the earth. This variation can affect the basic characteristics of the global climate and by implication droughts. If we zoom in to the phenomena of the globe itself, the movements of the earth (global tectonics and the associated volcanic eruptions), the circulation of the atmosphere and the associated circulations of the world's oceans, and the variations in surface reflectivity (albedo) between the existing deserts, tropical forests and snow and icefields are all thought to play a role in the creation of drought events. And finally, if we acknowledge the ability of humans to modify the surface and atmosphere of the earth, through changing the land cover, modifying the composition of the surface waters and the atmosphere itself, or conversely managing the environment in a way which increases demands for water, we need to recognize that humans themselves can create an increased vulnerability to droughts (Menzhulin, Savvateyev, Cracknell and Boken 2005). That said, meteorological drought seems to be the main trigger for a series of drought conditions.

Meteorological Drought

Meteorological drought is normally defined as the result of natural forces which lead to the reduction of moisture in the atmosphere, and which reduce the chances of that limited moisture being precipitated (as rainfall, hail or even snow) and reaching the earth's surface. There seem to be three interacting systems: the regional scale – with local air pressure and winds – the larger scale conditions at the lower boundary of the atmosphere, and finally the surface characteristics of the earth:

The immediate drought producing mechanism almost always involves persistent and persistently recurrent subsidence of air (approximately a few hundred meters per day) which results in compressional warming and lowered relative humidity … [while soil moisture deficits] through radiative and thermodynamic processes, may lead to further sinking motion of the air columns aloft. (Namias 1983: 30)

Thus, explaining droughts which last for months or a single season is fairly well understood, but when they last longer the explanations become more complex and conjectural (Woodhouse and Overpeck 1998: 2708).

In the case of the USA, the usual onset of drought begins in the late spring to early summer with a:

semi-permanent mid- to upper-tropospheric anticyclone over the plains, sustained by anticyclones in both the eastern central pacific and eastern central Atlantic and accompanied by intervening troughs … that can persist throughout the summer [thus blocking the entry of any moist air from the Gulf of Mexico] … Once a drought-inducing circulation pattern is set up, dry conditions can be perpetuated or amplified by persistent recurrent subsidence [of air masses] leading to heat waves, clear skies, and soil moisture deficits. (Ibid.: 2707)

Specific droughts, however, have distinctive combinations of triggers as witness the 1977 drought on the west coast of the USA. In an attempt to explain it, the following was suggested. An El Niño event (a situation where the contrast of sea surface temperatures and air pressures in the Pacific Ocean creates stable high pressure systems and accompanying droughts in Indonesia and Australia) helped create a 'strong Aleutian Low [Pressure] with teleconnections [links] to the United States western ridge [of high air pressure] and eastern trough [of low air pressure]'. This, combined with snow cover over eastern and mid-west USA, strengthened the East Coast low pressure system, which may 'have helped strengthen the western ridge', together with a pattern of sea surface temperatures in the northern Pacific. All of which hindered moist air flows into the area and set up the drought conditions (Namias 1983: 35–6). So, it can get pretty complicated, and involve a fair amount of guesswork as to the detail of the processes for specific events.

A further complication is that more than the contemporary atmospheric conditions may be responsible for the current events. In the case of the 'record-setting El Niño of 1997–98' the phenomenon could have been the result of a combination of processes on different time scales. Some 'coupled processes on the international time scale … [together with] a warm phase in the decadal [temperature/solar?] oscillation and … a longer-term warming trend in the eastern Pacific … Thus, we could think of attributing El Niño or La Niña occurrences to a combination of processes acting on different time scales' (Glantz 1998: 17–18).

Consequently, while drought may usually be easily ascribed to regional pressure and wind systems, a broader view must look to larger longer distance atmospheric linkages (teleconnections) with variations of sea surface temperatures, snow and ice cover, and, particularly in the warm seasons, the character of the soil, including moisture content (Namias 1983: 30).

The implication so far is that these are purely natural forces, unaffected by human actions. However, given the current recognition of human activities impacting upon the global atmosphere, this assumption cannot be maintained in the twenty-first century, although it may be accepted for the global environment prior to the onset of the Industrial Revolution. For the purposes of this discussion at this stage in the argument, I shall suggest that meteorological drought should be seen as the initial phenomenon, derived mainly from changes in the global atmospheric characteristics at regional scales, in the sequence of phenomena recognized as various forms of drought.

With this as a proviso we can suggest that meteorological drought occurs when either precipitation is significantly reduced for a significant period of time, or the evaporation or evapo-transpiration of moisture from a land surface, or usually a combination of both, results in a reduction in local soil moisture for long enough to have an adverse impact on the local ecology.

Droughts Induced from Extra-Terrestrial Causes?

> It is not surprising … that flood and drought records in many parts of the world contain periodic lunar and solar elements. (Huggett 1997, Table 5.7)

Because the earth's atmosphere is powered literally by the sun, any modification of the rate of that energy input would have significant impacts on the atmosphere. The Hungarian scientist, Milutin Milankovich, suggested in 1941 that variations in the earth's distance from, and inclination of its axis towards, the sun led to variations in the intensity of solar radiation. Three variations, or cycles, seemed to be relevant. First were variations in the earth's elliptical orbit around the sun over a 100,000 year period. When the orbit was strongly elliptical we were alternately closer to or further away from the sun with corresponding maximum variation in solar energy inputs, but when the ellipse was less precise the seasonal variation in received solar energy was smaller. That has been proven and currently the variation between the seasons is only about 6 per cent, but it could reach as high as 20–30 per cent. The second cycle stems from the tilt of the earth's axis towards the sun, which varies from 21.8 to 24.4 degrees over a 42,000 year cycle, and affects the seasonal temperature extremes. The third cycle stems from the earth's wobble on its axis over a 22,000 year cycle, which can similarly determine the seasonal extremes (Flannery 2006).

While these cycles were recognized to be particularly significant in early explanations of climate change over long time periods, as Flannery shows, they are also relevant to explanations of shorter term climate variations such as drought. Even the apparently minor modifications of solar energy input may be important in explaining the incidence of meteorological drought. A study of solar activity from tree-ring radio-carbon data over the last 75,000 years suggested:

> In every case where long-term solar activity falls, mid-latitude glaciers advance and climate cools; at times of high solar activity glaciers recede and climate warms. We propose that changes in the level of solar activity and in climate may have a common cause: slow changes in the solar constant, of about 1% amplitude. (Eddy 1977: 173)

From this we can suspect that droughts would be associated with the warming of the climate.

Immediately recognizable variation in solar energy outputs is usually measured in the incidence of solar flares and sunspots. What seems to be their relationship to drought occurrences? For the United States:

> [A] minimum of area under drought in the rising phase of the Hale cycle [22 year sunspot cycle] and a minimum in the declining phase, come primarily from the drought episodes of the 1930's and 1950's ... [However] there also appears to be an association between drought area and the 11-year sunspot cycles. Drought area for both the western and contiguous United States tends to be near maximum during and in the year immediately after the 11-year sunspot minimum. (Diaz 1983: 12)

There are critics, however, of the relevance of sunspot cycles. 'With as many cycles as have been found in the sunspot data [11 year, 22 year, Bruckner c.33 year, and 80–90 year], it is not difficult to find one that fits, and thus explains, a certain weather occurrence' (Warrick et al. 1975: 12).

Can droughts be caused by meteorite dust in the atmosphere reducing solar radiation? No evidence seems to be forthcoming yet, but that dust is claimed to cause heavy rainfalls which 'fall with surprising regularity thirty days after earth passes through particularly dense meteor showers like the Geminids, Ursids, and Quadrantids ... A fall of meteorite dust would provide ample opportunity for condensation and the 30-day delay is very close to the time it takes for such particles to drift 100 kilometres down to the weather zone'. Would droughts be possible between the dust periods? The author did not think so as 'the air near earth will never become too pure for condensation' (Watson 1985: 60, quoting Bowen 1964), but the absence of heavy rains would provide other opportunities for drought to be stimulated.

Droughts as Part of the Earth's Environment?

The Clash of the Air Masses?

What of the circulation in the earth's atmosphere that seems to lead to meteorological drought? Such a process usually results in air masses dominated by dry air in high pressure systems where the entry of moisture-bearing low pressure systems is blocked and the inherent descending air is warmed, its capacity to absorb moisture increased and moisture on the land surface rapidly evaporated. Similar effects result from air movement over mountainous terrain, where the lee-side of the mountains experience the 'rain shadow' effect with, again, descending air being warmed and able to soak up surface moisture. This is at relatively low levels in the atmosphere; are movements at higher levels relevant?

Changes in the direction of stratospheric winds may increase the possibility of meteorological droughts. These air movements, more than ten miles above the earth's surface, have been noted as changing direction (east to west, then west to east) over the Atlantic Ocean and these changes have been associated with droughts in West Africa and hurricanes in the Caribbean. Thus when these stratospheric winds were blowing east to west the Sahel had droughts in the 1970s, 1980s and 1990s, and when they were blowing west to east more hurricanes were experienced in the Caribbean (Davies 2000: 199). At the level of the ocean, a decrease in sea surface temperature off West Africa increases drought risk, whereas an increase in temperature of 0.5–1.0°C brings rain (Ibid.). This may well be associated with the shift in the Intertropical Convergence Zone (ITCZ) – the low pressure zone of warm almost saturated air masses originating near the equator and moving up to 30 degrees north or south with the seasonal overhead sun. When the movement south is less than normal, drought conditions are usually found in western Africa (Rao et al. 2006).

The Movements of the Oceans?

The possible links between oceanic circulations and drought have been emphasized in at least the last 20 years. Research has shown the linkages between extremes of wet and dry conditions in the tropics and adjacent temperate zones with the El Niño-Southern Oscillation (ENSO) phenomenon of the central Pacific Ocean. An unusual upwelling of cold water in the eastern Pacific is associated with high pressure systems in the eastern and compensating low pressure systems in the western Pacific. This gives rise to a negative Southern Oscillation Index (SOI) or El Niño – the difference between sea level air pressures in the east and the western Pacific – bringing wetter than normal seasons to the western edge and drier than normal seasons to the eastern shores. A reversal to upwelling warmer water reverses the pressure systems and the rainfall regimes, and is recognized as a positive SOI or warm phase – labelled La Niña.

Beginning with the West African droughts and famines of the early 1970s the search for explanations first raised the possibilities of linkages to the ENSO phenomena, and the 1982–1983 droughts affecting Brazil, southern Africa, India and Australia were seen to reinforce the thinking (Glantz et al. 1987). By 1997 it could be claimed that 'differences in SST [sea surface temperature] anomalies between the northern and southern hemispheres, most marked in the Atlantic sector, were related to Sahelian rainfall' (Hulme and Kelly 1997: 223). And by 2003 the impacts were broadened: 'extreme droughts in southern Africa are the result of a number of atmospheric circulation interactions ... with some droughts and periods of rainfall, for example, connected with ENSO events' (Vogel 2003: 839).

Increasingly, evidence is accumulating that this phenomenon has been present for some time – certainly over the last 150 years, with additional links to Indonesia, South America, and possibly China. For Indonesia the records of droughts between 1844 and 1897 and between 1902 and 1998 show that only six of the 43 recorded droughts were not associated with El Niño events (Boer and Subbiah 2005: 333)! This has led to claims that 'it is the next feature that explains a large amount of climatic variability after the seasonal cycle and the monsoon system' (Allan 2000: 4). Similar oscillations of SST and associated El Niño conditions have been noted for the Indian Ocean (Shannon 1994) and the links between the conditions in the Pacific and Atlantic Oceans have been suggested as relevant drought triggers for the American Southwest:

> A comparison of historic climate data and tree-ring based reconstructions of precipitation in the Four Corners region [Arizona, New Mexico, Colorado and Utah] with tree-ring based reconstruction of the Pacific Decadal Oscillation (PDO) and the Atlantic Multidecadal Oscillation (AMO) indicate that severe and persistent drought in the Four Corners region occurs when the PDO is negative and the AMO is positive. (Benson et al. 2007: 340)

Another movement of the oceans which has been suggested as relevant is what is termed the 'Global Conveyor'. This is the cold and salty deep ocean water which originates in the Arctic areas of the north Atlantic and which flows south under the warmer Gulf Stream and ultimately to the south Atlantic and the other oceans. Past global warming periods, with relevance to potential droughts, have been attributed to the rapid increase in volume of the conveyor as a result of the breakup of the continental ice sheets over North America in the seventh century glacial retreat and the draining of the glacially-dammed prehistoric Lake Agassiz (covering part of central Canada) to the south and east at about 12,000 BP. The massive influx of this slightly warmer water may have raised the temperature of the Conveyor sufficiently to affect local climates (Broecker 1995).

The most recent evidence claiming a link between Indian Ocean temperatures and the occurrence of droughts in Australia has suggested that the appearance of a warm Indian Ocean compared to a cool Timor Sea (an oscillation labeled the

Indian Ocean Dipole-IOD) was a major cause of the 'Big Dry' of 1995–2008 in southeastern Australia (Ummenhofer and Gupta 2009). The jury is still out on these links, but the evidence is accumulating.

A perhaps more bizarre way in which the movement of the oceans can create drought comes from the fears of rising sea levels associated with the threat of global warming. Tuvalu is a group of coral atolls in the Pacific Ocean, with a maximum height above sea level of 5 metres but with most of the land less than 2 metres. The corals are porous, absorbing rainfall which sits as a freshwater lens above those corals saturated by sea water. Sea levels have been claimed to be rising on the atolls since the 1990s and, if continued, will threaten potable water supplies and even the future of the principal root crop *pulaka*, which is grown in pits in the coral (Corlett 2009). Kiribati's atoll population of over 97,000 is in a similar predicament and is reported to be negotiating with the New Zealand government on possible assistance and potential evacuation if continued sea level rise exacerbates the 'drought' problem and the islands become uninhabitable (Encyclopaedia Britannica Year Book 2009: 424).

The Volcanic Effects?

Included in the array of natural factors increasing the chance of drought occurrence is the presence of atmospheric pollutants which prevent normal solar energy levels reaching the earth. Such a reduction would reduce air temperatures and photosynthesis and adversely affect plant growth and survival. In addition, it would affect atmospheric circulation patterns and thus the possibility of droughts. That the ejecta from volcanic eruptions might affect solar energy inputs and thus cause changes in climate was suggested as early as 1913, after several major eruptions (Humphrys 1913), and the creation of the so-called 'dust veils' and subsequent associated droughts have been documented from evidence of extreme volcanic eruptions.

That such dust veils could contain enormous amounts of materials blocking the sun's rays has been shown by estimates from several major volcanic events. 'The gigantic eruption of the Tambora volcano on the island of Sumbawa in Indonesia in 1815, of the volcano Krakatoa … in 1883, and Katmai in Alaska in 1912, hurled several cubic kilometers of matter into the air' (Flohn 1969: 207). The dust clouds reached well over 30 kilometres above the earth, remaining in the upper stratosphere for up to 10 years, and resulted in a reduction of overall solar radiation of at least 5 per cent with measurable reductions of air temperatures. Vulcanologists estimate the output of such explosions as a Volcanic Explosivity Index (VEI). The index for Mount Tambora was 7, while Krakatoa was 6, and these have been amongst the biggest eruptions measured (Friedrich 2000: 68).

Deep ice cores from drill sites in the Greenland Ice Cap have provided some of the strongest circumstantial evidence of the links between volcanism and earth temperatures and thus droughts. Reporting on the records for 1873 to 2001, it was suggested that 'The largest negative temperature anomalies in Greenland records are preceded by large volcanic eruptions. Results ... indicate that volcanism explains roughly 15–30 per cent of the variability in global temperatures ... [and] [t]he longest period of warming (1912 to 1963) is also the longest period within the instrumental temperature record that is uninterrupted by volcanism' (Box 2002: 1840 and 1842).

The earliest specific evidence of the links of such events with droughts seems to come from the effects of the volcanic explosion of the island of Thera (now Santorini) in 1640 BC and whose VEI was the same as Krakatoa, namely six. Thera's ash appeared as a sulphuric acid concentration in the 'Dye 3' core of a Greenland ice sample and was dated at 1645 BC (+/- 7 years). Coincidentally droughts in the eastern Mediterranean were claimed to affect the Minoan civilization and the influence of the associated ash cloud was claimed to extend to Anatolia, Palestine and China:

> Several effects could have been caused by the eruption. Dry fogs and dim sun were recorded during the reign of King Chieh [in China]. The weather was colder and more irregular. Crop failures brought famine, and heavy rains caused flooding and destruction in towns. The floods were followed by seven years of severe drought ... [T]he eruption occurred 24 generations before 841 B.C., which would place it during the reign of King Chieh in 1600 plus or minus 30 years B.C. (Friedrich 2000: 81 and 93)

Similar claims for the effects of volcanic eruptions on Chinese history have been made recently and drought has been at the core of the arguments. An eruption in Iceland in 626 AD has been claimed to have resulted in a cooling of the climate in north China for several years, leading to large livestock losses and famine over the period 627–629 AD and the fall of the eastern Turki Empire (Fei et al. 2007). An earlier paper suggested that the eruption of the Icelandic volcano Eldgja in 934 AD led to heatwaves and loss of life, along with starvation, over the following years, and to the collapse of the Later Jin Dynasty in China (Fei and Zhou 2006).

One more recent volcanic explosion to have created widespread drought conditions was that of Mount Tambora in the East Indies in 1815. Described by researchers as causing the 'Year without a summer' (Stommel 1979) it brought crop failure to Europe and northeast North America and may even have affected the outcome of the Battle of Waterloo (Simons 2005). The eruption of Mount St Helens in the USA in 1980 put 2.7 km³ of ash into the atmosphere and produced a similarly cooler-than-average summer in the Northern Hemisphere with associated crop losses. The 1982–1983 drought in Australia and Indonesia was linked in part to the eruption of El Chichon in Mexico which put 16 million tonnes of debris 42 km up into the atmosphere on 4 April 1982 (Watson 1985: 71). In 1991 Mount Pinatubo erupted in the Philippines and the resulting dust veil was claimed to have been responsible for an average reduction in global temperature of 0.5°C (Crutzen 2006), while there were even claims that this was linked to the coincidental doubling of the size of the ozone hole in 1992 (Johnson 1993).

More recent evidence of a super volcano, Mount Toba in Indonesia, which erupted c.75,000 BP, has suggested that the associated climatic distortions may have been even more significant and may 'have completely altered the genetic evolution of human kind' (Savino and Jones 2007, quoted in the *Natural Hazards Observer*, 32(3): 18, 2008). Drought may have been only one of many volcanic effects.

Hydrological Drought

Meteorological drought is the logical precursor to a hydrological drought, when surface waters and reservoirs dry up and rivers cease to flow and eventually even aquifers are affected. However, disruptions to the surface water supplies which thus encourage the onset of a hydrological drought may arise from other causes. Volcanic eruptions and earthquakes can destroy aqueducts, surface canal systems and underground conduits or pipelines, and may also fracture impervious rock structures and leach out the contents of aquifers. In these cases the drought is a water shortage completely separate from any atmospheric condition, a feature of many earthquake prone regions such as the Middle East or southwest Asia where irrigation systems are the basis of societies.

Agricultural Drought

'I'd have a couple of inches' [of rain] he would say, 'in November, and another inch three weeks later. That would give me summer feed. Then I'd have the rest as a good slow, steady fall in June or July when the weather was cool, and I'd be set like a jelly for the rest of the year'. (An Australian stockman on his ideal season, quoted in Ratcliffe 1963: 196)

The agricultural droughts are the ones that bring drought initially to the attention of the media and the public, regardless of where that public is. Photographs of wilting crops and the corpses of livestock littering the desiccated rangelands can be guaranteed to sell newspapers and claim prime time television. But the causes of this type of drought can be very convoluted. Crop yields, as an example, can be affected by an impressive variety of factors, with drought being only one of many. The list is, frankly, depressing:

1. abiotic factors, such as soil water, soil fertility, soil texture, soil taxonomy class, and weather;
2. farm management factors, such as soil tillage, soil depth, planting diversity, sowing date, weeding intensity, manuring rate, crop protection against pests and diseases, harvesting techniques, post harvest loss, and degree of mechanization;

3. land development factors, such as field size, terracing, drainage, and
    irrigation;
4. socio-economic factors, such as the distance to markets, population
    pressure, investments, costs of inputs, prices of output, education levels,
    skills, and infrastructure; and
5. catastrophic factors that include warfare, flooding, earthquakes, hailstorms,
    and frost (Boken 2005: 5).

What is immediately obvious from this list is that human activity is intimately
bound up in the processes noted. It is not only the moisture falling from the clouds
or available at the surface or accessible underground which is the most important
factor, rather it is the ability of the demand for that moisture, as created by human
management of the environment, to be met, which is crucial. Which leads us to
what might be called socio-economic drought.

Socio-Economic Drought and the Human Factor

Mankind itself is an actor in drought stimulation. Certainly reckless deforestation
can create a land surface which sheds rather than absorbs precipitation and an
agricultural drought can result from inadequate soil moisture for plant growth
(Illustration 4.1). In the eighteenth century the islands of Ascension, St Helena and
Mauritius were showing signs of desiccation from excessive land clearance and
cultivation. As a result of these and other examples: 'deforestation and consequent
aridity was one of the great "lessons of history" that every literate person [in the
nineteenth century] knew about' (Williams 2003: 346 and 430). More recent
evidence from western Australia showed that land clearance for new farmlands
(some 1 million acres per year in the 1960s), by removing the trees which had
previously caused air turbulence and increased the chance of precipitation, had led
to reduced rainfall. 'In Western Australia's wheat belt and coastal strip over the
past 30 years [since 1974] ... [t]he average rainfall in that period is 15–20% below
rates prior to the mid-1970s ... the change has been too fast to be explained by
conventional [global warming] climate models' (Luntz 2004: 8). The researchers
thought that half of the decline resulted from land clearance and the rest from
human-induced greenhouse gas emissions. This line of enquiry leads logically into
the problem of desertification which we shall examine later.

But not all human mismanagement of natural resources is derived from
ignorance of natural systems, as there are many examples where human
management is constrained by a rapidly growing human or animal population,
putting pressure upon existing management systems, or where management is
confined within archaic political or economic structures which limit the ability of
the population to absorb the stresses posed by seasonal droughts.

The northeast corner of Brazil, the Nordeste, is an area with a long familiarity
with droughts. Rainfall in the region seems to be linked to the location of the

Illustration 4.1   Aerial View of the Victorian/South Australian Border
The cleared farmland of South Australia contrasts with the uncleared original Mallee
scrublands on the Victorian side. Considerable variations in albedo (solar reflectivity
values) and surface temperatures would be measurable either side of the border.

Intertropical Convergence Zone (where air masses from south and north of the
equator meet) and seem to be negatively related to rainfalls over sub-Saharan
Africa (Rao et al. 2006). But droughts are not only the result of the lack of rainfall.

In 1936 the Brazilian government recognized that the area had droughts every
five years or so and began a series of policies to provide relief. However, the
policies have had limited success and a major barrier seems to be the ownership
of the land. By the 1990s most of the area was still held in large properties, 44 per
cent being held in properties larger than 500 hectares and a further 28 per cent
held in properties of from 100 to 500 hectares. Thus, for most of the population
the effective economy was either subsistence farming on blocks barely enough
to provide a living in good years, or poorly-paid part time work on the large
land holdings. In fact part of the government policies has been to encourage out
migration from the area which has a high rate of population increase.

A recent comment suggested that the benefits from government policies
'cannot hide the fact that overpopulation and economic overexploitation, and a
resultant incorporation of marginal lands into agricultural production, has made
poor people in the semi-arid region more vulnerable to the impacts of future
droughts ... Societal and technological fixes have in the past provided only "band-
aids" for what are clearly more basic underlying social (and political) problems'
(Magalhaes and Magee 1994: 72–3). In other words meteorological drought has

been only one side of the equation. On the other side was inequality in access to the resources of the area which exacerbated the vulnerability to agricultural drought.

Changing human demands upon the environment have a role to play also. The European settlement of Australia in the latter half of the nineteenth century provides just one example of a process going on at the same time in the Americas and North Africa:

> The most pressing problem [for the rural settlers] ... was the unreliability of production caused by the recurrence of droughts. Drought had also affected the pastoral industry before the advent of closer settlement [government policies to subdivide the large pastoral properties into smaller sizes to encourage a denser occupation of the land], but the light stocking techniques of the graziers and the mobility of the sheep flocks were some insurance against disaster. Once cropping and intensive livestock farming became the dominant forms of agriculture, the effect of drought was felt more acutely. The small farmer seldom had any reserves of capital and one or two years of drought were sufficient to drive him from the land. (Davidson 1969: 49)

Basically it was a combination of settlement invading areas where rainfall was not only lower but also more variable, together with new resource management needing higher and more regular rainfalls to survive. As one commentator put it more crudely: 'It might be said then, that drought came to Australia only with the arrival of the European settlement, since it is the European who has demanded more of the land and the climate than they are willing in some years to give. Given its inflexible structures, it may be that our civilization in its present form is incapable of coming to terms with a problem such as drought' (Coughlan 1985: 148).

On the Great Plains of the United States a similar situation had developed:

> As the limits of productive agricultural land have been reached, more marginally arable lands have been put into agricultural production in times of favourable climatic conditions and through the use of irrigation. This practice has resulted in an increasing vulnerability to drought in many areas of the Great Plains. (Woodhouse and Overpeck 1998: 2710)

The pressure to expand irrigation from deep wells, in particular, had led to the 'mining' of the Ogallala Aquifer, a massive fresh water aquifer underlying the Great Plains. The resultant draw-down of the water table by 30 metres locally had led to abandonment of local irrigation schemes.

Even traditional resource use systems have received criticism. Traditional nomads in the semi-arid lands have been accused of courting droughts. Nomadic pastoralism, so the argument goes, 'is inherently self-destructive [because of the system] ... of keeping as many animals as possible alive, without regard to the long-term conservation of land resources' (Allan 1965: 321). One such attitude was illustrated after the Sahel drought of 1968–1972, when the reaction of a Fulani herder in the spring of 1973 was: 'I had 100 head of cattle and lost 50 in the drought. Next time I will have 200' (Glantz 1976). The camel itself has even been labelled an environmental disaster 'by enabling man to lead a nomadic existence … [it] has made it possible for him to exploit the vegetation over ever vaster and more varied stretches of land; so the camel is responsible for the creation of the desert' (Charles Sauvage, botanist, quoted in Monteil 1959: 573).

But don't put all the blame on the animals. A report into the 2002 drought in Sicily stated that the island's water scarcity was largely determined by the 'many unfinished hydraulic infrastructures dotting the Sicilian landscape', the poor state of repair of the functioning structures, the high level of leakage, water theft, and the 'excessive number of different public bodies responsible for water distribution, whose responsibilities regularly overlapped or were unclear', and whose staff often had little technical competence (Giglioli and Swyngedouw 2008: 401–2). Also implicated in the drought was the Mafia, whose 'economic aim was not the revenue deriving from the completion of works to commoditize and deliver water, but the possibility of achieving immediate gain in the process of their construction … a situation which led to many infrastructural works never actually finishing' (Ibid.: 399).

Some also put the blame upon governments and talk about a 'government drought'. This was the response in the Mallee region of South Australia in 1970–1971, when the Commonwealth Government, because of a glut in the market, imposed wheat production quotas for the current year based upon the average of the previous five years. Unfortunately for the farmers, two of those five years had been severe droughts. These pulled down the admissible quota for 1970–1971 and led to grumbles about an effective 'government drought' constraint on production and farm incomes (Heathcote 1980, Whitwell and Sydenham 1991).

In some cases grumbling citizens take the law into their own hands, particularly when there are conflicting uses for water. In Klamath Falls, Oregon, USA, by July 2001 some 1,400 farmers in the Klamath River Basin had been cut off from the normally available Federal Bureau of Reclamation irrigation water since April 'and watched their land dry up because a federal court had said the water must be preserved for the [endangered] suckerfish, protected under the controversial ESA [Environmental Sustainability Act]'. By July 'feelings were running so high … that even the local sheriff … decided not to intervene as the protesters opened the head gates from the Upper Klamath Lake, with a chainsaw', and the waters flowed again (McCarthy 2001: 10).

Just how much the impacts of drought can be blamed upon nature and how much can be blamed upon humans is debated still. 'The drought literature is, in fact, seriously split between those emphasizing natural, physical forms of reduced water input or availability in relation to crop and other losses, and those looking to food security or humanly magnified desertification hazards for answers' (Hewitt 1997: 80). There is no doubt however that the increasing global human population will continue to increase its demands for water and without an increase in nature's supply, future drought seems inevitable. And the impacts of those droughts will probably closely follow the historical pattern, as we will suggest in the next chapter.

Chapter 5
The Scope of Drought Impacts

Drought impacts are non-structural and spread over a larger geographical area than damages that result from other natural hazards.

(Wilhite 2000a: 6)

While drought is usually associated with distinctive patterns of weather and may be associated with distinctive climatic regions of the world, its impacts reach beyond the short-term time periods of the weather and the spatial boundaries of current climatic types. Today's droughts have impacts upon the global environment and human societies, while past droughts have played a role in shaping that environment and influencing previous human societies.

Droughts have had and still have a major influence, firstly upon the accelerated weathering of the earth's surface, secondly upon the periodic extensions of the boundaries of global arid zones, and thirdly upon the sustainability of patterns of plants and animals. Each of these might be examined as evidence of the significance of drought upon all life forms. However, when drought occurs at the interface between human activities and the physical environments, then society begins to take notice.

Because of this link a change in the characteristics of either could affect the pattern of drought occurrence. Yet, to be recognized, drought impacts usually have to affect society either directly or indirectly – the death of plants or wildlife in remote locations does not usually make the headlines, unless someone claims to own them or to be interested in their welfare.

The Changing Scales of Drought Impacts

To assess the nature of drought impacts, we need to know something about the scale of the climatic component and the size of the area affected. It has been suggested that significant droughts could last from months up to years, and affect from relatively small areas (less than 100 sq.km) up to virtually the whole area of a nation (more than 1 million sq.km).

> Average-size (say $10^5 km^2$) droughts, for example, pose only fine-grain environmental fluctuations for the world as a whole, for continental masses, or for the largest nations. Global ecological or economic patterns and even truly national properties (e.g. gross domestic economic production) of large nations will be relatively little affected by such droughts. Conversely, however, individual vegetational associations and farms, all but the largest drainage

basins, and small nations experience droughts as coarse-grain, all-concerning environmental fluctuations. Important ecological and social properties at these scales may be greatly perturbed by the occurrence of a drought. (Clark 1985: 20)

As a rough estimate he suggested that short duration droughts lasting less than one year would have ecological effects – on crops, livestock and vegetation, but for a limited area. Droughts lasting more than one year would have both ecological and socio-economic effects, possibly including population movements/migrations associated with sharp economic downturns. The impacts of droughts lasting say two years or more would depend upon the proportion of the national area affected, and more pertinently the proportion of the nation's economic productive area affected. In effect, even for long droughts, the larger nations would appear to be less vulnerable than their smaller brethren.

A Hierarchy of Impacts?

Any attempt to assess the significance of drought impacts needs therefore to differentiate between three broad areas of those impacts:

1. Impacts upon the physical environment and ecosystems.
2. Impacts upon human life-support systems.
3. Impacts upon human society itself.

Such a broad division needs to be elaborated to demonstrate the complexity of drought influences (Table 5.1).

Table 5.1   The Scope of Drought Impacts

Major area of drought impact	Subsidiary areas of impacts
Physical environment/Ecosystems	Ecosystems; wildfires; desertification
Human life-support systems	Economic impacts
Human society itself	Political impacts; famine; disease; scientific thought

Source: Much simplified from Table 1 in Wilhite 2003: 751–2.

A hierarchy of drought impacts in general might be identified. Primary impacts might be seen as the reduction of accessible moisture from reduced precipitation or hydrological storages; secondary impacts might be seen as the resultant reduction in soil moisture levels affecting plant growth and livestock wellbeing leading to crop failures, livestock deaths, accelerated soil erosion and ecosystem modification; while tertiary impacts might be identified as the changes in ecological conditions and human activities which result from the experience of the secondary impacts (Parry 1990).

Seen from another viewpoint, the general impacts of drought might be classified according to the phenomena affected. Thus environmental effects might include damage to wildlife habitats, deterioration in water quality (as increased salinity), increased levels of air pollution from drought-accelerated soil erosion, increased frequency of devastating wildfires, and finally the devastation of all life forms through the process of desertification.

The economic effects of drought would include loss of livelihood from crop failures, livestock deaths, cessation of river navigation, property damage from wildfires, reduced hydro-electricity generation, loss of tax revenues to government, and the opportunity costs of drought relief aid diverted from other public funding activities. Should the economic impacts prove to be irreversible, land abandonment and out-migration of the population might result from or parallel local famines. Some sectors of society, however, might benefit from drought. Examples are transport firms moving relief supplies, drought-free areas selling emergency food or supplies to victims, and even the affected areas might receive some long-term benefit from public spending on infrastructure improvements such as roads, which might not otherwise have been provided.

The social impacts of drought might include disruptions resulting from reduced family incomes, and increased social tensions as livelihoods become threatened; traditional support systems are stretched and eventually collapse, anti-social behaviour such as brigandage and prostitution increases, and in the absence of effective relief, famine might result.

Thus, there are obviously a variety of possible impacts of droughts and I wish only to indicate the scope of those impacts by focusing upon six themes:

1. droughts and the ecosystems;
2. droughts and wildfires;
3. the role of droughts in desertification;
4. droughts and famine;
5. droughts and diseases;
6. droughts and their socio-economic and political impacts.

Drought and the Ecosystems

One of the most influential classifications of the earth's climate and associated vegetation zones was by the German geographer W. Köppen (*Die Klimat der Erde*, 1931). He specifically identified a group of global plants as Xerophitic or drought resistant and suggested that these occupied a significant proportion (some 26.3 per cent) of the global land area. In fact, when the United Nations considered the future development of the world's arid lands as part of the post World War II rehabilitation program, their map of the world's arid lands (UN, *World*

Distribution of Arid Regions, 1977) owed much to his prior work and revised the area of extremely arid, arid and semi-arid lands up to 32.8 per cent (Heathcote 1983: 16). By implication, approximately one third of the global land consists of ecosystems which are claimed to be at risk of drought stresses. What are those stresses, and how do the various life forms cope, and what is the relevance of their coping strategies to human management of these drought-prone environments?

Coping with the Droughts: A Normal and Long-standing Problem

Coping with water stress at some time in their lives is a normal part of life for most plants and need not always be fatal (Pitman 1986). Indeed the occurrence of droughts as part of the process of the drying out of global ecosystems in past geological time has been seen as an integral component in the evolution of humanity: 'Were it not for a trend towards cooling and aridification during Tertiary times, modern humans might not exist' (Huggett 1997: 6). The argument was that the aridification in Africa from about 25 million years ago led to the retreat of the woodlands (C4 plants) and their replacement by grasslands (C3 plants), which encouraged the spread of humans and an associated complex of plants and animals. The process was less developmental (i.e. through a steady sequence of environmental states) than evolutionary (i.e. through a series of rapidly changing rather than constant environmental conditions, which would include the challenges of extensive droughts).

Plants and animals cope with drought in a variety of ways, some behavioural, some morphological, some physiological, and often by a combination of all three. They maximize the search for moisture from either the atmosphere or soil, minimize their use and loss of moisture, tolerate high air temperatures, and have carefully controlled reproductive cycles.

Back in 1927 these plant adaptations to drought stress were seen as falling into four categories (Shantz 1927). First were the *drought escapers*, those plants growing only in non-drought conditions and escaping drought by lying dormant as seeds *in situ* in the soil. These would include the ephemerals and some of the grasses and herbs which had rapid reproductive cycles (six weeks for the Californian Grama grass *Bouteloua aristidoides*). A second group were the *drought evaders*, the perennial trees and shrubs which had access through their deep root systems to permanent ground water and were thus independent of surface moisture conditions (the mesquite (*Prosopis* spp.) has root systems which reach down to 80m). Tolerance of saline water, exudation of toxic substances to 'poison' the surface underneath their canopies to eliminate any competing plants (e.g. Creosote bush, *Larrea* spp.), and tolerance of considerable 'dwarfing' during water stress periods were also characteristic of this group. Next were the *drought resisters* – the succulents and cacti which stored water in the root and stem tissues from periods of surplus for the periods of deficiency. Some plants such as the 'rootless' cacti *Tillandiia* seem to survive even on dewfall and coastal fog-drip in Peru. Finally, the *drought endurers* represented those perennial shrubs which might not be tapping deep ground water but which, by their very efficient control of moisture loss by transpiration, through leaf-shedding, protective coverings to leaf surfaces and ability to orient leaves away from direct sunlight, together with delayed seed germination by thick protective seed covers (saltbushes *Atriplex spp.* and some *Acacias*) and capacity to remain dormant for long periods of drought, were able to revive when the drought broke (Heathcote 1983: 80–81) (Illustration 5.1).

However, as one commentator has suggested, developing characteristics which ensured survival through droughts was 'at the expense of those [characteristics] that favour high productivity'; you could not have both in the business of survival (Cocks 1992: 51).

A similar classification could be applied to animal and insect strategies to cope with drought stresses. Thus, there were the *drought escapers* – fleeing the affected area when drought came, as their progeny was safely stored as eggs or juveniles buried in the soil, or fleeing and only returning when the drought broke to reproduce quickly several times to build up population numbers. Examples would be the zebra finch's (*Taenopygia castanotis*) multiple clutches and the female kangaroo's ability in times of abundant forage to 'carry an embryo, a

Illustration 5.1   Weltwitschia Photographed c.32km South of Swakopmund, Namibia

The leathery foliage and 'lifeless' appearance of this desert survivor are clear enough.

small joey in the pouch and a suckling juvenile at foot' (Ludwig et al. 1997: 42). Secondly, the *drought evaders* survive by retreating to moister micro-climates in shady locations or burrows in the day and having mainly nocturnal life styles, as do many insect species such as the Algerian woodlouse (*Hemilepistus*), scorpions (*Leirus quinquestriatus*) and camel spiders (*Galeodes granti*). Thirdly were the *drought resisters* whose ability to survive high daily temperatures and maintain body water by limiting sweating and panting, and concentrating urine and faeces (kangaroo rats, jerboas *Jaculus* genera), was combined with the ability to rapidly take up any available moisture, even through the skin as with the Australian lizard *Moloch horridus*. Finally, there are the *drought endurers* – surviving by aestivating or in a dormant state in burrows or in the mud of dried up lakes. Examples would be gophers (*Citellus* spp.), lung-fish or mud-hoppers (*Neoceralodus* spp.) and some frogs (e.g. the Australian desert frog *Cyclorana alboguttatus*). In the case of snails there are many instances of periods of dormancy for many years, sealed safely in their shells (some four years in the British Museum for a famous *Eremina desertorum* from the Sudan). In the case of both plants and animals, the onset of a severe drought might see the apparent devastation and removal of all life, but the environment nonetheless would contain life in various forms, while beyond the drought's perimeter other less tolerant life forms await the drought-breaking rains to reinvade and recolonize.

For some, even indirect impacts can be disastrous, as a study of toad populations in the Cascade Mountains, USA showed. During increasingly frequent El Niño [drought] years, less precipitation falls in the Cascade Mountains, so the toads lay their eggs in shallower lakes and pools. The shallow water column above the egg clutches does not screen ultraviolet rays as well as deeper water, and the UV-B wavelength makes embryos more susceptible to infection, causing more than 50 per cent mortality during these seasons (Kiesecker et al. 2004: 141).

Drought might pose a threat to the existence of a species through human actions. Eggs of the Chinook salmon are laid in the cold headwaters of the Sierra Nevada in California before the progeny hatch to swim downstream to the Pacific Ocean to mature. Normally, mature fish at about 3–4 years old and weighing 20–50 pounds move back up the rivers to the traditional spawning grounds. In 1988, in the second year of a drought and consequent low water, three-quarters of all salmon moving upstream, some 1.56 million fish, were caught (USACE 1993). If such catches had been maintained, the future of the species would have been at risk.

Historical Role of Past Droughts

The ability of life forms to survive droughts or at least leave progeny for the return of non-drought conditions may be significant in the short term, but what if the drought continues for several years and a new more arid climate appears? How significant does drought seem to have been in the past, the remote past before human appearance on earth? Recent scientific research has suggested that the history of ecosystems and their evolution on earth appears to have been marked by 'inconstancy, not constancy, of environmental conditions … [thus] environmental stability is fleeting, environmental change is perduring' (Huggett 1997: 315–16). How important then has been drought, as a factor in environmental change, in these mutable ecosystems? One commentator suggested that 'Drought will lead to major mortality of individuals of many species in the driest parts of their ranges. However, it rarely seems to directly cause major changes in the distribution of species, largely because any such adjustments would have taken place in the past. (Kirkpatrick 1994: 93).

The periodic occurrence of drought then may cause temporary changes, but extensive droughts over many years in the past may have led to the creation of arid climates and significant modifications of those environments. The transition from no-drought to drought conditions may take months, but rainfall patterns can show step-like trends, that is rapid change from one condition to another, and this could prove critical for some species.

Ironically, one of the governmental drought relief measures for graziers with starving livestock in drought in Queensland (Australia) in 1899–1903 was to distribute chopped-up Prickly Pear cactus *(Opuntia spp.)* as emergency cattle and sheep feed. When the drought broke, the uneaten dried pads sprouted and grew again and the relief of one menace led to the spread of the weed which was not controlled until the release of cochineal insects (which fed on the cacti) in the 1920s (Nix 1994).

The Australian Experience of the Role of Drought in the Ecosystem

As an example of the impacts of long past droughts and associated aridity, sequences of arid conditions in Australia seem to have occurred within the last 1.2 million years, with major dune formations developing in extreme aridity c.700,000 years ago and, after more humid intervening periods, again in moderate aridity c.18,000 years ago. But the stress of past aridity and the continuing appearances of droughts have made permanent marks upon the present Australian ecosystems:

> This selective pressure led to the evolution in animals of a variety of behavioural and reproductive adaptations including long lives (allows longer time to reproduce), ensociality (individuals cooperate in caring for the young), polyoestry (ovulating more than once each year enables rapid reproduction following losses and r-strategy reproduction [many offspring]), parthogenesis (in some lizards; enables reproduction without males when they are hard to find), aestivation (more or less similar to hibernation but occurs during the dry summer months), and omnivory (opportunistic feeding). Comparable adaptations have occurred in Australian dry country plants. (Archer et al. 1998: 18)

Commenting on the history of droughts in Australia one observer noted that the impacts depended upon whether 'redundant' or 'keystone' species were affected

– the keystone species being those 'which, if removed, induce the loss of other species'. He went on to suggest that over most of Australia 'ecosystem activity is driven in stop-go fashion by irregular rainfall events ... the spatial extensiveness of this ... [being] unusual ... Not unexpectedly [as a result], plants have developed characteristics which ensure survival through drought at the expense of those that favour high productivity' (Cocks 1992: 50–51).

Survival among Australian bird species depends upon their inherent nomadic seasonal movements, with normal breeding in spring, but in the arid interior suitable conditions depend upon the irregular rainfall and 'breeding follows rain irrespective of time [of] year' (Keast 1959: 89). Droughts stop breeding and reduce the number of eggs laid or young reared. The break of the drought triggers a rapid buildup of populations with multiple broods and larger clutch sizes. As an example, even a one year drought in the interior in 1957 caused 'almost complete exodus [of bird species] ... [F]or some species practically no residue remained inland ... Almost the whole teal population in Australia perished and the species was rare for five years until the inland plains were flooded again in 1962/63' (Frith 1971: 1–2). This high incidence of bird nomadism has been reaffirmed recently with the ability to search out opportune occurrence of resources seen as a major benefit and aid to survival (Flannery 1994: 90–91).

Any survival in times of drought in the interior depended upon the existence of micro-climates which functioned as drought refuges, especially for smaller mammals (0.1–5.5 kg), and competition from the imported rabbits and domestic cattle, sheep and feral camels put enormous pressures upon such refuges. In effect, at such times each successive drought saw fewer refuges containing native species left and the native species as well as the invaders suffered (Morton and Price 1994: 155–6).

Droughts and the Disappearance of Australian Megafauna?

Have any species been exterminated by past droughts? A considerable debate exists on the role of drought in the disappearance of the global megafauna, i.e. the large prehistoric versions of the larger land mammals. That the megafauna disappeared, and rather rapidly at various locations around the world, seems not to be disputed; the debate rages about why they disappeared.

In Australia, megafauna such as giant lizards, snakes, birds, kangaroos and wombats, mostly weighing in at over 45 kilograms, existed from at least 500,000 BP. By c.46,000 BP 90 per cent of the larger terrestrial animals had vanished, and drought and gathering aridity have been touted as a major factor. Whether drought destroyed their habitat and/or food supplies; whether drought drew them to shrinking surface waters where they could be relatively easily despatched by human hunters; or whether their habitat was destroyed by climate change from an asteroid impact, the debates are so far inconclusive (Williams 1993, Archer 1998, Luntz 2007, Cupper 2007). A recent commentator has discounted climate and drought altogether, the demise claimed to have taken place 'during a time of reasonably comfortable climate and well before the onset of aridity in the lead up to the most recent glacial expansion' (Wells 2007).

While the North American megafauna (except for the bison and the bears) are claimed to have been wiped out by human predation c.11,000 BP, a similar argument but with a different date line has been put forward for Australia. The last megafauna died out between 46,000 and 50,000 BP and there are claims that for 3,910 years there was an overlap, with human occupation of the continent thought to begin c.45,000–50,000 BP. 'Such a short overlap period between humans and megafauna, at a time when climate wasn't dramatically severe, indicates that human interaction was the most logical explanation for the megafauna extinctions' (Long 2007: 19). However, as yet no massive kill sites such as were found in North America, nor skeletons with embedded spear points, have been produced, so the climate protagonists remain unconvinced. Research goes on and drought may yet be proved to have had a role, however minor, in the disappearance of the Australia megafauna.

Ecological Benefits from Drought?

Droughts, however, have had their advantages. G.C. Watson, a government surveyor who prospected a railway route into the interior from southwestern Queensland in 1882, summed up his experience thus: 'What by some people and those who do not study the physical laws of Australia may be considered a drought is but provision of nature for the improvement and increased fertility of the soil by pulverization from exposure as well as suspension of vegetation'. In 1893, James Tyson, a millionaire pastoralist with properties in southwestern Queensland, made a rather pointed comparison between the merits of Australian and English soils, which echoed Watson's comments: 'England is little more than a manure heap whose soil is poisoned by the animal droppings of centuries, and whose vegetation is at least less wholesome for stock than the health giving native grasses of inland Australia, whose pastures are constantly purified by a tropical sun and occasionally renovated by bush fires' (Heathcote 1969: 187–8).

A similar argument appeared in 1904, when the Commercial Agent for New South Wales told the Royal Colonial Institute [London] that 'There is this to be said in favour of drought – lessons may be learnt – it rests the ground' (quoted in Chatterton and Chatterton 1996: 37). In effect, drought decimated the feeding opportunities for wild or domesticated livestock and for herbivores, and the results could be a disastrous collapse of the population. But that collapse could act as 'an evolutionary means by which [surviving] individual plants escape grazing long enough to recover, build up root reserves, grow and produce seeds' (Ludwig et al. 1997: 44). It also meant that the rabbit population (which had been successfully competing with native herbivores as well as domestic cattle and sheep since the introduction of the first rabbits with the first European settlers in 1788) had been substantially reduced and the annual costs of their control were saved for several years.

The Commercial Agent was putting a positive gloss on the end of a major drought affecting eastern Australia from c.1895 to 1902, which virtually bankrupted Tyson, but there had been earlier droughts which had had similar, but even more convoluted benefits. A long drought in the 1870s 'beggared everything: rabbits, kangaroos, native cats, rat-kangaroos, farmers, graziers, the soil itself', but when the drought broke, the scrub vegetation regrowth, although of little economic value to the farmers and graziers, was 'cast like a protective net over plants, birds, mammals, insects that would have disappeared for ever in the disasters of the next seventy years' (Rolls 1994: 30). In this case the drought encouraged economically useless vegetation, i.e. relatively inedible forage, which slowed the continuing onslaught of domestic livestock on the sustainability of the ecosystem.

A more recent example of the complexity of assessing drought's impacts comes from Western Australia, where declining rainfalls over the southwest of the state since the 1970s have raised fears for future human water supplies and stimulated construction of a desalination plant for the state capital, Perth. In contrast, however, the decline of between 10 and 15 per cent in annual rainfalls does not seem to have significantly reduced grain production and may well have slowed the pace of a previous problem, namely increasing salinization of farmlands, by slowing the movement of saline ground water through the subsoil (Ludwig et al. 2009).

Out of Africa?

For Africa extensive droughts in geological time, related to the creation of arid climates, are certainly claimed to have been ecologically significant.

> The trend towards drier climates (aridification) over the last 25 million years led to several adaptive radiations [of life forms] … The radiations displayed a relay effect … started at the food web base with the diversification of grasses, herbs and weeds diversified during the spread of grasses. Many songbirds and Old World rats and mice, which fed on the seeds of the expanding plant groups, then underwent a radiation. Lastly, modern snakes … expanded as the rats and mice, and eggs and chicks of songbirds, appeared on the menu. (Huggett 1997: 284–5)

The relics of past droughts may still be evident in contemporary environments. For West Africa one study found that:

> The scale of the temporal dimension to West Africa's forest-climate relationship may be properly measurable not in thousands but in hundreds of years. It would also not be inconceivable for today's vegetation to be responding simultaneously to recovery from drier conditions at each of these   time-scales – i.e. from protracted and deep aridity around 12,000, from a short deep aridity 3,500 and from a relatively arid phase 700–200 years ago. Whether expressed through effects on soil, soil fauna and flora, or vegetation distribution, lag effects from

each dry phase might remain relevant, interfering with present responses to more recent climatic variation. (Fairhead and Leach 1996: 188)

In this case it would seem that the ecological impact of past droughts may be long lasting.

Drought and Wildfires

> After so long a drought even the stones were ready to burst into flame. (Samuel Pepys in his diary, commenting on the precursor to the Great Fire of London 1666. Quoted in Fagan 2000: 129)

Apart from the obvious impact of drought on plant survival, a more indirect impact of drought upon vegetation is through the encouragement drought gives to the outbreak of wildfires. The role of wildfires in the evolution of global environments seems to be integral and drought creates the fuel by which the fires can transform the environment. For example, most of the major devastating wildfires, bushfires as they are locally termed, which have affected Australia have been either during or towards the end of major droughts, when the natural vegetation has been desiccated and formed the fuel to feed any ignition, whether from natural or human causes (Pyne 1991).

The creation of the fuel load, as the professional firefighters term it, is the combined result of the desiccation of the surface vegetation, the accumulation of that dead plant material and the removal of natural barriers (surface waters) to fire spread. When the high temperatures and low relative humidity of most drought events are joined by high winds, only an ignition point is needed to create a potential environmental disaster. Indeed, in most fire-fighting handbooks, drought conditions are built into the indices of fire risk. In the USA:

> The fuel moisture level is factored into an index that includes weather conditions, wind speeds, fuel loads (the total oven-dry weight, per acre, of all the fuels in the area), and drought conditions. The information is processed by the National Fire Rating System, which determines the fire risk for every climatic region in the country. (Junger 2001: 22–3)

In Australia a researcher suggested in 1947 that if rainfall had been only 30 per cent or less for three months then there was a serious fire risk, and by 1978 experts admitted that the hypothesis 'has stood the test of time well' (Luke and McArthur 1978: 66).

As an experienced American fire-fighter acknowledged, the professionals can see a fire threat whenever drought occurs. In July 1989 the American west 'was well into one of the worst droughts of the century, and I was out there to see the wildfires that it was sure to produce' (Junger 2001: 4–5). In 1988 in the western

USA the Palmer Drought Index already stood at more than −4.0 showing that the entire west was experiencing the worst drought since records began in the 1870s:

> Barges were running aground on the Mississippi River; railroad tracks were warping in New Jersey. The drought culminated in the fires in Yellowstone National Park. From late June until early November fires burned a total of 1.2 million acres across the west, more than half of which was in Yellowstone itself. (Junger 2001: 23)

It was to prove one of the worst drought induced wildfires in the history of western USA.

Essentially the longer the drought continues, the greater the risk of a major conflagration. In the Amazon Basin the frequency of major fires in the rainforest has been linked to ENSO events, with fire frequencies now 'intra-decadal with measurable ENSO events in 1972–1973, 1977–1978, 1982–1983, 1986–1987, 1991–1991, 1997–1998, and 2002–2003'. Such has been the increased frequency associated with these droughts that the researchers claimed that these unintended fires may have been destroying more of the rainforest than the previously assumed culprit – intentional burning to create grazing lands (Moran et al. 2006: 344). In 1997–1998 over 5 million hectares were burned in Rondonda State alone, mainly to create pasture lands from the rainforest. In Indonesia similar ENSO events have been associated with extensive forest fires, although here the human input has been more obvious. In the swidden [shifting cultivation] agricultural systems on the edge of the Indonesian rainforests, 'fire … is a natural ally of the farmer' and the 1982–1983 ENSO and associated drought event 'was seen as a unique opportunity to engage in a drastic land clearing campaign' to prepare future agricultural lands (Malingreau 1984). In more recent years fires have prepared lands for the profitable oil palm plantations (Stigter et al. 2005).

Given the preconditions for fires, where are they most likely to occur on a global scale? The most hazardous regions seem to be those possessing either a Mediterranean or a continental climate. In the former, most rain falls in the winter and forests and grasslands become desiccated during the annual summer drought. In the latter, the air is likely to be dry throughout most of the year and the fire season may persist for a long period. Hence we have the regular wildfires of southern France and Spain, California and the southeast of Australia, together with the widespread conflagrations which periodically affect western USA and central Australia, the latter showing a high correlation between annual wildfire occurrences and rainfalls in the previous three years (Griffin and Friedel 1985: 66).

Ignition can come from a variety of causes. Any tourist flying over central Australia cannot help but be amazed at the criss-crossed fire scars of lightning strike induced fires covering hundreds of hectares. On 19 March 1999 'more than 7000 lightning strikes were recorded within one hour on this day' (Watson 1999: 105). Rural fire ignition certainly can include lightning strikes from summer thunderstorms, but can also be caused by ill-considered human burn-offs of dead vegetation and the exhaust from farm machinery, as well of course by arsonists. In central Australia, however, humans are few and far between, so nature must carry the responsibility here.

Apart from the increased risk of wildfires, what other impacts have droughts on the ecosystems? The broader impacts are complex:

> There can be stem death and resprouting of plants; there can be whole plant death and a flush of germinants; there can be increased predation of animals; and, preferential grazing on a 'green pick'. There will be less fuel, shorter plant heights; fewer vertebrate animals and insects. These consequences may mean that there are depletions in seed stores in the soil or on the plant … [and] there may be species with no reproductive backing if a cohort is eliminated. (Gill 1999: 53)

Historically there have been suggestions that fire, mainly drought induced, has encouraged the take-over of grassland from previous woodlands and forests. If droughts and associated fires occur often enough, it is argued, forest might give way to grasslands or cropland. In fact, humans are now using this fire tool in Brazil to reduce the Amazonian rainforest to pasture for cattle, but did it happen 'naturally' in the past? Pyne's study of the role of wildfires in Australian history claimed that the border between the tropical savanna (grassland with scattered woodland) and pure rainforest could be shifted when grass-fueled fires, combined with drought, might break through the microclimatic barrier that separates grassland from forest and raze even rainforest into prairie (Pyne 1991: 61). A similar argument has been applied to the origin of the prairies of North America, although the deliberate use of fire by humans has been claimed to be a stronger factor (Malin 1956).

The continuation of any vegetation change seems to depend upon the frequency of drought and fires. Currently for northern Australia the savanna experiences fires every one to two years, whereas guesstimates for the rainforest suggest fires possibly every 300 years. For the mountain ash forests of southeast Australia the fire interval for tree kills seems to be about every 100 years, with the under-storey devastated every 50 years or so (Gill 1999).

Although hard to believe, droughts can adversely affect tropical rainforests even without human help. Research into the effects of the 1997–1998 El Niño associated drought over Sarawak, Malaysia, showed that tree mortality overall in the pristine rainforest was four to six times the normal non-drought attrition rate and for the most common tree species, *Dipterocarpaceae*, could be twelve to thirty times the normal rate (Nakagawa et al. 2000). Similar results were found in the Lambar Hills National Park, Borneo, where the same drought produced tree mortality rates three times the pre-drought period, with the common species suffering kills over six times the normal rate (Potts 2003). Thus not only overall tree kill but also species composition was affected by the drought, illustrating that not even one of the earth's most humid and high rainfall environments is free from drought threat.

Apart from the impacts on vegetation and loss of human property in drought induced wildfires, drought leads to accelerated soil erosion as witness the 'Dust Bowl' of USA in the 1930s. Wildfires can even cause the break-up of granite rock surfaces through exfoliation as surface temperatures reach 500°C (*Australasian Science* 2003, 24(5): 10). On 6 February 1983 a series of bushfires broke out in the drought-affected countryside of South Australia and Victoria. Half a million hectares of country burned, 71 people died along with over 300,000 domestic animals and property damage amounted to approximately $A400 million. The link with drought was clear enough. Less clear was the link between drought and soil erosion, with up to 80 tonnes of soil lost per hectare from some of the, now bare, burned areas in the drought-breaking rains of March 1983 (Allan and Heathcote 1987). On 7 February 2009, in drought conditions, similarly devastating bushfires erupted in Victoria, killing 173 people, destroying 2000 homes, burning 430,000 hectares of woods and farmland, and estimated to have caused the deaths of over a million native animals. In this case, there were no drought-breaking rains, but the effects on soils must have been considerable.

Droughts also stimulate air pollution as bared soil is eroded and the particulates encircle the globe: 'In another look at forest fires and El Niño, scientists using NASA Satellite data have found that the most intense global pollution from fires occurred during droughts exacerbated by El Niño. The most intense fires took place in 1997–1998 in association with the strong El Niño. (*ENSO Signal and Network Newsletter*, 2003, 18(2): 2).

Wildfire risks from drought threats even affect tourism. On 4 May 2007 *The Guardian* newspaper in UK headlined 'Fire risk restricts Peak District access', as threat of fire from a 'mini-drought' which had already been associated with over 1,500 heath and grass fires was deemed sufficient to advise visitors to keep to marked paths in the national park (*The Guardian*, 4 May 2007: 11). Such was but a slight example by comparison with the major disruption caused by the fires in the Yellowstone National Park in the USA in 1988, when c.45 per cent of the park's area was burned and of that c.8000ha virtually destroyed (*Encyclopaedia Britannia Year Book 1989*: 199).

Drought and Desertification

The (Apparent) Paradoxes of Desertification

Desertification is both absolute and relative.
Desertification is both natural and social.
Desertification is both an ecological and a moral problem.
Desertification is Inevitable but Unpredictable.
Desertification is Everyone's business and No One's (Spooner 1987)

The link between drought and desertification is, at first glance, obvious. Assuming a definition of desertification as the creation of desert-like conditions, the recognized role of drought in removing environmental moisture and thus creating the aridification of the environment would suggest that drought is a major player in the process of desertification. As usual, however, in all matters relating to the state of the global environment, apparently obvious relationships have to be qualified by closer inspection of the complex realities.

To establish the possible role of drought in desertification we need to outline first the history of the concern over desertification with the nuances of its definition and documentation, and then confront the discourse about its causes. In the process we will find all of Spooner's paradoxes.

The History of Concerns for Desertification

Desertification, commonly understood to mean the spread of desert-like conditions, came to international attention with the environmental devastation and loss of human and animal life accompanying extensive droughts in West Africa from 1968–1973, which were captured on televised documentaries. This media coverage may have played a part in subsequent extensive United Nations sponsored aid programs and associated scientific investigations into what appeared to be evidence of long-term environmental deterioration, culminating in the United Nations Conference on Desertification in 1977 (UNCOD).

In fact, concern for evidence of adverse environmental change can be found in the accounts of European explorers in the deserts of Africa, Arabia and Asia from the late eighteenth century onwards. A Danish Expedition to Arabia of 1761–1767 (Hansen 1964), and the activities of explorers such as Charles Doughty (1843–1926), Sven Hedin (1865–1952), Baron von Richthofen (1833–1905) and Ellsworth Huntington (1876–1947) commented upon the evidence of past civilizations amid the desert sands. Huntington's 1907 *The Pulse of Asia* drew upon his own and earlier explorations to suggest that the evidence of climatic variation and associated environmental deterioration and reduced capacity to support the population in central Asia might have caused the folk migrations which led to the Mongol invasions of southwest Asia and Europe in the thirteenth century.

In the areas of nineteenth century European colonization, droughts on the Great Plains of North America in the 1880–1890s (Brown 1948), eastern Australia in 1895–1902 (Heathcote 1965), and southern Africa in 1918 and the early 1920s (Kokot 1955), raised concerns for the long-term viability of human occupation. The 1930s saw further concerns for what appeared to be a combination of climatic variability and human mismanagement of the land's resources, with the evidence of the Dust Bowl on the southern Great Plains in 1935 and the first global study of soil erosion (Jacks and Whyte 1938). In West Africa, Stebbing (1935 and 1937) anticipated an advance of the southern edge of the Sahara as a result of over zealous clearance of the forest and savannah vegetation, and Ratcliffe (1963) studied the link between overgrazing and the expansion of the Australian deserts.

Post Second World War reconstruction efforts reflected continuing popular concerns such as Sear's *Deserts on the March* (1949) for the Great Plains, Calder's *Men Against the Desert* (1951) and Aubreville's *Climats Forets et Desertification de l'Afrique Tropicale* (1949) for the North African deserts. Pick and Aldis published concerns for *Australia's Dying Heart* (1944), and Lowdermilk's (1953) global review of soil mismanagement repeated earlier calls to reverse the loss of resources.

In this context, the traditional nomads of the desert got short shrift. There were claims that 'nomadic pastoralism is inherently self-destructive, since systems of management are based on the short-term objective of keeping as many animals as possible alive, without regard to the long-term conservation of land resources' (Allan 1965: 321). Even the camel got a serve, since by enabling man to lead a nomadic existence, '[it] has made it possible to exploit the vegetation over ever vaster and more varied stretches of land; so that the camel is responsible for the creation of the desert' (Charles Sauvage, botanist, quoted in Monteil 1959: 573).

The new interests brought calls for a reversal of such environmental resources. In 1966 a retired British forester with experience in Kenya and Nigeria and the founder of the 'Men of the Trees' Society, published his *Sahara Conquest* (Baker 1966), a further scenario to reclaim the desert, in this case by massive tree planting schemes. Some such calls had a political context. In *The Struggle between the Desert and the Sown* (1955) the Israeli author Reifenberg documented the efforts of the new nation of Israel to reclaim the deserts as the culmination of a long history of desert advances and retreats in this part of southwest Asia. In the Israeli view the Arabs had been the 'father not the son of the deserts', and thus the Israeli claim to the Arab lands was environmentally justified.

International scientific interest culminated in the United Nations Arid Zone Research Program which ran from 1951–1971. This brought together international scientists to study the global arid areas in an effort to understand both the physical characteristics and the past and present land uses, in order to better plan for future management and possible desert reclamation. In effect this program laid the foundations for future United Nations interest in the desertification phenomenon.

This concern over the increasing evidence of environmental deterioration seems to have reflected in part not only the increased scientific evidence of that deterioration, but also a growing moral concern for the global environment and human relations with it. The 'Environmental Movements' of the 1970s and the subsequent 'Green Movements', by focusing upon evidences of resource mismanagement through human ignorance or greed, saw desertification as but one example of the adverse impact of humans upon the environment. This impact had been of concern earlier (Glacken 1967) and had gained an added ethical dimension (Nash 1990, Passmore 1980, White 1967). Thus by the 1990s desertification had been consolidated as an international catch cry, a political force and a global scientific problem for some of the reasons noted above, but also possibly as a result of the end of the Cold War removing competing political issues (Driver and Chapman eds 1996).

The Context for Defining Desertification

> Specific definitions of desertification have been many and varied, partly reflecting the fact that it has been seen on the one hand as a *process of environmental deterioration* and on the other as the *product of environmental deterioration* – the devastated environment itself. (Glantz and Orlovsky 1986)

All definitions agree upon certain basic criteria: explicit or implicit evidence of changes in the characteristics of the present environment by comparison with previous characteristics; those changes have resulted in a reduced capacity of the environment to support human life; the resultant degraded environment has a desert-like appearance; and, unless rectified, those changes may continue to further reduce the capacity of the environment to support human life in the future.

To adequately and convincingly identify desertification in those terms, however, scientists needed compatible objective data for a considerable time period for specific areas of the world. Identifying either the processes at work or the end product – the desertified landscape – has proved to be extremely difficult.

The first problem has been the lack of the requisite historical data sets and the assumption of linear trends between those data sets which do exist. The degradation of the capacity of the soil to sustain vegetation or crops (through removal by wind or water erosion, by modification of chemical content through salinization, alkalization or acidification, or by modification of the soil texture and thus moisture retention capacity by compaction) is recognized to be a significant indicator of desertification. Global descriptions of soil characteristics, however, are still sparse and of varying quality, although formulae for estimates of soil erosion rates are available and have been partly used in the Global Assessment of Human-Induced Soil Degradation (GLASOD) commissioned by the United Nations Environmental Program and incorporated in the World Atlas of Desertification as the most scientifically acceptable measure of desertification (UNEP 1992).

The other main indicator of desertification is change in vegetation cover, whether of natural or domesticated plants. Change here may be indicated by reduced vegetation quantity and/or quality (in terms of reduced bio-diversity and/or reduced biomass and/or productivity), but these indicators are much more difficult to measure. The problem here is the inter-seasonal and inter-annual fluctuations in that cover, and the extent to which the cover at any one time reflects a linear trend towards increasing or decreasing density, or whether it reflects merely cyclical changes.

In west Africa, where the contemporary concern for desertification originated, there have been serious disagreements among scientists as to the extent of recent vegetation change (Table 5.2). The differences reflect contrasting assumptions on the historical extent of natural vegetation and the relationships between climate and natural vegetation; differences in the classification of 'natural' vegetation, and a failure to recognize historical human-induced deliberate or accidental revegetation of previously sparsely vegetated areas.

The global satellite coverage of the 1970s onwards has improved upon the earlier estimates of vegetation conditions, which had been based upon subjective assessments by individual explorers/observers. This satellite coverage has provided the base for a Global Vegetation Index – the global vegetation cover averaged over the 1983–1990 period as a base line for subsequent documentation of trends (UNEP 1992), but of course this does not help the debates on changes prior to the 1980s.

The second problem has been the interpretation of the data. Faced with what appears to be a desertified landscape in terms of apparently degraded vegetation and/or soils, the scientist must assess whether the landscape is in decline from a more productive past or in process of rehabilitation to a more productive future.

A third problem is the time scale chosen for the analysis of the significance of the environmental changes. While significance on a human scale may be set on scales ranging from inter-seasonal variations to trends over decades or a generation (usually seen as 20–30 years), significance in ecological terms may range from decades to centuries or millenia, and while these latter scales may be seen as irrelevant for human planning purposes, they may be relevant to any attempt to explain the desertification processes. The basic difficulty remains, however, that for the vegetation cover, as for the soil degradation observations, the implied trend will depend upon the length of time between observations. Oscillations in conditions from whatever cause, which may occur within those observations in time, may not be noticed.

Table 5.2   Conflicting Views of West African Desertification

Country	Forest loss during the twentieth century					
	Orthodox view[2]			Revisionist view[2]		
	Forest cover[1]			Forest cover[1]		
	1900	1990s	Loss/ Gain	1900	1990s	Loss/ Gain
Ghana	9.9	1.6	−8.3	2.5	1.6	−0.9
Cote d'Ivoire	14.5	2.7	−11.8	6.0	2.7	−3.3
Benin	1.1	0.4	−0.7	0.5	0.4	−0.1
Sierra Leone	5.0	0.5	−4.5	0.1	0.5	+0.4
Liberia	6.5	2.0	−4.5	5.5	4.8	−0.7

Notes: 1. Forest cover areas in millions of hectares. 2. Sources noted in Fairhead and Leach 1996: 189.
Source: Fairhead and Leach 1996.

A fourth problem facing definitions of desertification is the fact that there are several interested parties concerned with the phenomenon, some of whom may have an interest in stressing the extent of, and dangers from, desertification, and others more interested in playing down its significance. For scientists seeking research funding for projects of personal interest, for self-identified victims of the process and their political leaders, and for political groups who have identified particular organizations or ideologies as contributing to the process, there may be an incentive to exaggerate the significance of the phenomenon, or at least its natural as opposed to human causes (Glantz 1977, Heathcote 1986b). For scientists seeking funding for other research areas, for resource managers accused of exacerbating the desertification processes or seeking to gain time to complete exploitative management strategies, and for political groups anxious to defend specific resource management orthodoxies, there may be an incentive to play down the dangers. There seems to be evidence of both approaches to the phenomenon in the literature (Beinart 1996, Driver and Chapman eds. 1996, Garcia 1981, Heathcote 1980, Watts 1983).

A fifth and final problem in defining and explaining desertification is that it is studied by both natural and social scientists, and the investigations of one group may underestimate the significance of factors of interest to the other. This division has been most evident in the debates about whether desertification stems from natural fluctuations, or from the effects of human activities producing environmental stresses which cannot be sustained in particular locations (Rhodes 1991, Thomas 1993).

Desertification Defined

Bearing in mind such difficulties, the variability of definitions is not surprising. However, the consensus view as adopted by the United Nations has evolved from the definition adopted by the UNCOD in 1977:

> Desertification: The intensification or extension of desert conditions; it is a process leading to reduced biological productivity, with consequent reduction in plant biomass, in the land's carrying capacity for livestock, in crop yields and human wellbeing. (UNCOD 1977: 3)

to that used in the World Atlas of Desertification:

> Desertification/land degradation is defined as: land degradation in arid, semiarid and dry subhumid areas resulting mainly from adverse human impact. (UNEP 1992: vii)

and further modified as:

> Desertification is land degradation in arid, semi-arid and dry sub-humid areas resulting from various factors including climatic variations and human activities.

(UN Convention to Combat Desertification in Countries experiencing serious
Drought and/or Desertification, particularly in Africa, June 1994)

Thus consensus has defined the phenomenon as limited to the drier areas of the
world where highly variable climatic conditions, particularly drought, combined
with the adverse impacts of human activities, appear to constitute a major threat to
the long term sustainability of support for human occupation in those areas.

Desertification Documented

Since UNCOD in 1977 there have been various estimates of the area desertified.
The World Map of Desertification, showing areas at risk and prepared for the
conference, identified 4561 million hectares as 'affected or likely to be affected'
(UNCOD 1977, Annex 1). If the extreme deserts were excluded, on the grounds that
their condition could not worsen, the area was reduced to 3762 million hectares. In
1984 when the progress of the UN Plan of Action to Combat Desertification was
reviewed the area was reduced to 3475 million hectares (Tolba 1984). Dregne and
colleagues estimated 3562 million hectares in 1991 (Dregne et al. 1991), while
a UN estimate was only 1035 million hectares (UNEP 1992). More recently, it
has been claimed that while some experts estimate that 70 per cent of the world's
drylands are desertified, others suggest only 17 per cent (Reynolds et al. 2003).

In effect these varying figures are not strictly compatible as they refer to
drylands defined in different ways at different times, and to the inclusion of areas
suffering vegetation degradation but which may not have been suffering soil
degradation. At least the United Nations figures (UNEP 1992) do include estimates
of soil degradation based upon GLASOD.

The 1992 estimates (Table 5.3) provide basic information on soil rather than
vegetation degradation and while, therefore, marginally more acceptable, must
still be viewed with caution. The message is that over 15 per cent of global soils
appear to be degraded, and that over half of that area (53 per cent, or 1035 m.ha.)
is located within the global drylands. The largest degraded areas are in Asia and
Africa, with 38 per cent and 25 per cent of the global total respectively, and the
proportions of the global degraded drylands are similar – 36 per cent and 31
per cent respectively. Interestingly, the largest proportion of regional drylands
desertified is in Europe (33 per cent), but all regions have at least 10 per cent of
their drylands showing desertification.

The most serious dryland soil degradation and implied desertification is
identified in Table 5.4. Using the worst two categories of soil degradation (strong
and extreme degradation) Asia and Africa again dominate the scene, although it
is Africa's turn to lead with almost 54 per cent of the world's seriously degraded
dryland areas compared to Asia's 32 per cent. By comparison, the other regions
each contain less than 6 per cent of the global areas.

Table 5.3   Global Soil Degradation c.1992

Region	Soil Degradation (m.ha.)					
	In susceptible drylands %[1]		In other areas %[2]		Total degraded area %[3]	
Africa	319.4	24.8	174.8	10.4	494.2	25.2
Asia	370.3	22.1	376.6	14.6	746.9	38.0
Australasia	87.5	13.2	15.4	7.0	102.9	5.2
Europe	99.4	33.1	119.4	18.3	218.8	11.1
North America	79.5	10.9	78.7	5.4	158.2	8.1
South America	79.1	15.3	164.3	13.1	243.4	12.4
Total global	1035.2	20.0	929.2	11.9	1964.4	100.0

Notes: 1. Percentage of total area of regional susceptible drylands. 2. Percentage of total of regional other areas. 3. Percentage of global total degraded area. Total degraded area of 1964.4m.ha is 15.1 per cent of global land area.
Source: UNEP 1992: 25.

Table 5.4   The Most Serious Soil Degradation Areas in Drylands c.1992

Region	Soil degradation areas (strong and extreme) m.ha.	Percentage of global total areas (strong and extreme) %
Africa	74.0	53.8
Asia	43.7	31.8
Australasia	1.6	1.1
Europe	4.9	3.5
North America	7.1	5.2
South America	6.3	4.6
Total	137.6	100.0

Source: UNEP 1992: 13.

Maps of the desertified areas of the world, whichever definitions were used, all show the most intensive areas of desertification to lie on the humid edge of the drylands (Dregne et al. 1991, UNCOD 1977, UNEP 1992). This is usually explained as the result of the encroachment of agriculture, with its implied soil disturbance, into areas traditionally given over to purely livestock grazing, together with increasing pressure upon vegetation resources for fuel and building materials from increased human populations. This somewhat simplistic view, however, has been criticized.

Desertification Explained? How Important is Drought?

There is an extensive literature devoted to alternative explanations for the desertification phenomenon. Essentially the arguments fall into three camps. First are those which claim that natural processes, specifically climate change or variability – mainly through drought and feedback linkages between the characteristics of the ground surface and air temperatures – are the cause (Bryson and Murray 1977). Second are those which place the blame squarely on human mismanagement of the environment, mainly through human resource demands which exceed the capacity of the environment to supply in the long term (Sinclair and Fryxell 1985). Third are those which see a mix of both natural processes and human activities combining and inter-acting to create an unstable environment (Hare et al. 1977).

In the *World Atlas of Desertification* (UNEP 1992) the causes of desertification were listed as entirely human derived (Table 5.5) and no specific listing of natural causes was provided. Implicit, however, were links between human activities and the natural ecosystems. As identified, deforestation and overgrazing imply the reduction of vegetative cover resulting in less protection for the soils from solar insolation and increased wind speeds, leading to increased evaporation and potential wind and water erosion, along with possible increases in ground water levels and soil waterlogging from reduced vegetation soil moisture requirements. Agricultural activities include the ploughing up of fragile soils, thus baring them for erosion; attempts to grow crops in areas with insufficent soil moisture leading to crop death and exposure of the soil to further erosion; and excessive application of irrigation water leading to build up of salts in the soil (salinization). Over-exploitation implies excessive use of vegetation for fuel or building materials, which reduces its capacity to reproduce or recycle essential nutrients to maintain soil fertility. Bio-industrial impacts imply contamination from pollution sources and are more usually associated with intensive land uses outside the drylands, hence the relatively small area shown.

Table 5.5   Causes of Desertification

Attributed cause	Area affected (m.ha)	Percentage of total desertified
Deforestation	578.6	29.4
Overgrazing	678.7	34.5
Agricultural	551.6	28.1
Over exploitation	132.8	6.8
Bio-industrial	22.7	1.2
Total	1964.4	100.0

Source: UNEP 1992: 25.

There is no doubt that climatic variability, particularly in terms of precipitation trends over decades, has had measurable impacts upon the success of human occupation of the drylands of the globe, some of the most telling evidence coming from the Sahel (Nicholson and Farrer 1994, Hare 1983, Kates 1981). In such locations seasonal climates resemble desert climates (low precipitation, high evaporation, and high solar insolation) so that any human resource management involving reduction of protective vegetation cover at such time would run the risk of permanently damaging the environment (Aubreville 1949, Bullock and Le Houe'rou 1996).

The specific time sequence and causal factors of landscape change has been suggested as:

> *Drought* is a natural factor, which occurs on a yearly or biennial time scale. *Aridification* [also called] desiccation … is a climatic trend lasting for some decades, centuries or millennia. Land degradation, which is anthropogenic and the result of inadequate land use, occurs on a human time scale of 25-year generations. (Mainguet 1999: 126)

Such damage might be interpreted as the result of longer-term climate change, such as a trend towards increasing aridity, and labelled desertification. However dryland ecosystems demonstrate considerable ability to recover from periods of desiccation, so that whether or not desertification is identified could depend upon the time period chosen for the study, as suggested earlier. Indeed the edge of the Sahara Desert defined in terms of vegetation cover has been identified from satellite imagery to have shifted 240 km south between 1980 and 1984, but after several annual retreats and advances by 1990 was only 130 km south of the 1980 location (Tucker, Dregne and Newcomb 1991).

Such variation suggests the importance of short-term rather than long-term climatic changes and specifically the role of drought (83 per cent of the oscillations), but may include the effects of human impacts (Hulme and Kelly 1997: 215). A similar argument has been raised recently for trends in Portugal, where desertification in the southeast and northeast of the country has been associated with a recent spate of drought years: 1980–1981, 1982–1983, 1991–1992, 1994–1995 and 1997 (Santo et al. 2005: 184). The links between drought and desertification have also been suggested for Ethiopia (Mersha and Boken 2005: 227).

Certainly, clearance of vegetation for cropping or through excessive livestock grazing may change the albedo (reflectivity of solar radiation) of the ground surface and increase the air temperatures, thus increasing evaporation and creating drought-like conditions, which encourage further soil erosion from increased wind speeds. The maximum reflectivity of snow is 0.95, of bare sand is 0.40, of dry soil is 0.35, whereas from grass or a grain crop the maximum is 0.25 and from a forest canopy 0.20 (Berger 1981: 764). In such a context a recent study in northwest China has suggested that 77 per cent of the 1.4 million square kilometres examined was desertified, with 54 per cent of that being severely or 'super severely' affected. Of the causes listed, water damage and salinization were claimed to have affected c.16 per cent of the area, but over 56 per cent of the damaged area was claimed to have been affected by wind erosion, which is of course stimulated by drought devastation of protecting vegetation (Li et al. 2004). An earlier study had claimed desertification of 2,100 square kilometres per year on 'the perimeters of China's deserts between 1975 and 1987 ... [which] was attributed to overgrazing, deforestation and less than optimal farming practices' (Griffin, Kellogg, Garrison and Shinn 2002: 234).

It is thus not surprising that most desert reclamation techniques involve attempts to establish new vegetation cover to directly protect the soil, to reduce reflectivity (albedo), and in part to act as wind breaks (Baker 1966, Zhu and Liu 1981). The role of increased albedo on drought impacts may have been overplayed, however, as it has been questioned even in the areas of West Africa where the original links were first claimed. Using satellite imagery and historical records, a study of vegetation change in West Africa did not find evidence 'of historical regional increase in albedo due to land-use changes on a scale likely to produce or intensify drought' (Gornitz 1985: 309).

Table 5.5, however, omits mention of some of the less obvious but possibly equally important factors, such as the history of colonialism and recent economic development, both of which have been identified as relevant to the spread of desertification. Most colonial powers in Africa, for example, introduced new tax systems, which forced a monetary economy upon a previously subsistence pastoral or agricultural community. This required extra numbers of livestock to be grazed for cash sale to pay the taxes, or food crops on the better soils to be replaced by cash crops and pushed into areas climatically more vulnerable or with poorer soils more prone to erosion (Baker 1984, Franke and Chasin 1981, Garcia and Escudero 1982, Macdonald 1986, Morgan and Solarz 1994, Watts 1983). Subsequently, improved veterinary services and permanent water supplies were developed with foreign aid, which replaced traditionally limited seasonal supplies and allowed larger herds to be carried all year, and this often led to overgrazing of the ranges (Glantz 1977).

In addition, the increase of human populations in the drylands threatened by desertification, from 57 millions in 1977 to 135 millions by 1984 (UNEP 1992: iv), and particularly the rising populations in Africa (Caldwell and Caldwell 1990), have brought increased pressure upon the environment and demands for increased access to water. As an illustration, Lake Chad in Nigeria covered an area of 25000^2 km in the early 1960s but by 2000 had been reduced to 1350^2 km by decreasing precipitation, allied with a massive increase in demand for irrigation.

The highly variable seasonal rainfalls of drylands have always offered marginal benefits for farmers and pastoralists, but nonetheless both historically and currently were and are seductive attractions. A variety of invaders have been attracted, from starving peasants desperately trying to provide food for their burgeoning families, and entrepreneurs or speculators anxious to exploit infrequent bonanza years or to experiment with the latest agricultural innovations, to political leaders determined to demonstrate personal leadership and national expertise.

To an Australian symposium on 'Cropping at the Margins' in 1981, an experienced farmer explained that:

> In the 20 years we have been farming [in central New South Wales] the success rate is only really 1 in 4. One year we expect total disaster. Two years we barely pay your way but the fourth year is a bonanza. It is the potential in farming that keeps them going because if you can secure a big one it is really worth it. (de Kantzow and Sutton 1981: 137)

Twenty-seven years earlier, in 1954, Nikita Krushchev (then First Party Secretary of the Communist Party of the Soviet Union, and subsequently national leader), probably had similar hopes when initiating the Russian 'Virgin Lands Scheme' in the semi-arid areas of Kazahkstan. This scheme was to plough up about 42 million hectares of traditional grazing lands and bring about 650,000 new settlers on to the 'new farms'. Quizzed some years later on the rationale for the scheme, its director Leonid Brezhnev (who was subsequently to replace Krushchev as national leader) claimed that, 'It was not just a matter of boosting grain production in one republic but of providing a cardinal solution to the grain problem for the whole Soviet Union' (Brezhnev 1978: 7). The massive soil erosion which followed scattered the top soils throughout central Asia as the regional droughts reappeared, and while crop yields were initially promising, their increased moisture demands brought increased vulnerability to those droughts. Brzehnev reminisced, 'Had we heard the warnings of experts that overall ploughing would turn the steppes into a desert? Of course we had, and had taken them into account', but they had learned the hard way (Ibid.: 162). Only by a combination of extensive soil conservation measures, crop rotations, shelter belts and retirement of some farms back into pastoral lands has occupation been maintained, and that with a renewed respect for the threats from drought (Glantz, Rubinstein and Zonn 1994).

The Russian story was just one example from a series of studies from Australia, the West African Sahel, the North American Great Plains, northeast Brazil, Kenya, Ethiopia, and northwest and south Africa, which were brought together in a hard-hitting review, appropriately titled *Drought Follows the Plow* (Glantz 1994). The idea may be a new one, but the historical evidence seems clear enough, that unwise exploitation or intensification of land use will result inevitably in the deterioration of the land's resources and an increasing vulnerability to hazards such as drought. Widening the argument of fears over the role of increasing human populations in desertification, the editor of the review commented that 'the stork will eventually outrun the plow, as originally suggested by Malthus', and that as a result 'droughts will continue to follow the plow'. This idea should be instructive in that 'it warns planners and policy makers (even those wearing rose-colored spectacles) that there are adverse consequences to developing new agricultural lands and rangelands, if those plans are not carefully done' (Ibid.: 174).

Finally, the periodic devastation from civil strife and warfare has no doubt contributed to the desertification process as local populations have been driven away or slaughtered, their lands abandoned, cultivation ceased and irrigation neglected (Glantz 1987).

The Future of Desertification and the Role of Drought

Despite over 20 years of international efforts, the complexity of the factors involved in desertification has meant that efforts to reverse the trend have had limited success. Reviews of the results of the 1977 UNCOD initiative were not particularly impressive (Mabbutt 1987, Odingo 1992, Rapp 1987, Spooner 1987) and despite ongoing research and publications by the United Nations Environment Program through its *Desertification Control Bulletin,* which documents attempts to halt desertification, the debate about definitions cannot hide the fact that the phenomenon continues to be extensive, and that its extent is locally increasing.

A new UN 'Convention to Combat Desertification in those Countries experiencing serious Drought and/or Desertification, particularly in Africa', was agreed in 1994 and came into force in December 1996. The new emphasis was to be upon support for local schemes to rehabilitate desertified areas or reduce the potential for their expansion. This shift in scale held better promise for the future, since the specific causes of desertification are more often related to local economic, social, and political events in the context of the seasonal weather, rather than to broad changes in climatic patterns.

Having said that, however, the forecasted global warming climate scenarios suggest an increased aridity in the drylands, with associated increases in natural soil erosion, even without human interference (Bullock and Le Houe'rou 1996). Such a scenario does not offer much hope for reversing desertification in the drylands in the future, unless the pressure of human resource uses can itself be reduced. In that context even periodic droughts will continue to be a major determining factor in the stimuli towards desertification. This trend was recently re-documented (Salvati et al. 2009).

Drought and Famine

> The elders [northern Uganda] argued that no rain or inadequate rain over three consecutive years meant either migration or starvation. (Webster 1979: 152)

> Many studies of drought and famine (and more generally, of society-environmental issues) have fallen into two camps: those that characterize such problems as arising out of social structures (perhaps aggravated by environmental vagaries); and those that view such problems as the result of some variant of environmental determinism. (Waterstone 1991)

Famine and drought have long been related. Writing one of the earliest studies of famine, Castro drew attention to the evidence from China:

> The most terrible famines in China, according to the historical record, have resulted from the great droughts. A statistical tally of these disasters has been compiled by Alexandre Hosie; out of the thousand years from A.D. 620 to 1620, 610 were drought years in one province or another and 203 of these were years of serious famine. (Castro 1953: 131)

In a similarly pioneering study, but this time of drought, Tannehill reinforced Castro's argument:

> History shows that drought lies at the bottom of most famines. Men who have studied the famines of India say that there is no doubt that these famines have been caused directly by failure of the annual rains. In China there are many famines that are due almost solely to natural causes, chiefly droughts ... There are many contributing factors, but failure of the rains is the principal cause in China and India. The same has been true in Europe. (Tannehill 1947: 23–4)

Other studies have tried to highlight the link between droughts and famines. Discussing the relationships between regional droughts and food production, Brown and Finsterbusch suggested that for the period 436 BC to 1970 AD famines where drought was specifically mentioned as a cause included African, Indian, Chinese and Russian examples (Table 5.6). A report on famine in the Indian State of Jodhpur in 1943 commented that 'the vagaries of the monsoon [i.e. drought] can cause a fodder famine, or a grain famine or a water famine ... [W]hen the failure occurs on all fronts simultaneously, the situation becomes acute and is known as a "*trikal*" or treble famine'. The conclusion was that 'drought [was seen] to be invariably associated with famine as its most frequent cause' (Mathur and Jayal 1993: 18 and 27).

Table 5.6   Reasonably Authenticated Major Famines Associated with Droughts

Date	Location	Effects
1064–1072	Egypt	Nile flood failed for 7 years
1660–1661	India	'No rain for two years'
1769–1770	India (Bengal)	Drought, 10 million died
1803–1804	West India	'Drought, locust, and war', thousands died
1876–1879	North China	'Almost no rain for 3 years', 9–13 million died
1920–1921	Russia	Drought, millions died

Source: Adapted from Brown and Finsterbusch 1972: 6–7.

But as a disaster, how important is famine? In a study of human deaths from disasters from 1900 to 1990, famine was second after civil strife as a cause, responsible for 39.1 per cent of the total as opposed to civil strife's 48.6 per cent (Blaikie, Cannon, Davis and Wisner 1994: 4). Drought as such did not rate a mention, but a comment was made that 'continuing war in northeastern and southwestern Africa has made the rebuilding of lives shattered by drought virtually impossible' (Ibid.: 5). So perhaps drought as a distinct disaster cause is difficult to separate out?

Hewitt, some three years later, discussing hazards, placed drought impacts into a final fifth category of 'Complex Disasters', but in considering major disasters around the world from 1963 to 1992 put drought as first in terms of population affected, third in terms of significant economic damage and seventh in terms of deaths (Table 1.1). He went on to suggest that:

> Drought events vie with or exceed the others in total number of persons affected while, in the course of this century, death tolls in drought-related famines exceed any of the others, except epidemics … Drought-related losses are magnified by the increasing spread and productivity of commercial agriculture when exposed to water shortage, and the marginalisation and impoverishment of populous rural societies in the tropical world. (Hewitt 1997: 59)

He further commented that 'the special and severest association of drought and disaster concerns food security and famines' (Ibid.: 86).

But what is the more recent evidence of such linkages? In the early studies of the famines in the Sahel of West Africa in the 1970s the finger was clearly pointed at drought. United Nations Disaster Relief Reports of the various emergency situations arising on the African continent have appeared in the 'Situation in Africa and Overview' published since 1984. The enclosed table of countries affected suggested, for example, that 81 per cent of the 32 disasters in 1984 were the result mainly of drought. In 1985 the percentage was 73 per cent of the 22 emergencies and in 1986 the percentage related to drought was 68 per cent of the 19 emergencies. The scenario appears to have continued into the twenty-first century, with the addition of AIDS as a major contributor, for in mid-2002 the influential journal *The Economist* published a report entitled 'With the Wolf at the Door', commenting upon the origins of the current famines in Africa. Various factors were listed (Table 5.7) but drought was second only to AIDS. A more recent assessment continued to highlight drought as a persistent factor in the incidence of hunger, although whether drought was included in 'adverse weather' was not clear (Table 5.8).

Under a similarly outspoken headline 'Is this the end of the world? Earthquakes. Hurricanes. Floods. What is happening to our planet?' *The Independent on Sunday* on 16 October 2005 provided a graph drawn from an international World Disaster Report for 2004, showing that of deaths from natural disasters, drought and famine together topped the list with 46 per cent (Moreton 2005).

Table 5.7   Causes of famine in Africa in mid-2002

Political unit/ Causes	Political unrest	War	AIDS	Flood	Drought	Population affected
Angola		x				Over 1 million
Zambia			x		x	Over 2.5 million
Zimbabwe	x				x	Over 6 million
Malawi			x	x	x	Over 3 million
Mozambique			x	x	x	0.4 million
Swaziland			x		x	0.14 million
Lesotho			x	x		?

Source: *The Economist*, 1 June 2002: 47–8.

Table 5.8   Persons affected by Hunger in Southern Africa and Causes 1999–2003

Nation	1999	2000	2001	2002	2003
Angola	Civil strife	Civil strife	Civil strife	Civil strife	Civil strife
Lesotho				Adverse weather	Drought, frost
Madagascar		Floods, cyclones		Drought, economic problems	Drought, economic problems
Mozambique	Drought in parts	Floods, cyclones	Drought in parts	Drought in parts	Adverse weather
Malawi				Adverse weather	
Swaziland				Drought	Drought
Zambia			Adverse weather	Adverse weather	
Zimbabwe				Drought, economic disruption	Drought, economic disruption
Population affected (000)	1825	2350	4425	16,700	9500

Source: Linnerooth-Bayer, Hochrainer, Mechler and Suarez 2007: 16.

In recent years, in fact, there has been considerable debate as to the relevance of climate as a whole, and specifically drought, to famine. The debates provide a spectrum of opinions which range from allocating to drought the all-powerful role in the creation of famines to one where drought has no relevance at all.

One of the earliest themes of the relationship between humans and the environment held that human activity was directly and completely influenced by environmental conditions such as droughts, as we have seen in the context of desertification above. In the 1970s Bryson and Murray, using archaeological evidence from around the world, suggested the existence of *Climates of Hunger*, those 'changed climates that no longer support the crops and herds, berries, fruits, and game they once did. Climates change: a culture closely tied to a particular climate finds itself in danger' (Bryson and Murray 1977: 3).

The collapse of the Mycenaean Empire c.1230 BC and the eclipse of the Mesa Verde culture in southwest USA in late 1200 AD were among the examples cited. Two years later Roberts and Lansford published their book *The Climate Mandate*, claiming that:

> One, two, or even more consecutive years of bad weather for food production will almost certainly come along sooner or later, and they will have serious adverse impacts on human life that are beyond the ability of technology alone to prevent or remedy ... Clearly we must recognize the climate mandate which dictates that the earth's bounty will rise and fall from time to time and place to place in response to climatic fluctuations. If we heed the climate mandate ... humanity should be able not only to survive but to prevail over the hunger and starvation that have threatened so many people for so many centuries. (Roberts and Lansford 1979: 165 and 188–9)

A second hypothesis suggests that while drought alone may not be responsible for famines, in association with other factors it may contribute to the vulnerability of a society to famine, or may be the actual trigger which precipitates the onset of famine. Relevant here is the argued link between drought and desertification noted above. As put by researchers in Africa, 'it is not so much the drought that destroys the land, but the destroyed land that brings on the drought' (Franke and Chasin 1981: 125). The significance of drought as affecting food supplies reflects of course the extent to which agriculture contributes to a society's economy, and since 'agricultural productivity is closely dependent upon rainfall conditions, climatic factors also become important exogenous variables in economic growth and development' (Winstanley 1983: 209–10). Even in Italy as late as the eighteenth century, prior to the agricultural revolution, farmers usually could only produce surpluses of 25–30 per cent above personal food needs each year. With 85–90 per cent of the population bound to agriculture, 'A few unfavourable years in a row or even one unfavourable year could cause famine and death to the population ... [because] the supplies of a good year could not last for more than one year' (Rossini 1981: 731–2).

A third hypothesis is that drought is essentially irrelevant as a cause of famine: other factors are more significant. This is particularly the view of the political economists who see famine as the result of inequalities in access to resources within societies – inequalities as the legacy of a commercial economy imposed through colonial governments (Susman, O'Keefe and Wisner 1983), or of a capitalist system imposed upon traditional subsistence economies (Watts 1983), or of a breakdown of human food entitlements (Sen 1981), or as the result of the greed and folly of individuals or faceless multinational companies (Tudge 1977). A more recent example is Davis's *Late Victorian Holocausts,* where the claim is made that in droughts in Asia in the nineteenth century:

> Although crop failures and water shortages were of epic proportion – often the worst in centuries – there were almost always grain surpluses somewhere else in the nation or empire that could have potentially rescued drought victims. Absolute scarcity, except perhaps Ethiopia in 1889, was never the issue. Standing between life and death instead were new-fangled commodity markets and price speculation, on the one side, and the will of the state (as inflected by popular protest), on the other. (Davis 2001: 11)

In these views it is the mismanagement of the political economy which is the origin of the famine, not the failure of the rains. Cash crops have replaced essential food crops in agriculture; traditional societal buffering mechanisms against drought-induced shortages have been forgotten in the scramble for gain by individual or company capitalists; super-power politics have diverted scant resources into military budgets; power plays by rival tribal groups have brought a breakdown of law and order and the collapse of the basic food distribution systems. Nonetheless even Davis has to admit that drought has played a significant political role through its links with famine, noting that 'modern historians have clearly established the contributory role played by drought-famine in the Boxer Rebellion, the Korean Tonghak movement, the rise of Indian extremism and the Brazilian War of Canudos, as well as innumerable revolts in eastern and southern Africa' (Davis 2001: 13). Subsequent research has provided more examples of drought stimulated civil conflicts as existing divisions within the societies were re-emphasized (Levy et al. 2005, Righarts 2009).

One region of the world which has had a long history of droughts and associated famines is the northeast of Brazil, the Ceara Province. With almost a third of Brazil's population in the 1970s, the region had endured 51 droughts over the period 1692 to 1970, some 18.3 per cent of the years. A researcher suggested that the drought impacts had been enhanced by a combination of:

> the *social psychology of the rural peasantry*, a profound and mystical attachment to the arid backlands which – in spite of drought recurrence – blinds the agriculturalist to many of its inadequacies ... [and] a relict feudalism [which] permits the majority of the benefits and privileges in life to pass into the hands of the few who own the land, while the great peasant populace faces hunger and bankruptcy in the best of years. (Brooks 1975: 30–31)

Apart from famine and death, responses had included rural rebellion on three occasions, as we shall see in Chapter 9.

In the 1993 drought in Bihar province, India, lower caste tribal groups were suffering the worst. A senior government official told a visiting reporter for the *Times of India* that the situation was: 'tremendously fragile. The maize crop keeps them going for three months and the rice crop for two. *Lac mahua* and *tendu* leaves see them through another two months at the most. Then for five months they have to really struggle. They migrate, live off roots, berries, anything they can gather. So, if crisis strikes in that period, we reach the famine stage quite swiftly' (Sainath 1996: 292). The major problem in both Bihar and Ceara seemed to be that the groups had inadequate access to lands from which to obtain sufficient food.

But virtually every famine may provide a different interpretation of the relevance of drought as a contributary variable. The causes of famine are many and varied:

> In famine settings ... innumerable indicators signal actual or potential distress, including crop loss; food reserve depletion; irregularities in interstate grain transfers, population movements, and death rates; soaring prices in major food markets; hoarding; and liquidation of farm holdings. The attention various cues receive, the interpretation they are given, and the sense of urgency imparted to them within the highest centres of control (e.g. central government) depend on such sundry factors as the potential strength of the afflicted population, the other pressing national problems, philosophical tenets dictating the state's position on welfare funding, proximity to the next elections, availability of foreign assistance, and the time that has elapsed since such a disaster last visited. (Torry 1979: 536)

Yet the claims continue: 'In the quarter of a century since 1967, droughts have affected 50 per cent of the 2.8 billion people who suffer from all natural disasters. Because of drought's direct and indirect impacts, 1.3 million lives were lost, out of a total number of 3.5 million people killed by disasters' (Kogan 2000: 196). For Asia there were claims that 'severe drought over the past three years (1998–2001), in combination with the effects of protracted socio-political disruption, has led to widespread famine affecting over 60 million people in central and southwest Asia' (Barlow et al. 2002: 697).

Is there any evidence of the links between drought and famine becoming stronger over time? One study attempted to compare the effects of meteorological droughts in the Sahel of West Africa which occurred in 1910–1915 and 1968–1974 and which were of similar characteristics and covered similar areas. The comparison was summarized (Table 5.9). The differences between the impacts seems to have reflected the fact that the increased human population along with the increased livestock population and smaller area under food crops had made the areas more vulnerable to potential famine, but that stronger government controls on food prices and more available money from the cash economy and improved agricultural yields had blunted the drought impacts somewhat. Yet the potential for more serious impact remained:

another *minor* climatic stress like that of 1910–15 or 1968–74, in conjunction with the population growth postulated in the Matlock and Cockrum (1974) model and with the multifarious social, political, and economic changes could precipitate a system collapse in the Sahel, as could the continued and related exploitation of an already fragile resource base. (Bowden et al. 1981: 509)

Rainfall in the Sahel had been periodically rising since the 1980s but a drought in 2004 'combined with suitable meteorological conditions for locusts, led to the devastation of crops and resulted in food shortages on the western Sahel' (UK Meteorological Office 2005: 12).

Table 5.9   Comparison of the Impacts of Droughts in the Sahel in 1910–1915 versus 1968–1974

Factors	1910–1915	1968–1974
Population Index	100	300+
Mortality	'enormous'	'minimal'
Population migration	large scale and permanent	small scale and temporary
Livestock deaths	smaller	greater because more at risk
Grain prices	more inflated (millet x30)	less inflated (millet x2–3)
Food available	often none	scarce but some usually available
Crop area in foods	larger	? smaller, more cash crop areas
Vulnerability to famine:		
Nomads	less	most vulnerable
Farmers	worse off	better off

Source: Kates 1981.

Does drought pose a threat to world food supplies? There were fears in the 1970s, stimulated by the Sahel experiences. These earlier concerns were dismissed – commentators suggesting that 'certainly the world-wide famine predicted a few years ago is not going to take place unless there are cataclysmic changes in the weather' (Mayer 1977: 7). Yet, fears arose again in the early 1980s and 1990s.

Recognizing that cereal grains dominate some 75–90 per cent of the global agricultural area, certain production areas were more vulnerable than others to the drought threat: 'The FAO rates the West Asia/North Africa region as most critical among factors causing future food-deficit areas for rain-fed crop production' (Oram 1985: 141). While the same source also noted that the USA usually held c.80 per cent of marketable grain stocks, reports in March 1989 saw the drought in 1988 as responsible for major grain losses. 'US bread basket turns to dust' reported *The Independent* (23 March 1989: 12) and the global food reserve was under threat.

Commenting upon global food grain stocks in the mid-1990s a claim was made that:

> With worldwide drought damage in Canada, China, and the United States, the globe had only a 54-day supply of stock-piled grains. This was down from 89 days earlier in 1988, and below the 60-day global minimum set by the United Nations Food and Agriculture Organization. (Opie 1994: 283)

This was the lowest since the concerns in 1972–1973. The writer went on that despite the improved efficiency of grain production as a result of the Green Revolution of the late 1970s and 1980s, 'if a worldwide drought hit for several consecutive years ... [the FAO] concluded that it would collapse food surpluses and threaten widespread hunger on a global basis' (Ibid.). Bearing in mind the caveat of the commentators in 1977 above, recent concerns for climate change in the twenty-first century may require such optimism to be modified in the future. Certainly, the various contributors to a recent volume on the management and monitoring of agricultural drought all recognized that drought could and did have continuing serious impacts on national food productions, despite the best of available technological aids to its anticipation and management (Boken and Subbiah 2005).

In conclusion, therefore, while the significance of drought as a factor in human loss of life through famine is clearly debateable, a link cannot be denied, and the bulk of the arguments illustrate the fact that drought as a threat to the well-being of a society cannot be ignored.

Droughts and Diseases

> If we don't have more water, cholera, gastro-intestinal diseases, diarrhea, dysentery, malaria, skin disease, eye disease, and the epidemics we already have will all become more severe. There will be social and political incidents. (B.G. Verghese of the Centre for Policy Research New Delhi, quoted in Ward 2002: 5)

While drought may directly affect the incidence of human malnutrition and subsequent deaths by the destruction of food crops, there appears to be increasing evidence of indirect links between drought and outbreaks of disease, which are debilitating, if not always fatal to humans. The linkages between droughts and diseases affecting plants, animals or humans may at first appear tenuous, but recent research has forced a re-examination of old, and a careful appraisal of new, evidence.

The impact of bubonic plague upon human populations has long been documented (Zeigler 1969), but the role of drought in creating the environmental conditions which would favour the spread of the disease and its impacts on humans has only recently been suggested.

> Modern research on surviving wild-animal reservoirs of plague [in East Africa] ... has concluded that most plague outbreaks are caused by sudden and severe climatic change. Massively excessive rainfall is the most likely cause of plague spread, especially if it follows a drought, although a severe drought followed by normal weather could, theoretically, also spark an outbreak. (Keys 2000: 25)

Massive rainfalls increase rodent food supplies and plague immune rodents breed faster and need to spread into new feeding areas to maintain their enlarged populations, taking with them of course the plague bacteria/bacilla which infect non-immune rodents and eventually humans. Similar effects follow normal seasons replacing prior droughts as food supplies rapidly increase and the accompanying population explosion occurs. Keys, however, suggests that the most dramatic scenario would be where a severe drought was followed by massively increased rainfall and claimed: 'That, or something very much like it, is almost certainly what took place in east Africa during the world-wide climatic chaos of the 530s [AD]' (Ibid.: 26).

He was referring to the droughts produced by the dust veil from the Proto-Krakatoan volcanic explosion of c.530 AD, and which seem to have affected China, Japan, Mongolia, parts of Europe, Arabia and South America as well as East Africa. In his view: 'It was this failure [of the monsoon rains in East Africa] that almost certainly caused bubonic plague to break out of its naturally immune wild-rodent pool and spread to the Mediterranean and Europe, changing the region's history for ever' (Ibid.: 401).

Hemorrhagic fever epidemics associated with a mega-drought in the sixteenth century are believed to have killed up to 80 per cent of the native population of Mexico and 'a comparable long-lasting drought is thought to have affected much of the western United States a few centuries prior to that' (Meehl and Hu 2006: 1605).

A similar argument but linked more broadly to the El Niño phenomenon has been raised recently. From Peru came a 2002 report that 'the worst outbreak of Bartonellosis, an insect-borne disease fatal to humans, appears to be closely related to El Niño events' (http://www.gsfc.nasa.gov/newsrelease/releases/2002/02-017. htm). Similar concerns have been voiced in Australia about the increased risk of Ross River virus outbreaks, again associated with El Niño-influenced drought conditions (Nicholls 1986).

An even more complicated link has been suggested as a result of the invasion by seawater of coastal freshwater aquifers in California, which had been drawn down by excessive pumping during droughts. The drawdown in the delta of the Sacramento-San Joaquin rivers had shown the presence of TCMs [trihalomethane] in the aquifer:

> TCMs are potential cancer-causing chemicals formed when water containing
> certain precursors ['organic substances produced by decaying vegetation; the
> other is bromides, which are salts of seawater origin'] is chlorinated during the
> process of disinfecting drinking water. (California 1989: 36)

This is yet another example of the complexities of chemical interactions in our
environments and the link with the potential role of drought.

The association of droughts with accelerated soil erosion and the consequent
dust storms has also been linked to disease risks, in effect a separate phenomenon
from the volcanic induced 'dust veils' which themselves are thought to create
droughts. It is well known that significant volumes of dust are blown around the
globe in any one year with some estimates suggesting at least 1 billion tonnes.
Most comes from the larger deserts of Africa and central Asia, but the medical
significance of this movement has only recently been highlighted. 'The idea that
large scale disease outbreaks could be caused by dust clouds from other continents
has been floating around for several years. While it seemed far-fetched ... the
theory is now gaining acceptance as scientists find that it may explain many
previously mysterious disease outbreaks' (Pohl 2003: 18).

Drifting around the globe with the suspended dust particles are 'soil pollutants
such as herbicides and pesticides and a significant number of micro-organisms –
bacteria, viruses and fungi' (Griffin, Kellogg, Garrison and Shinn 2002: 228).
The two major source areas have been claimed to be 'The Sahara and the Sahel
regions of North Africa, which have been [periodically] in a drought since the
late 1960s' (Ibid.: 232). Dust originating in these regions and drifting across the
Atlantic Ocean landed in the Caribbean and the American mainland, bringing
'crop pathogens that make their way across the Atlantic [and amid them] are
those that cause sugar cane rust (*Puccinia melanoccphala*), coffee rust (*Hamilivia
vastatrix*), and banana leaf spot (*Mycospherolla musicola*)' (Ibid.: 233). Even the
Caribbean coral reefs were at risk, when 'a soil fungus (*Aspergillus sydowii*)
began to attack and kill seafan coral [in the Carribbean] from African dust' (Pohl
2003: 11). On the other side of the world, the outbreak of bovine foot and mouth
disease in South Korea in 2002 has been linked to dust blown from drought-
affected northern China and Mongolia (Ibid.).

As recent commentators have suggested, 'There is an air bridge that spans the
great oceans, and we haven't been paying much attention to it' (Griffin, Kellogg,
Garrison and Shinn 2002: 235) Interestingly, the Australian medical organization
the Menzies Foundation considered sponsoring a seminar on 'The Health Effects
of Drought' during the 1982–1983 drought, but the Executive Committee decided
in June 1983 'that the time for such a seminar was not opportune and the proposal
was therefore deferred' (personal letter from E.C. Wigglesworth, Executive
Officer, 29 June 1983). In the light of the above this would, in hindsight, appear to
have been a missed opportunity to focus attention on a developing global problem.

The Socio-Economic and Political Impacts of Drought

> Queensland Cotton is looking to diversify away from cotton production and
> marketing because of the drought ... Any alternative would have to produce a
> return on capital of more than 15 per cent before tax. Queensland Cotton's
> production in 2001 was a record 850,000 bales. In 2002 it fell to 530,000 bales
> and in 2003 to 225,000 bales. (*The Australian*, Business Section 29 July 2003: 18)

While we have examined the ecological impacts of drought, and its role in the
generation of wildfires, desertification, human diseases and famine, it is in their
socio-economic impacts that droughts first make media headlines and cause
governments to take notice. Warrick and colleagues summarized the scope of the
droughts in the USA in terms that have universal relevance:

> The range of possible effects of drought includes financial hardship, bankruptcy
> and geographic dislocation for the farmer; regional economic disruption and
> migration; massive government relief and rehabilitation; food shortages,
> rising prices and health effects at the national level; and severe health effects,
> disruption of social systems, international conflicts, and starvation and famine at
> the international level. (Warrick et al. 1975: 35)

In a simple but telling diagram (Figure 5.1) the sequential links of drought impacts
were traced from individual agriculturalists to the international political arena.

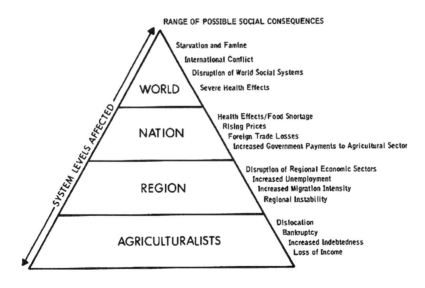

**Figure 5.1   The Range of Social Consequences of Drought vis-à-vis Social
System Levels**
Source: Warrick et al. 1975: 35.

Tracing those links, however, is not as easy as might first appear, and counting the costs of those links can be problematic.

In 2005 a researcher questioned some US official estimates of natural disaster damages. He noted that the Chicago Mercantile Exchange in 2000 had estimated that 20 per cent of the nation's $9 trillion economy was 'weather sensitive', without explaining how this value was derived or defining what was considered to be weather sensitive. Doing his own sums, the author considered that the losses from climatic disasters had been exaggerated: 'In summary, comparison of climate extreme losses and gains derived using various measures of the nation's economy reveals that [such] changes are extremely small' (Changnon 2005: 4–5).

In detail the economic costs of a drought may be difficult to differentiate. A study of *Australian Domestic Product, Investment and Foreign Borrowing, 1861–1938/9* had difficulty finding the real cost of droughts, even at the station-property level: 'In practice, stations tended to lump almost any exceptional outlays … into a drought account' (Butlin 1962: 85). At the national level, an anonymous commentator suggested that 'the mind may well boggle at estimates of the cost of a drought calculated to include wool shorn from sheep never conceived and wheat not harvested from crops never sown. Even if it be admitted that one can in some way lose something which never came into being, the calculation of values of the items becomes very involved' (Anon 1966: 52). Part of the involvement is that the decrease in production from drought raises the value of what was produced, upon which the value of the 'lost' production is usually calculated.

Taking into consideration the complexities of accounting noted in those original reviews – particularly, for example, the dubious economics of the inclusion of losses from crops not sown and livestock not conceived – there is continuing evidence of substantial economic losses from droughts. Over the period 1945 to 1975 I estimated the average cost of droughts in Australia as over $A100 million per year, roughly four times the cost of any other natural hazard (Heathcote 1979). The economic costs of the 1982–1983 drought were claimed to have multiplied up to $A5 billion, and this despite a decline in the share of GDP contributed by the agricultural and pastoral sectors from 15 per cent in the 1950s to c.4 per cent in the 1980s (Chisholm 1992, Lawrence, Vanclay and Furze 1992). The droughts in New South Wales and Queensland from 1992–1995 have been estimated to have cost $A4 billion and this with the agricultural and pastoral sectors contributing only 3 per cent to GDP (McLennan 1996: 117). Further declines in farm exports by 4 per cent were forecast for 1997–1998 (ABARE 1997: 5). Drought continued to stalk Australia in the early 2000s. In 2006 the Australian Bureau of Agricultural and Resource Economics (ABARE) reported:

> Taking into account the downward revision in crop and livestock earnings identified in this report, the gross value of farm production of these commodities in 2006–07 is forecast to be down by 35 per cent or $6.2 billion below the value in 2005–06 … [T]he drought is estimated to reduce economic growth

in Australia in 2006–07 by around 0.7 percentage points from what would otherwise have been achieved. (ABARE 2006: 1)

The social and demographic impacts of droughts have been harder to quantify. The farm sales and rural depopulation, for example of the western Eyre Peninsula (South Australia) in the late 1930s, seem to have been fairly typical of a process going on in most of the wheat farmlands in Australia at that time, namely a combination of poor commodity prices, high rural debts, the impacts of accelerated soil erosion and drought losses of crops and livestock (Heathcote 1993). The 1982–1983 drought appears to have had a similar effect on rural communities struggling with declining incomes and rising costs (Burnley 1986, Gregory 1984). In effect the drought, when combined with declining economic returns, seems to have speeded up the processes of social and demographic change already present in many rural communities.

Given the four types of drought we might expect a variety of socio-economic impacts. Meteorological drought is usually seen as the main driving force creating concerns about socio-economic impacts. After all it affects the productivity of rain-fed farming, withering crops and pastures, reducing the yields from water catchments and impacting directly upon regional economies. Thus Braudel, in his masterly review of *The Mediterranean and the Mediterranean World in the Age of Philip II*, felt the need to identify 'Drought: the scourge of the Mediterranean'. Here:

> A few changes in temperature and a shortage of rainfall were enough to endanger human life. Everything was affected accordingly, even politics. If there was no likelihood of a good barley crop on the borders of Hungary [barley was the oats of northern Europe], it could be assumed that the Grand Turk would not go to war there that year; for how would the horses of the *spahis* be fed? If wheat was also short … whatever the plans for war drawn up during winter or spring, there would be no major war at harvest time … Is it any wonder then, that the only detail of daily life that regularly finds its way into diplomatic correspondence concerns the harvest? … That Philip II should be kept minutely informed of the variations of the weather from seedtime onwards; that the price of bread should rise and fall depending on the rainfall … is very revealing of the state of the Mediterranean food supply in the sixteenth century. It was no mere 'economic' problem, but a matter of life and death. (Braudel 1992: 181–3)

But lack of local precipitation may not always produce a drought. Egypt, as 'the gift of the Nile', was, and still is, not really concerned about meteorological drought over Egypt, but is concerned about the Nile being affected by a hydrological drought resulting from meteorological drought over Ethiopia and Uganda, i.e. over the headwaters of the Nile. A similar argument could be applied to the irrigated areas of the world, some 200 million hectares in 1978, currently about 270 million, and expanding annually. Created to remove the local threat of drought, as one of the earliest measures taken by societies to meet the drought challenge, the irrigators are less concerned about their local annual rainfalls than precipitation over the catchments of the rivers or reservoirs providing their water supplies. For such areas a local reduction of precipitation is not really relevant; it is a hydrological drought threatening to affect the water coming from the river, pipe or aqueduct which is of concern.

Thus, a hydrological drought will create distress among irrigationists. Only rarely does no water at all reach the fields or reservoirs, but occasions when the farmers received as little as 10 per cent of their annual allocations are not unknown, and history can provide plenty of examples of empty reservoirs, crumbled irrigation channels, and reduced industrial production and hydro-electricity outputs. For the farms and pastoral operations the failure of irrigation supplies would compound the effects of an agricultural drought when reduced local precipitation had withered the crops and vegetation and local irrigation areas could not supply any shortfalls. It is in these situations of agricultural drought that governments are usually stirred to defensive actions, although it is often the case that the local situation may have already deteriorated into conditions of socio-economic drought before the bureaucracy is energized and effective relief measures are put in place. In this latter situation the loss of rural productivity will have led to loss of income or bankruptcy, with subsequent abandonment of properties, local lawlessness, the collapse of rural centres, and out migration of the population.

We should qualify the above generalization, however, by stating that limited access to water may create an effective socio-economic drought, even when strictly speaking a meteorological drought could not be recognized, even over the water catchment areas. That limited access might be created by a legal or physical barrier, a property boundary, or a political frontier. In every such case the 'drought' is a man-made fabrication and nature can plead not guilty. To further explain some of the complexities of the socio-economic impacts of such droughts, let us look at what happened in a highly industrialized nation, England and Wales, in 1975–1976.

The Drought in England and Wales 1975–1976

Over the period 1975 to 1976 reports accumulated of significantly reduced rainfalls over England and Wales. The results seemed to provide evidence of various types of drought. While the winter of 1974–1975 proved to be wetter than average (October 1974 to March 1975 precipitation was 107 per cent of the long term average over England and Wales), the summer was much drier than average (May to August precipitation was only 67 per cent of the average for 1916–1950), and this was followed by a dry winter with precipitation (October 1975 to March 1976) only 62 per cent of the 1916–1950 average. The drier conditions continued into the following summer with April to August 1976 receiving only 48 per cent of the 1916–1950 average. In the official *Atlas of Drought in Britain*, produced especially to document the phenomenon, the comment was made that: 'Parts of southern England received less than 40% of the 1941–1970 average precipitation over the 7 months from October 1975 to April 1976' (Doornkamp, Gregory and Burn 1980: 7). This meteorological drought was paralleled by hydrological impacts with streams failing, and river flows substantially reduced and reservoir levels affected. Rains over the period 27–29 August 1976 broke the hydrological drought from southwest Wales to Norfolk, but elsewhere the drought did not end until 10 September 1976. 'The period May 1975 to August 1976 was the driest six month period since records began in 1727 and the return period of drought has been estimated by some to be of the order of once in a 1,000 years' (Doornkamp, Gregory and Burn 1980: 7).

The desiccation effects appear to have intensified by a concurrent heat wave from 20 June to 6 July 1976, when tourists in Exeter were seeking baths and cool drinks from local hotels! However, rather typically, the drought overall was broken by heavy rains from 10 September onwards; the period of September and October was claimed to have had rains well in excess of the previous 250 years of record, which led to the official declaration of the end of the drought on 8 October 1976.

The effects of the meteorological and hydrological droughts were to create both agricultural and socio-economic drought impacts. Although overall drinking water supplies were down by 10 per cent, the broad community demands for drinking water were met, often by increased use of the accessible ground water aquifers, but water rates rose by around 20 per cent in 1977. Industry also seems to have been relatively little affected, but the main economic impact was on the agricultural sector of the economy. Crop yields were down 22 per cent compared with 1974, prices of vegetables rose and farm incomes were claimed to have dropped by 9 per cent in 1976 (Cox 1978: 143). The drought was estimated to have added £35 million to the cost of supplying Britain with water, while £185 million would be needed to improve future supplies (Cox 1978: 141).

Apart from the obvious change in the fields with crops withered or dying, there were other environmental impacts documented. Fish stocks in rivers and reservoirs declined, there were reports of beech tree deaths in the chalk country of the Cotswolds in August of 1977, increased incidence of wild fires in Amenity areas, especially moorlands where peat fires lasted several weeks, and suggestions of increased risks of soil erosion from soils bared by vegetation death or wildfires. Drying out of subsoil also caused structural damage to building foundations; £50–60 million was paid out in insurance claims in 1976 as a result (Cox 1978: 144). Surprisingly rabbits seem to have thrived and 'by 1977 there were more varieties of birds nesting in Britain than at any time since records began' (Cox 1978: 145).

Even more interesting were claims on the one hand that infant-survival rates improved during the drought because of the warmer weather, while some 700 of the 3658 human deaths in Greater London from 27 June to 10 July 1976 were from the heat wave associated with the drought. Even more intriguing was the suggestion that:

A little more than a year later a brief boom in boy babies in August 1977 may
have been a response to the drought of 1976 ... The surge of accumulated trace
elements into the water supply after the end of the drought [heavy rainfalls]
was linked to an adjustment in the sex ratio of babies born 320 days later.
(Doornkamp, Gregory and Burn 1980: 8)

So what can we conclude from the events in England and Wales over the summer
of 1976? The media and officialdom all agreed that the nation was in a drought
situation, with the media, as expected, probably exaggerating the effects. It was
claimed that 'The British Government viewed the drought as a major setback to
the agricultural industry, but it did not believe that special action was necessary
to improve returns to the industry as a whole' (Ibid.: 58), and in fact the impacts
upon agriculture were seen as less than expected (Roy, Hough and Starr 1978).
Yet another commentator, drawing upon first- hand experience of farming through
the drought, agreed that: 'The basic policy of the Government and of the water
authorities had been to gamble on getting through the drought without taking any
really drastic steps, and no doubt [by the official end of the drought on 8 October]
they could say that their gamble had been successful'. However, as a victim, he
commented that 'the very real hardship which a policy of rationing and wait-and-
see had imposed upon millions of people will remain graven in the memories of
all of us who were caught up in this struggle' (Cox 1978: 142).

The nation had survived a period of stress but at a cost and with moves for a
change in government policy as a result. 'Demand management is becoming a
basic part of water resource planning in Britain ... and the use of economic and
mandatory sanctions, coupled with publicity and education programmes such as
those necessarily imposed during the drought, will inevitably become standard
tools in the water industry in future years' (Doornkamp et al. 1980: 68). However,
other less scientific opinions were not so sure of a beneficial future outcome:

The drought provided, for those of us caught up in it, a common, shared
experience unlike any other in Britain since the blitz. What we learnt may be
of some value if another such dry spell hits us. If we are spared such a second
ordeal I hope it will be useful to have an immediate record of what the drought
was like, before we begin to forget it or before – which is more probable – we
begin to embroider our memories into myths. (Cox 1978: 9)

Another less impressed scientist, comparing the UK impacts with his experience
of drought impacts on Botswana, suggested that:

One's judgement as to the presence or absence of drought, or its intensity, is
specific to the production or consumption of particular economic goods. Britain,
in 1976, experienced rainfall which was variously described as being the lowest
for 200, 500 or even 1000 years. The consequent 'drought' led to the closure, or
restriction to part-time working, of a number of industrial concerns, as a result

of failure or shortages of urban water supplies ... British agriculture, in contrast, was relatively little affected ... I suggest that to many British farmers 1976 was not really a drought year at all. (Sandford 1979: 34)

Cox, as we have seen, would disagree, but the different views might reflect the fact that in Britain drought was not life threatening, whereas in Botswana it could be.

So it would seem that even in relatively modern times, even though the onset of drought is usually seen as a potential natural disaster, the details of the effects, the impacts and significance of drought can be debated. The reason seems to lie partly in the problem of defining the phenomenon, but also in the ways in which those impacts are tackled by the society at risk. As we shall see, societal attitudes to drought and their strategies to meet its challenges have varied considerably over time and these are the subject of the next part of our analysis.

Chapter 6
Society at Bay? Managing the Droughts

Human Attitudes to Disasters: An Overview

> Our forebears could manage with the instruction they received in their youth, but, as for us, we must begin our studies again every five years, if we do not wish to become out of date. (J.W. Goethe, quoted in *Journal of the Australian Institute of Agricultural Science*, 40(2), 1974: 148)

The long history of human occupation of the globe is pockmarked by the collapse of apparently thriving societies in the face of environmental disasters. In those disasters were included sharp and rapid changes of climate and associated ecosystems, and some of those disasters seem to suggest a role for droughts. To try to uncover human attitudes to droughts, therefore, we need to first scan human attitudes to disasters in general, to see whether the generalities may also be reflected in attitudes to drought.

Societies, of course, are complex mixes of people, institutions and organizations and as a result responses will be complex, occurring at different scales within the society, from those of individuals and groups to those of formalized institutions and governments, national and international. Responses therefore will be varied, but are there any common attitudes to the range of 'natural disasters' which might be found in most societies?

Scholars attempting to explain human attitudes and activities tend to use two contrasting explanatory strategies:

> At one extreme ... [are] those scholars who defend the positivist and most traditional social science theses ... [and] take for granted that there exists ... one single evaluation method – a single rationality parameter ... At the other extreme, there are ... the 'cultural relativists' – who assume that every risk assessment is merely a cultural construction [which] ... varies according to the cultural milieux in which it is employed ... [and therefore] objective evaluation is absolutely impossible. (Poli 1994: 83–4)

Another way of putting this dilemma is the debate between an 'etic' or an 'emic' approach to the evidence. The terms stem initially from linguistics, with etic describing linguistic features which do not serve to distinguish one sound or meaning from another, and emic describing features that do distinguish one language element from all others. In other words, etics are general principles that can be applied to any socio-cultural context while studies which stress the dominance of specific cultural constraints upon human actions would probably be seen as supporting emic arguments. I shall try to balance these contrasting strategies below.

What does research into human interactions with the broad phenomena of natural disasters tell us about human attitudes and responses, or lack of responses, and can those findings be applied to drought also? Based on the ample evidence of adverse drought impacts upon societies, contemporary western thought generally sees drought as a potential disaster for humanity as a whole, not only with severe adverse impacts on life support systems (agriculture and industrial production), but also with life threatening possibilities. These are in addition to significant impacts upon the environment itself. As a potential disaster, the onset of drought has required a societal response, usually to attempt to negate, or reduce, or offset any potential impact.

Studies of human responses to natural disasters in the social sciences have provided a general framework within which to place global responses and local case experiences. For the sociologists, a seminal compendium of established research provided evidence of a hierarchy of stakeholder responses to the whole range of so-called 'natural disasters', including drought, from the individual through groups and communities up to the United Nations Disaster Relief Organization (UNDRO) (Drabek 1986). In addition, the importance of prior experience in increasing risk awareness was stressed, particularly with the greater frequency of disasters nowadays, although it was recognized that the belief that 'lightning never strikes the same place twice' is still prevalent. Four phases of disaster management were suggested:

Preparedness – planning and warning mechanisms before the event occurs;

Response – pre-impact mobilisation and post-impact emergency actions;

Recovery – restoration and reconstruction after the event; and

Mitigation – hazard perception and subsequent adjustments to future strategies in an attempt to improve the response to the next event (Drabek 1986: 9)

Such a rigid framework however implies an institutional approach which may belie the realities of small communities faced with massive disruption and even complete destruction. It also implies that potential victims believe that they can respond satisfactorily to the impacts of any disaster, but the range of human perceptions of the risks from disasters suggests otherwise. In effect risks can be seen, as one commentator put it:

to be socially allocated, rather than [the] outcomes of the incidence of natural extremes, or equipment or operator failures. They depend primarily on the social order, rather than climate or, say, weapons potential.

As such, they 'express success or failure in the shared responsibilities and expectations of public life' (Hewitt 1997: 360).

Three studies in the 1950s–1960s attempted to identify the scope of human attitudes to nature, drawing upon a study of the varied colonial and indigenous attitudes to be found in southeast Asia (W.L. Thomas), a variety of communities in southwest USA (F.R. Kluckhohn and F.L. Strodtbeck)and a study of western philosophical works on human–nature relations up to the end of the nineteenth century (C.J. Glacken) (Table 6.1). While Thomas's study suggested a fivefold division, the other two favoured a broader threefold breakdown of the range of attitudes evident in the literature and from the surveys in the semi-arid southwest USA.

Kluckhohn and Strodtbeck's study was taken up by two geographers researching human attitudes to natural disasters and is worth examining in more detail. The original sociological study was of five separate cultural groups (two of native American Indians, alongside others from Spanish Mexican, Mormon, and Anglo–American communities), who were all facing the semi-arid environments of the southwest of the USA. There appeared to be a range of human attitudes to

Table 6.1   Human Relationships with Nature: Three Views

Land, Man and Culture in South East Asia W.L. Thomas (1957)	Variations in Value Orientations F.R. Kluckhohn and F.L. Strodtbeck (1961)	Traces on the Rhodian Shore C.J. Glacken (1967)
Five Themes	Three Themes	Three Themes
1) *Dominance of physical environment:* Forces of nature dominant/determining/influencing passive humans.	1) *Man subject to Nature.*	1) *Environmental influence on human activity*
2) *Landscape harmony:* Ecological balance through optimal strategies of resource use	2) *Man in harmony with Nature.*	
3) *Human adaptation to nature:* Broad physical limits exist within which humans act, respond, adjust.		2) *A Designed Earth*
4) *Human action changes the habitat:* Environment is altered deliberately/accidentally over short/ long term.	3) *Man master over Nature.*	3) *Man as a Geographical Agent*
5) *Cultural dominance:* Cultural ideas and society values including belief in technology providing increasing power over nature.		

Source: Adapted from Burton and Kates (1964).

the environment, encapsulated in the three concepts of 'Man under Nature', 'Man with Nature' and 'Man over Nature', as illustrated by the brief statements in Table 6.2. Responses to environmental stresses, it was argued, would reflect whichever belief was dominant in the society at risk.

Under these latter three categories we might expect societies with a 'Man under Nature' view to have a fatalistic attitude to the vagaries of weather, including drought if it were recognized. Such a view would accept drought as inevitable, possibly as a punishment from the gods for some past or current misdemeanors, and while prayers to the gods might, hopefully, provide some relief, on the whole the society would resign itself to tolerating and accepting its impacts, for other than prayer or flight there were no strategies possible. Societies with a 'Man with Nature' view would recognize the threat and the need for a response. That might require modifying activities to reduce drought impacts, implying possibly some significant inevitable losses to the society, and trying to return to the 'correct' response by working with Nature to negate the impacts of drought. For 'Man over Nature' societies, drought was a challenge to be faced and conquered, by human labour, ingenuity and technological innovation. Science in particular was to be marshalled to address the problem and to protect society against loss or damage.

Further geographical research which developed out of studies into natural hazard impacts identified a range of responses. At one end were traditional or 'folk' strategies; those of pre-industrial societies with a relatively low level of technology, minimal environmental manipulation, multiple and varied mitigation strategies, resulting in low property damage but high death rates. In contrast were 'industrial' strategies, found in those industrialized nations possessing a high level of technology, and relatively uniform disaster management strategies, resulting in

Table 6.2   Perceptions of Man–Nature Relationships

MAN SUBJECT TO NATURE
'My people have never controlled the rain, wind, and other natural conditions, and probably never will. There have always been good years and bad years. That is the way it is, and if you are wise you will take it as it comes and do the best you can.'

MAN WITH NATURE
'My people help conditions and keep things going by working to keep in close touch with all the forces which make the rain, the snow, and other conditions. It is when we do the right things – live in the proper way – and keep all we have – the land, the stock and the water – in good condition, that all goes along well.'

MAN OVER NATURE
'My people believe that it is man's job to find ways to overcome weather and other conditions just as they have overcome so many things. They believe they will one day succeed in doing this and may even overcome droughts and floods.'

Source: Adapted from Burton and Kates (1964). The discussion was based upon Kluckhohn and Strodtbeck's 1961 study.

high property damage but low death rates. A combination of both strategies was found in a third composite group (Burton, Kates and White 1978).

In these studies, to try to provide common base lines for comparison, an attempt was made to distinguish between 'adaptations' (long term biological and cultural responses) and 'adjustments' (short-term responses to specific events) in the context of a variety of coping strategies. This led to a 'Choice Tree of Adjustments' (Figure 6.1). In effect the hazards or disasters were seen as at the interface between human activity and natural events, with safe human activity limited by both natural and human constraints.

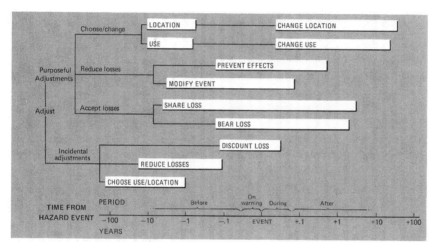

Figure 6.1   A Choice Tree of Adjustments to Natural Hazards
Adjustment begins with an initial choice of a resource use, livelihood system, and location. For that choice various incidental and purposeful adjustments are available, at somewhat different time scales for initiation. The most radical choice is to change the original use or location.
Source: Burton, Kates and White 1978: 46.

More recently the range of human responses to natural hazards has been set out as 'purposeful actions', with associated 'incidental adjustments' which elaborated on the original choice tree (Table 6.3).

Psychological studies and a long history of human trial and error have added many dimensions and further insights into human attitudes to natural disasters, drought included. People seem to have an inherent wishful hope that things will not change, and attitudes to drought seem to reflect that. Over the centuries, farmers faced by drought have been ever hopeful that it is but a temporary feature and that normal seasons will return. In 1894 drought devastated two-thirds of Nebraska, USA, but in 1895 when the drought had abated, the State Relief Commission

Table 6.3   Human Responses to the Challenges of Natural Hazards

Purposeful actions	Incidental adjustments
Bear the losses	Discount the losses
Share the losses	
Modify the hazards	Reduce vulnerability
Prevent the effects	Improve the warnings/forecasts
Change the resource use	Choose different resources
Move the location	Choose a new location

Source: Kates 1995.

which had been set up to coordinate collection and distribution of voluntary aid to stricken communities reported that:

> The unusual atmospheric digression of last year should in no sense be considered
> an example, and not more to be anticipated than the expectation of a recurrence
> of the insect [grasshopper] visitation of 1870. The years of anomaly are believed
> to be in the past. (Ludden 1895: 10)

Human Attitudes to Drought: A Wide Spectrum?

So far the discussion has made the assumption that drought is a measurable phenomenon and has adverse impacts upon human societies either directly or indirectly. How have those societies responded to the challenge? In attempting to illustrate those responses we have to recognize that there is a wide spectrum of attitudes to be considered. At one extreme there might be no specific response at all; the society might have no concept of drought either because their environment was not exposed to meteorological drought, or that if it were they did not recognize it as a distinct phase of the weather or as having any adverse impacts upon them. On such grounds dwellers in a tropical rainforest might not be expected to have any knowledge or concern about drought, whereas occupiers of a variable climate such as the semi-arid environments fringing the world's deserts might be expected to have a high level of concern and experience of droughts in all forms.

We might suggest a series of questions to be borne in mind as the evidence of human responses to drought is examined:

1. Is drought recognized as a phenomenon? If so, is it seen as a physical threat to life or to livelihood; and can its impacts be mitigated or must they be tolerated?
2. If recognized, is drought seen as purely a natural phenomenon or can it be seen, at least in part if not completely, as the result of human actions?
3. If drought is not recognized, why not?

Setting the Scene; the Interplay of the Natural Event System and the Human Resource Use System

One way of trying to show the way that human activities are affected by and themselves affect the natural event system is by a series of graphs (Figure 6.2).

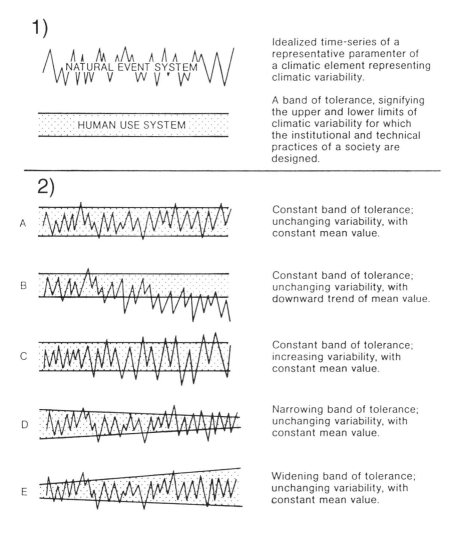

1)

NATURAL EVENT SYSTEM

HUMAN USE SYSTEM

Idealized time-series of a representative paramenter of a climatic element representing climatic variability.

A band of tolerance, signifying the upper and lower limits of climatic variability for which the institutional and technical practices of a society are designed.

2)

A — Constant band of tolerance; unchanging variability, with constant mean value.

B — Constant band of tolerance; unchanging variability, with downward trend of mean value.

C — Constant band of tolerance; increasing variability, with constant mean value.

D — Narrowing band of tolerance; unchanging variability, with constant mean value.

E — Widening band of tolerance; unchanging variability, with constant mean value.

Figure 6.2   Some Possible Associations between Natural Event Systems and Human Use Systems over Time

1. The two elements in the associations; 2. The range of possible associations and their implications. For explanation of the scenarios please see text.
Source: Heathcote 1990: 55.

The natural event system is depicted on the vertical scale as a sequence of varying levels of the event: for our purposes in looking at drought, the amounts of annual rainfall could be used. The human resource use system is depicted on the vertical scale as a range of values within which the events of the natural event system (in this case drought) could be accommodated and would represent resources to be exploited. The range would represent the collective knowledge and technical skills of a community available to face the challenge of the fluctuating quality/quantity of the events – say fluctuating annual rainfalls.

With reference to the graphs: **A** shows a time over which the human resource use system appears to have adapted to the unchanging climatic regime, with only minor instances when excessive or insufficient rainfall caused stress. **B** shows the human resource use system facing an increasing challenge as the annual rainfalls began to decline and the human management was unable to adapt. Such a scenario might be found in the semi-arid areas of the world where an invading 'pioneer' community, attracted by an apparently benign initial climate, found itself unable to cope with the declining rainfall. **C** shows a different kind of challenge where the variability of the climate/rainfall increases over time, resulting in greater exposure to increased frequencies of both droughts and floods, which pose increasing threats to the community. In this case it is not the gradual reduction of the rainfall but the erratic swings from 'drought to deluge' which cause the stress.

So far the only variability has been the natural event system, but graph **D** introduces the other major variable, namely the capacity of the human resource system to cope with, and benefit from, natural events. Here, over time the resource management becomes more difficult because it is itself flawed or based upon resource expectations which no longer exist. Cultivation techniques may lead to excessive soil erosion over time or excessive irrigation may lead to water-logging and soil salinity. In this case the natural pattern of rainfall has not changed but human mismanagement has reduced the available resources to be exploited.

Finally, graph **E** suggests that the band of tolerance of the human resource use system may itself be expanded to bring into use resources previously unknown or inaccessible. The scenario here would be illustrated by the impact of the Agricultural and Industrial Revolutions upon the capacity of humanity to access global resources since the eighteenth century. In effect it is demonstrating graphically the expansion of the technological resource management capacity of societies over time.

Not illustrated is a final graph **F** which would have tried to combine the natural event variability of graph **C** with the technological expansion of graph **E**. Here would be an attempt to show the effects of the expansion of technological innovations upon the natural event system, most specifically of course being the implied links to global atmospheric warming. As we shall see in the final chapter there could be several possible graphs.

Traditional Drought Management?

In trying to understand the incredibly long evolution of human attitudes to drought, a convenient division between what might be called traditional attitudes and those of more contemporary origin might be suggested. Convenient it may be, but helpful to understanding it is not. Harking back to the divisions suggested in the research into attitudes to natural hazards, namely the contrast between the traditional or 'folk' attitudes and those of the industrialized nations, separating the two was the third group of attitudes incorporating elements of both, in other words a transitional group. This, at least, was recognition that in the world of the twentieth century and most obviously in the world of the twenty-first century, change is implicit – in political alignments and powers, in the globalization of markets for produce and in the overwhelming surge of information and communication between the global communities. In recognition of this complexity I have suggested, first, a brief review of what might be called attitudes to drought in human mythology, then a look at the historical evidence of the responses of traditional societies to the drought threat, before a final coverage of the complex patterns of transition going on around the globe at the present time.

But let us consider a reasonably recent news item to illustrate the difficulties. *The Independent* newspaper on 17 January 2006 carried the following headline: 'Tribesmen drive cattle into streets of Nairobi as drought hits pastures'. An accompanying photograph showed scrawny long horn cattle tended by brightly robed Masai teenagers padding along a suburban street amid cars and trucks, heading hopefully for the grass verges and parklands of the capital, but destined to be denied access to the Presidential Palace grounds and their verdant lawns. Carrying out a traditional pastoral drought strategy, the Masai were merely shifting their starving livestock from the droughty areas to more attractive grazing lands, but which in January 2006 happened to be the President's lawns in the capital city … and he was not amused.

Traditional Strategies: Drought in Mythology?

> More important, we must check our tendency to assume that all people need are scientific assessments of the perils they face. Myth and ritual also help them to cope with disaster, albeit in a very different way. (Dvorak 2007: 9)

Dvorak was reporting on his collection of myths and rituals associated with volcanoes which showed that not all cultures viewed volcanic eruptions as destructive. For the African Masai their volcano Oldonyo-Lengai was the 'Mountain of God', in the Philippines the volcanoes are seen as fertilizing the soils with their ash, whereas in Hawaii the eruptions were seen as the acts of creation, with the lava flows as the menstruations of the Goddess Pele (Ibid.). Are there parallels with attitudes to droughts?

A recent study (Armstrong 2005) has pointed out the importance of mythology as a timeless strategy for coping with what Shakespeare termed the 'slings and arrows of outrageous fortune' – life's problems and challenges. From the Greek, both *mythos* and *logos* were equal paths to truth: *logos* as the objective facts, *mythos* as explanations of pain, sorrow and the value of life. Thus 'if a beloved friend died or if people witnessed an appalling natural disaster, they found that they did not simply want a rational explanation [*logos*]. Instead they developed mythical narratives which, like poetry or music, brought comfort that could not be expressed in purely logical terms ... A myth was not true because it was factual but because it was psychologically effective', i.e. it gave hope and comfort to the mind.

From the nineteenth century onwards *logos* came to dominate human thinking through the rise of science, and *mythos* receded from the human memory, so that to discover the significance of drought perception in mythology we need to go back beyond the onset of *logos*. Perhaps the most influential early source on mythology was J.G. Frazer's classic study in magic and religion *The Golden Bough*, first published in 1922 and drawing on an impressive global search of ethnography. From his initial interest into the murderous rule which governed the accession to the priesthood of Diana at Aricia in Italy, his quest for understanding led into the broad fields linking magic and religion with human attempts to control or at least explain the forces of nature.

Thus, Frazer noted that while for Aricia each priest officiated after killing the previous incumbent, this was mirrored by events in the 'medieval kingdom of the Khazars in southern Russia, where the kings were liable to be put to death either on the expiry of a set term or whenever some public calamity, such as drought, death or defeat in war, seemed to indicate a failure of their natural powers' (Frazer 1957: vi). This led to the traditional role of the king, namely 'to give rain and sunshine in due season, to make the crops grow and so on' (Ibid.). Later in the book Frazer was to provide a chapter on 'The Magical Control of the Weather', with sections on 'The Public Magician', 'The Magical Control of Rain', 'The Magical Control of the Sun' and 'The Magical Control of the Wind'. He credited public magicians as 'the direct predecessors, not merely of our physicians and surgeons, but of our investigators and discoverers in every branch of natural science' (Frazer 1957: 62). Thereafter for each of the three sections, examples were provided of the techniques by which magic was invoked in order to manipulate rainfall, affect the duration and power of sunlight, and enhance or reduce the power of the wind. Following on from Frazer what evidence can we find in mythology of human attempts to identify the causes of drought, its impacts and of strategies to cope with it?

Various causes of drought can be found in folk literature and religious texts. Thompson (1958) noted that drought was assumed to be produced by magic in early Irish and Indian oral histories, in some cases by a goddess, in others by sacred weapons thrown into the sky, in others by magic drying up the rivers. In Buddhist and South American Indian mythology the end of the world was seen as a continuous drought. In Aryan mythology the Vedic drought was caused by a demon and in Indian mythology an arch-demon Vritua had hoarded up all the world's water to create a wasteland (Campbell 1991). Australian Aboriginal mythology adapted knowledge of drought-escaping burrowing frogs (*Cycloruna albuguttata*) to recognize a monstrous frog called *Tiddalick* which had swallowed all the water in the land (Illustration 6.1). 'Legend has it that the drought ended only when *Noyang* the eel performed a comical dance that caused the amused frog to burst out laughing and release the water' (Hudson 2002: 32).

The reasons provided for drought seemed to range from the failure of the ruler's powers (Frazer 1957), as God's punishment for misdeeds (in Semitic myths) and Seth as 'God of chaos and evil, the personification of desert drought' in early Egyptian religion, to malevolent acts by demons (Campbell 1991, Schama 1995). On the side of the humans, however, were the drought breakers. In the southwest of North America the Thunderbird or Phoenix spirit brought storms to break the drought, while in China the appearance of the Dragon star constellation heralded the onset of the rainy season, and prayers to the Dragon King were common if the rains failed.

That societies struggled to understand droughts is illustrated from the Christian story of the 'Flight from Egypt':

> To learn the secrets of rainfall irrigation, as those lowland farmers [from Egypt] came to realize, they would have learn the wider secrets of the skies. And we hear their questions, asked with monotonous urgency in the books of Enoch and Job ('Hath the rain a father?'; 'Who can number the clouds in wisdom?'; 'Can any understand the spreading of the clouds?'). (Hamblyn 2001: 19)

Illustration 6.1   Tiddalick the Frog

Such actions are not merely history, however, for recent research into ethno-climatology among Andean Indians has shown a detailed knowledge of star patterns which are assumed to have significance for future seasonal conditions. Observing the brightness of the Pleiades star cluster on 24 June (i.e. prior to the normal planting time for their potato crop) decides whether planting will be on time or delayed:

> In the years when the Pleiades are bright, large, numerous or otherwise favorable, they plant potatoes at the usual time. However, when the Pleiades are dim, small, scanty or otherwise unfavorable, they anticipate that the rains will arrive late and be sparse, so they postpone planting by several weeks. (Orlove, Chiang and Cane 2002: 431)

Clear observations reflect a potentially normal season, whereas when the star cluster was partially obscured by high cirrus clouds, this presaged El Niño dry conditions and poor harvests. Rather than fatalistically accepting climatic variability, the Indians seek information they can use to adapt to that variability.

Traditional Strategies: The Historical Records

Learning from the Locals

> There is no presumption that local understanding will prove to be scientifically correct; but the local understanding of hazards is important in its own right, and more frequently than is generally believed, local knowledge and practice prove to have scientific and social validity as well. (Porter 1978: 6)

> I think – indeed I know – that sometimes they [the people most affected by drought] have quite different models of drought causation than those of us versed in western science do … [H]e who wants to take local participation into account has not yet given it one quarter of the thought that has been given to the interpretation of satellite imagery. (Sandford 1979: 280)

If we adopt the broad responses suggested in Kates's overview (Table 6.3), what does the historical evidence tell us to complement Frazer's monumental study? What of 'bearing' and 'sharing' the losses? Studying three subsistence villages practising shifting agriculture in northern Nigeria in the early 1970s, the researchers interviewed 150 farmers, two-thirds of whom were Muslim, but 91 per cent were still illiterate in a community with no written records nor apparently any collective memory (Dupree and Roder 1974). Virtually all saw drought as a problem, but three-quarters of them thought it would not occur again, but if it did the basic strategy was to bear the loss (which in the later years of the 1970s with the Sahel drought in full swing, probably meant starvation). In addition there was prayer and, hopefully, help from relatives, together with a search for alternative jobs and sale of firewood and any craft items. Interestingly, two-thirds of the folk would not store any grain, because relatives traditionally had rights to any such storage!

Bearing and sharing the impacts of drought was implicit in the strategies of the Saharan Kel Adrar Tuareg at a similar time. Animals could be shared with distressed relatives as gifts or loans: a gift of an animal to a poor relative was seen as a moral act; a lactating camel could be loaned and incur no charge, but non-lactating camels would incur a 'rent', and there could be reciprocal gifts of livestock. In addition the herds were deliberately mixed (camels, cattle, sheep, goats, donkeys) to maximize the usage of available feed between the browsers (goats and camels), and grazers (cattle and sheep and donkeys), and if the drought persisted the tribe could fall back on hunting and gathering, even raiding ants' nests for their stored grains and, if all else failed, robbing their neighbours (Hollhuber 1974, Swift 1973).

For the nomads of West Africa in the 1970s–1980s, bearing and sharing was complemented by 'opportunistic stocking' (taking maximum advantage of the available forage, as and when it occurred) as standard responses, and was praised by a sympathetic field researcher who was not impressed by the economic benefits of any conservative stocking practices which could not take advantage of good seasons (Mortimore 1989: 207). Ten years later, having worked in East Africa, the same researcher found that this was still the preferred strategy of Turkana nomads in Kenya. But the raiding of neighbours' livestock was also useful, not only in drought years – for the grazing resources – but also possibly in the good years for extra livestock to cope with the flush of grazing! (Mortimore 1998). By the beginning of the twenty-first century, those methods were still present. The bearing and sharing still used the staple food base – livestock – in reciprocal borrowing and repaying agreements and through the bride wealth system (livestock payments by the groom's family), but the droughts of 1980–1981 and 1984 had caused a temporary sedentarization of the nomads around the food relief camps. This disrupted traditional alignments, and concentrated the livestock around the camps causing overgrazing and death of some animals (McCabe 2004 and 2009).

By contrast William Thesiger, summarizing his long experience of the Bedouin tribes in Arabia, considered that they had a fatalistic attitude to life's challenges:

> As I listened [to the Bedu] I thought once again how precarious was the existence of the Bedu. Their way of life naturally made them fatalistic: so much was beyond their control. It was impossible for them to provide for a morrow when everything depended on a chance fall of rain or when raiders, sickness or any one of a hundred chance happenings might at any time leave them destitute, or end their lives. They did what they could, and no people were more self-reliant, but if things went wrong they accepted their fate, without bitterness, and with dignity as the will of God. (Thesiger 1964: 226–7)

In fact, nomadism – the harvesting of naturally occurring surpluses as and when they occur, either directly by human labour or indirectly through the use of domesticated livestock – is one ancient strategy to cope with drought-withered crops or grazing. Move on to pastures new! Norman Tindale, an anthropologist who travelled in central Australia in the 1920s, meeting Aboriginal groups living basically in traditional modes, put his observations thus:

> Drought is the great and extraordinary hazard of desert life; one that comes with such frequency as to engrave its pattern on the lives of the next generations, if not on every one in these tribes. It is possible to speak from personal experience of several examples of groups forced to flee their normal harvesting territories because of drought. (Tindale 1974: 69)

Flight did not always save lives however, for in 1929 he had met groups of about 200 starving Aborigines as they arrived at Hermannsburg, a Lutheran Mission near Alice Springs. Despite the help at the mission, 'There were deaths of children and the crippling of others' (Ibid.).

Recently, evidence has suggested that such distress and forced migrations were not unusual, but might have not been compensated by movements in more favourable seasons. The usual explanation of the oscillating human occupation of the desert margins, which was that occupation was by encroachment from better-watered fringes, may need to be amended. Archaeological evidence from the central deserts of Australia suggested that there was a period of relative humidity prior to 25,000 BP, during which the bulk of the area of what became later known as the 'Western Desert' was occupied by Aboriginal groups. However, this was followed by a period of increasing aridity, the result of more frequent droughts. This required more nomadic strategies. In other words, the desert came to the settlers, not the reverse.

> In Australia Pleistocene foragers [50,000–35,000 BP] did not move into deserts fully equipped with a modern desert adaptation, but rather that climate change created deserts in areas where hunter-gatherers already lived and these human groups adapted their existing strategies to the new situations or else abandoned the landscape – a 'desert transformation' model. A review of deserts around the world suggests that similar colonization models may have broad applicability. (Hiscock and Wallis 2005: 35)

With the onset of drought groups would divide into smaller bands which either concentrated around permanent water sources or abandoned the area until conditions improved. During major droughts:

> [D]esert groups would fission and disperse in small groups and the bulk of the population would fall back on better watered country. It is not difficult to see this process acting as a sort of 'cultural pump', drawing people into the desert during good seasons and forcing people out towards the margins of the region, or towards key waters within the desert, during drought periods. (Smith 2005: 235)

Letting nature take its course and living with the consequences seems to have been the usual strategy. It certainly seems to have been a Navaho Indian pastoral strategy in Arizona from the 1870s to 1975. Comparing livestock numbers over the years with the estimates of available feed from an environmental model showed that droughts brought sharp livestock losses. The researchers commented that 'in other words in the drought on the range 1892–1893, 1910–1917, 1924 and 1928–1934, the Navaho were prepared to let their livestock take their chance' (Richmond and Baron 1989: 222). African pastoral experience during droughts could be very similar as Mortimore concluded: '"Breed, feed, and milk, for tomorrow they die" may look like the antithesis of sustainable livestock management – but it may approach the harsh reality for pastoralists in the arid zone' (Mortimore 1998: 73).

Social and kinship networks as with the Tuareg allowed transfers of surpluses in good years and the calling in of 'credits' in the drought years (Hollhuber 1974). In agricultural societies food needs were reduced by the fostering out or sale of children or other dependants. Mary Renault's opening sentence to the second of her Alexander Trilogy, *The Persian Boy*, sets the scene:

> Lest anyone should suppose I am a son of nobody, sold off by some peasant father in a drought year, I may say our line is an old one, though it ends with me. (Renault 1972: 323)

In the West African reality the wife went as well. 'The drought of 1940–1949 … was called *wanda wassu* by the Songhai, i.e. "send your wife away" with the meaning of "free yourself of a mouth to feed"' (Mainguet 1999: 36). And for the children in a drought-threatened Botswana in the early years of the twenty-first century, the fear was still there:

> When there is a drought I am scared that we will be separated and taken to different houses. I want always to be with my brother and sister. We love each other and we want to stay together even if there is a drought (girl 14). (Babugara 2009: 9)

Coping, in fact, involved a complex of strategies:

> Mobility; migration; extensive kinship ties; *hxaro* of Sarwo in Ngamiland giving of gifts for future credit; sharing of game; *tswana* compulsory tillage of the chief's plot [for common use] first; harvest tribute in good years to chiefs [for storage] *dilegafela*; sharing of cattle *mafisa* (herded, milked and draught, but on loan); communal jobs by age-requirements *maphato* (e.g. clean out dams); rainmakers. (Hitchcock 1979: 95–6)

The complexity of traditional strategies to cope with drought stresses seems to have been reasonably successful, as claimed by researchers checking on traditional water harvesting methods in both Africa and India, with traditional villages apparently better able to cope (Stigter et al. 2005).

But not all the strategies were effective. Near the site of a new–old strategy to cope with drought, the Lake Assad Dam on the Euphrates River in Syria, an archaeological excavation of the site of a village dated c.11,500 BC found evidence of over 150 edible plants and open forest providing deer and nut harvests. The subsequent virtually 900 years of droughts emptied the site, with the remains showing: 'That the people turned in desperation to less palatable foods, to drought-resistant clovers and medics that were far from nutritious and required much more processing to detoxify before consumption' (Fagan 2004: 92). More recently droughts and warfare in Mozambique in the nineteenth century led to famines and mass migrations of populations for better-watered lands. Quite apart from warfare, the farmers were facing drought well above the manageable two years of water shortages (Newitt 1988).

Nonetheless, African drought strategies have died hard:

> Adam Marra told us how during one [of the droughts] he had worked for a foreign aid agency, touring the area in one of its vehicles. Along every track babies had been left by their starving parents in the hope that some aid agency vehicle would pick them up and take away any that were still alive for treatment somewhere. It was, he said, the most awful experience he had ever had. (Balfour 2006: 293)

Human history has been battered by the periodic scourging of droughts, but has irrigation, one of the oldest defensive strategies, provided protection?

Irrigation: Panacea or Pandora's Box?

Irrigation systems appear to have been developed independently in at least three major global locations: southwest Asia (including the Nile and Tigris-Euphrates valleys), China and the Americas (Peru). Sites in Iran have been tentatively dated to 550 BC. Basin systems were in operation along the Nile by 3000 BC. Canals were mentioned in Mesapotamia in the Code of Hammurabi at 2300 BC, in Peru by 1200 BC and China by at least 300 BC. And in India, recent archaeological work has found a complex system of over 16 dams associated with 35 Buddhist monastery sites dating back to the second and first centuries BC (Shaw and Sutcliffe 2003). The historical record however does not always indicate that the systems have continued down to the present time. In the Indian case the complex seems to have been destroyed and the sites abandoned by the tenth century AD, probably as a result of the Muslim conquest. Irrigation systems are delicate infrastructures and need regular maintenance and protection as we shall see.

Irrigation itself usually takes two forms: diversion by gravity of the natural flows out of the watercourses, or reticulation which implies attempts to completely control natural flows by storage and controlled application to fields. While gravity flows were the initial technique, human abilities soon created a variety of lifting devices, with different potentials for water application by reticulation to different areas (Figures 6.3 and 6.4). The sowing of fields or swamps drying out after natural flooding is suggested as the predecessor to human-induced flooding or irrigation – a kind of 'geological opportunism' (Vita-Finzi 1971). The opportunistic effort involved in creating a diversionary water channel or water storage could obviously vary from a single human's work to that of a community or maybe even a nation, while the type of water allocation seems to have depended upon the size of the

System	Lift (m)	Energy	Area irrigated per day (ha)
1. Archimedes Screw 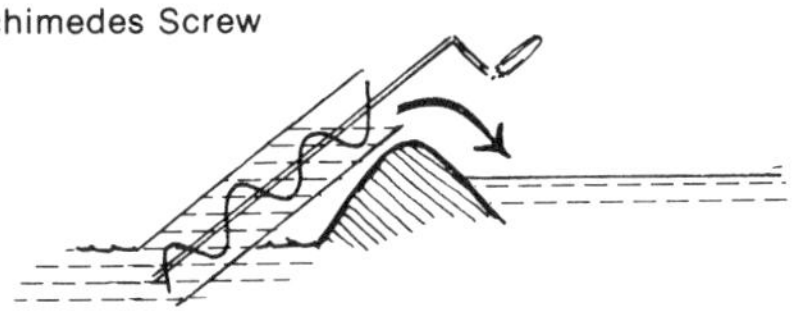	1	1 man	0.3
2. Shadouf 	2-5	1 man	0.1-0.3
3. Noria 	c.20	Water power	>0.8
4. Sakiya 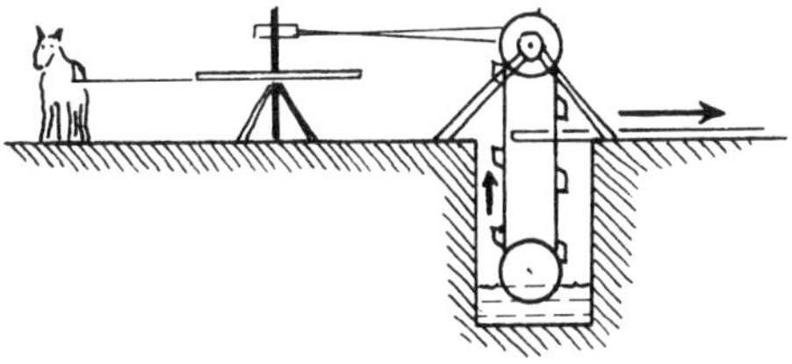	100?	Animal power	2-4.8

Figure 6.3　Traditional Water Lifting Systems
For each system the lift in metres, energy source, and approximate area irrigated per day in hectares is indicated.
Source: Heathcote 1983: 195.

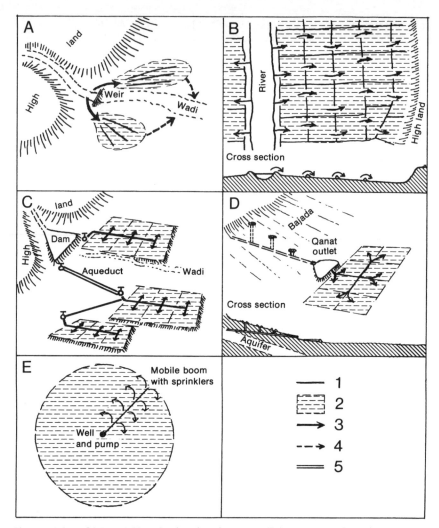

Figure 6.4   Old and New Irrigation Systems (Diversion and Reticulation)
Key: A = Diversion–Seil (flood) system; B = Diversion–Basin system; C = Reticulation–Dam and canals; D = reticulation–tunnels (*qanat* or *foggara)* and dams; E = Centre Pivot sprinkler system. 1 = Dikes or walls; 2 = Irrigated areas; 3 = Water movement; 4 = Drainage of excess water; 5 = Canals.
Source: Heathcote 1983: 192.

human demand and the technological skills at hand. In effect, they have ranged from a single farm through village fields to national projects producing and maintaining the agriculture of 'hydraulic societies'. At the smaller size level were, and still are, the tank systems of south and southeast Asia, with good examples in Sri Lanka.

The Tank Systems of Sri Lanka

The Dry Zone of Sri Lanka, almost two thirds of the nation's 70,000^2 km, is dotted by a tank system which was created to provide irrigation water at the village level. The climate of the Dry Zone ranges from arid to semi-arid with deceptively high annual rainfalls from 635–1,900mm. However, this rainfall is highly seasonal – 65–75 per cent of the rainfall coming from the northeast monsoons from October to early January. Significant nation-wide meteorological droughts have occurred roughly every 20 years, following two or three successive failures of the northeast monsoon, and 'Generally, there is a drought once in every 4–5 years at a given locality' (Tennakoon 1980: 8). Large water stores have been dated back to at least the fourth century AD and one of the most famous Sinhalese Kings of the twelfth century, Parakramabahu I, while further encouraging tank irrigation, is credited with the saying 'Not even a little of the water that comes from the rain must flow into the ocean without being made useful to man'. Abandonment of parts of the Dry Zone followed from political conflicts in the thirteenth century onwards, and the tanks were neglected.

The build up of population in the wetter south of the nation by the early twentieth century, however, forced a reappraisal of the traditional systems; a new wave of government sponsored settlers and tank builders began to re-colonize the abandoned areas. By the 1970s, the population had increased fivefold and an average of three tanks and associated settlements per square kilometre dotted the Dry Zone. The plan of the typical settlements is set out in Figure 6.5, with the tank central to the valley and the gravity-fed irrigated paddy rice fields below it divided into the oldest and more recent (field blocks) paddy areas. Beyond was the so called 'parkland' bordering the surrounding dry forest on the intervening ridges. Over time, as the population has increased, the pressure for more land has led to attempts to clear parts of the parklands, which have had limited success as some are above the water levels in the tanks and can only be used for dry farming *(chena)*. The latter has even been extended into the forest with the resulting clearance leading to soil erosion. Extension of the paddy blocks down valley has been more successful, but will eventually intrude into the tank catchment of the next village and has already caused disputes.

Thus, the return to the traditional tank system has provided limited defence against drought. That defence, however, has been eroded by the increased population and its demands for access to more water supplies. To this basic demand has been added the effects of the traditional subdivision of family lands among the children, thus creating smaller areas to support the new families and the limited areas of paddy which can be gravity fed from the tanks. In addition, 'desertification' through over-clearing of the forest lands for *chena* cultivation has led to dwindling timber supplies and soil erosion with sediments filling the tanks and devastating the dry fields. There are grounds for hope with more efficient use of the waters, private initiatives in agriculture, crop breeding for drought resistance and consolidation of land holdings, but drought continues to hover over all.

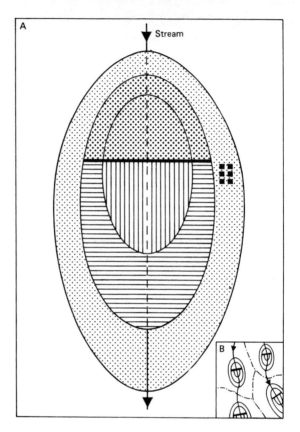

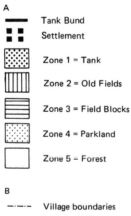

Figure 6.5   Schematic Diagram of a Typical Dry Zone Village, Sri Lanka
A–shows basic land use centred around the water storage tank. B–shows the general pattern
of adjacent village boundaries and similar land uses.
Source: Tennakoon 1980: 12.

The Larger Hydraulic Societies

In the Euphrates Valley the natural camber of the floodplain allowed water to flood artificially bounded basins up to 5 km from the main river channel, and similar terrain aided irrigation in the Nile Valley. With this 'control' of the waters there could be a significant boost to the people-carrying-capacity of the land. For Iran at 3000 BC, irrigated land could support up to six persons per square kilometre, whereas rain-fed agriculture could support only 1.2 persons, and hunter gathering only about 0.1 per square kilometre (Flannery 1971). As a hedge against drought this seemed to be the best strategy and irrigation has been universally associated with the rise of civilizations around the globe.

The surplus production afforded by irrigation affected the social organization of the groups practising this technique. As the scale of water manipulation increased, so the community activity necessary to establish and maintain the system had to be increased. At the largest scale the result was the creation of what was labelled 'hydraulic societies' (Wittfogel 1970). Such societies had to be centrally organized to control the labour needed to develop and maintain the canals and, in Wittfogel's view, this led to the creation of a 'totalitarian despotism' centered upon a king/ emperor (often deified as a water god), supported by hierarchies of priests who interceded with the god for the people and who provided the bureaucrats to organize the labour on semi-military lines. In this society importance was laid on the ability to predict changes in the water supply, hence calendars and astronomical observations were developed; the allocation of water required measuring devices, the most usual being the earliest types of clocks; canal construction encouraged the development of mathematics and geometry; and the increased potential value of the agricultural lands made land surveys essential.

Fully operational, this type of society functioned efficiently and supported a complex civilization. But it was vulnerable to any malfunction in the organization. As long as it kept the flows of water moving, the canals cleaned out and their banks repaired, and the water allocations on schedule, all appeared to be well. However, warfare could disrupt the complex organization, as happened in Mesapotamia with the thirteenth century Mongol invasion, and inadequate drainage could lead to the salting of soils, which has been claimed to have led to the collapse of the Sumerian civilization (Jacobsen and Adams 1958). But the ultimate threat was the failure of the water supply – drought! In the Nile Valley the relationship between success and failure was quite clear (Figure 6.6). The level of the normally bountiful Nile could fluctuate disastrously. Reduced runoff from drought in the water catchments of the upper valley meant difficulty in getting the water from the low river level to the fields and a smaller area able to be irrigated, which brought suffering and hunger. Excessive flows, with the banks over-topped and fields and villages flooded could destroy both land and people. Irrigation could both be panacea and Pandora's Box (Heathcote 1983), yet, as we shall see, it is still seen as a major defense against the threat of drought.

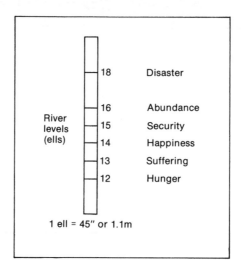

Figure 6.6   The Nile Gauge or Nilometer
The gauge measured the height of the annual flood and indicated the impact of the various levels upon Egyptian society.
Source: Heathcote 1983: 196.

The Rainmakers

> While socioeconomic strategies may have been somewhat more effective in coping with drought than were ideological responses, it cannot be doubted that rainmaking ceremonies provided people with a sense of security which was sorely needed in drought periods ... Far from being quaint tribal customs, then, the traditional responses to drought in Botswana must instead be viewed as important mechanisms for alleviating social stresses in an unpredictable environment. (Hitchcock 1979: 96–9)

Traditionally, modifying the nature of the drought or modifying its impacts might be attempted by general prayers or invocations to the gods or ancestors. A Somali prayer for rain is quite specific for what is needed of God:

> You who give sustenance to your creatures, oh God,
> Put water for us in the nipples of the rain!
> You who put water into the oceans, oh God,
> Make this land of ours fertile once more!
> Accepter of penance, you who are wealthy, oh God,
> Gather water in rivers whose beds have run dry!
>   (From 'The roobd'oon of Sheikh Aqib Abdullahi Jama:
>   a Somali prayer for rain', quoted in Andrzejewski 1970)

In fact, Boorstin claimed that one function of the Muslim Hadj, along with that of welcoming the renewed year and the bonfires to persuade the sun to rise, was 'to work charms to prevent drought' (Boorstin 1985: 120).

As Frazer illustrated for the early cthnographic literature, this was but one of many methods employed by rain-makers. In the southwest of North America various Indian groups seem to have made use of tools to induce rain – from 'bull-roarers' (wood or bone object whirled around on a string, giving off a low moaning sound) which were 'a vital part of the rainmaking ceremony amongst Apache, Navaho, Zuni, Ute and Kwakiutl people', to rain dances (Hopi). In Uganda in East Africa, the rain-maker had to manipulate the local hill as 'home of the winds' for the Bakitara tribe:

> All the local rainmaker had to do when plying his trade, was to wait for a likely-looking cloud to arrive overhead, and then cover the holes [from which the winds normally blew] with red backcloth, carefully weighted down with stones to stop the winds from blowing the cloud away. (Watson 1985: 261, 301–2)

Rain magic and the mobilization of large groups for rainmaking ceremonies 'serve[d] also as signals to the society at large that drought existed and additional measures should be taken to counter it' (Fleuret 1986: 227). Durkheim, in his study of religions in Africa, suggested that:

> Whenever the drought is very great, the great council assembles and summons the whole tribe. It is really a tribal event. Women are sent out in every direction to notify men to assemble at a given place and time. After they have assembled, they groan and cry in a piercing voice about the miserable state of the land, and they beg the Mura-Mura (the mythical ancestors) to give them the power of making an abundant rainfall. (Durkheim 1965 quoted in West and Smith 1996, frontispiece)

Quite apart from the calls on the rain-makers, however, the locals in Africa had their own ways of forecasting the weather, 'interpreting wind speed and direction, cloud formations, vegetation, and insect and bird migrations, for example – to predict weather patterns and the advent or cessation of precipitation' (Pratt, Cerda, Boulayha and Sponberg 2005: 276).

For the Aztec in central America, the 'god of rain Tlaloc ... [was] worshipped to ensure that rainfall would be timely and abundant, astronomical observatories were built to help predict the weather, and extensive irrigation works and reservoirs were constructed to store and transport water' (Liverman 2000: 37). At the summit of the main temple were symbolized two sacred mountains: *Tonacatepetl* 'where the Aztecs kept the maize that fed them (associated with Tlaloc, god of rain), and *Coatepec* ('Serpent Hill')' the god of war. The two shrines of the gods of rain and of war 'represent the supreme duality of the Pre-Hispanic world: life and death' (Moctezuma and Olguin 2002: 276). This concept of a rain god, in fact, was shared with other cultures in the region. For the Olmec culture of La Venta (800–400BC) he was also Tlaloc, but the Maya called him Chac, the Zapotecs termed him Cocijo and to the Totonacs he was Tajin (Ibid.: 459).

Multiple Strategies

To what extent drought was recognized separately from the spectrum of Nature's phenomena is not always clear, however. A recent contributor to a review of drought mitigation strategies and policies in Australia suggested that the Indigenous Australians did not recognize drought as such:

> There wasn't really a concept of drought in Indigenous culture – I'm not sure if there's even a concept of uncertainty … It's more a matter of going with the flow rather than trying to control it. (quoted in *ANU Reporter* 2004/5)

Yet in his study of Australian Aboriginal tribal cultures, as we have seen above, Tindale had noted several examples of groups forced to flee their normal hunting territories because of drought. These flights were in spite of detailed environmental knowledge incorporated into tribal mythology of:

> Springs, water places, pools of water, and streams … [as well as] prominent hills, especially those that provide a lookout or watching-place from which the movements of game, the activities of birds in the distance, and the smoke from fires in other parts of their country and others, may be seen and activities deduced. (Tindale 1974: 64)

In effect, drought was so integral to their lifestyle that it did not need to be separated out, it was 'normal'. The knowledge had even become part of the mythology surrounding the travels of the ancestors along the 'song lines' linking water points and sites associated with specific mythical events. That concept of linkages through space and time inspired Bruce Chatwin to give a wider, global, meaning in his 1987 book *The Song Lines*.

The longevity in human experience of the hazard of drought, in fact, had generated an equally long sequence of management strategies long before the Industrial Revolution encouraged humanity to challenge nature more directly. Indeed, one commentator considered the pre-historic human survival strategies through the alternating Ice Ages and warming periods of the last 15 millenia to be an illustration of 'Opportunism, flexibility, and mobility' (Fagan 2004: 68). From the hunter-gatherers to rain-fed farming activities an impressive array of coping mechanisms had evolved. Rain water was stored, channelled and concentrated onto favourable cropping sites; a variety of livestock with different capacities to cope with drought were raised and moved across a broad grazing range in search of forage resources; multiple seeds were sown to ensure that some germinated; plants with varying moisture requirements were used together for the same reason; and plots were scattered among different terrains to ensure that fleeting rain showers would be caught. A variety of methods for food storage and preservation (drying, salting, parching, fermenting, smoking, curing) were developed and might be combined with dietary changes when the drought stress was high and wild edible plants (often known as famine foods) were tapped (Fleuret 1986, Subbiah 2000).

Historically, diversification was the name of the game, with the stakes being the survival of the society. But the 'traditional' merges with the 'modern', and there is no sharp divide in time: humanity is on a constant learning curve, as a recent comment illustrated:

> In Kenya, nearly twenty years ago, one of the authors documented a wide range of more than seventy coping mechanisms that rural people used to survive drought. They ranged from boarding small children away in the home of a more fortunate member of the extended family to reliance on non-farm income and the use of wild famine foods ... Returning to those villages in 1990, he found many of those mechanisms still in place. But, in addition, coping now included neighbourhood-based women's self-help activity, and highly developed knowledge of how to 'play' the aid and relief systems. (Blaikie, Cannon, Davis and Wisner 1994: 205)

Sometimes, however, particularly when technology and overlapping bureaucracies and cultures are involved, the learning is difficult. A recent examination of the attempt by the Hualapai Tribe of northwest Arizona, USA (a community of c.2000 folk), to incorporate up-to-date scientific measurements and indices of drought into their traditional drought management strategies pointed out the difficulties. The tribe's economic basis rested on a combination of Federal and State government jobs (in government and education) with tourism, cattle ranching and some timber sales as additional income. Traditional methods of recognizing drought had not been able to foresee the impacts of droughts from 2003–2005, and a new official plan to use available scientific monitoring systems had been introduced.

By 2006, however, the results were still unclear. Despite the new data, the tribe had difficulty recognizing meteorological drought because of 'a good deal of confusion about how to calculate and interpret drought indices, such as the Standardized Precipitation Index and the Palmer Drought Index, which are necessary for the activation of the Hualapai Drought Plan'. The result was a compromise with 'drought conditions assessed by consensus in weekly meetings instead of strictly following the protocol outlined in the ... Plan'. There were also logistical problems: 'It may be difficult to get all tribal, federal and state drought representatives together on short notice to hold management meetings during times of drought' (Knutson, Svoboda and Hayes 2006).

Ironically, the initial success of the Green Revolution in the 1970s, with its concentration upon high yielding grains based upon irrigation and heavy mineral fertilization, increased the vulnerability of many societies to drought. It removed the diversity of crop types by concentrating upon new varieties and restricted production to areas which could be irrigated. This increased dependence upon a more limited range of crops and potentially unreliable, but vital, irrigation. In Kenya by the time of the 1982–1983 drought the new cash crop of maize had replaced the more drought-resistant traditional crops of millet and sorghum, while the decline in farm sizes and loss of traditional common land for extra livestock, along with the thriving market economy, had reduced the ability of the society to cope with droughts. Harvests were only a fifth of normal and cattle numbers were halved, but famine was prevented by official and international food aid. In Botswana a prolonged drought (1979–1987) on the rangeland:

> made legitimate a shift of dependency from the [traditional] extended family to the state and subsequently a greater dependency on the state. Traditional patterns of food security, which permitted some semi-independent production on the part of the poorer majority, were significantly eroded during the drought. They have not been revived. (Downing and Stowell 2003: 722–3)

In a 1972 drought in Papua-New Guinea, one of the results of the extensive official drought relief efforts was a similar reversion to dependence upon the central government for assistance, along with a new taste for tinned sardines, which had been part of the emergency drought food aid! (Waddell 1983).

On the opposite side of the world, in the Solomon Islands where ground water contamination from the invasion of salts from rising sea levels is causing droughts of a different type, 'many atoll populations have developed coping mechanisms to help them through these periods. Using less brackish beach springs or drinking coconut milk are typical examples' (Barr 1999: 31). Necessity is still the mother of invention.

Chapter 7

Drought as a Technological Hazard? The Mitigation Strategies of the Industrialized Countries in the Nineteenth and Twentieth Centuries

The nineteenth and twentieth centuries saw the rapid expansion of European political and strategic control of an increasing proportion of the globe. That expansion brought not only political control and exploitation of the indigenous people and resources of those new colonies, but also fostered the shift of a new wave, or rather a succession of waves, of new European settlers, bringing not only their families and crops and livestock, but also their experiences of land management and expectations of economic advantages to be obtained from the newly settled areas.

The newly colonized areas of the African, Australian and North and South American continents offered a wide range of environments, from tropical jungles, through temperate woodlands and the broad savannahs of subtropical grasslands and woodlands, to the hot semi-deserts and deserts of the plains and the cold deserts of the high mountains and plateaux. Through a combination of the patterns of advance and exploitation and the spheres of influence of the competing colonial powers, the bulk of the invasion by the new settlers tended to occupy first the most attractive and least extreme peripheral environments in terms of climate. Thus, it was from these relatively benign bases that the further spread of the 'pioneer settlers' (as they came to be known, despite the prior presence of the indigenous inhabitants) pushed into the interiors of the continents. That penetration brought not only new people, with their cultural baggage of how to exploit the environments to their advantage, but also a new dimension to life in the interiors. Increasingly over the years life in the interiors became linked, perhaps tied would be a better word, to the homelands of the settlers and particularly to the monetary economies that were established there and the techniques of resource use which they brought with them. In effect it was the initiation of what we now think of as the process of globalization, the creation of economic links between all parts of the world (Powell 1986).

Pioneering and its Problems

> My people believe that it is man's job to find ways to overcome weather and
> other conditions just as they have overcome so many things. They believe they
> will one day succeed in doing this and may even overcome droughts and floods.
> (Burton and Kates 1964 and noted in Table 6.2 above)

The pioneers may have believed the claims above, but they faced considerable
problems. In the case of the French settlers in North Africa, the Canadian and
American settlers in North America, the Spanish and Portuguese in South America,
the Dutch and British in southern Africa, and the British in Australia, that push of
people and cultural baggage was, for the most part, a shift from the relatively
benign coastal environments of their first bases down a gradient of decreasing
rainfalls to the continental interiors. Not only was it a push into regions where the
average rainfall was decreasing, but also into regions where the variability of that
rainfall was increasing. In effect, the chances of drier than 'normal' seasons were
more likely the further one went. This admittedly very broad generalization can
be faulted in detail, but it remains, in my view, a vitally important explanation for
at least part of the apparent increasing importance of the recognition given to the
drought hazard facing the varied global societies of the nineteenth and twentieth
centuries and its continuing relevance in the twenty-first century.

These pioneer settlers were much more potentially vulnerable to hardships
from increasingly hostile environments than they could have anticipated. In most
cases they were coming from a Europe very different in character from these
new lands, their experiences were of different relationships between the seasons
and the fruitful earth, so that much of their knowledge was not appropriate to the
new lands. In addition, their learning experience was hampered in part by the
efforts of private developers, 'boomers' of the attractions and potentials of the new
lands, and by colonial governments themselves acting as developers. These were
all hoping for successful settlement. Both were unwilling to admit that the new
lands might provide a hazardous investment for the settlers and their economic
capital, and so were relatively unwilling to broadcast any evidence of settlement
problems, losses or failures. Speaking of the chances of the success of settlement
in the late twentieth century, the suggestion was made that 'Even the most capable
managers can fail if they start at the onset of a period of low prices, low output
[from drought] or both' (Oliver and Tobin 1989). Also implied was the role of
mistaken assumptions in the fortunes of that settlement process.

It is not surprising, however, when one considers the successful transition of the
European economy through the Industrial and associated Scientific Revolutions
of the eighteenth and nineteenth centuries, which witnessed the spectacular
technological developments in the harvesting and processing of agricultural
produce, and its transport and distribution over the globe, that the new settlers
had a supreme confidence in their inherent ability to make their mark upon the
'new' lands. When they began to feel the new threats from drought it was quite

logical that they should attempt to draw upon the evolving technical and scientific knowledge of the time. Drought, it was assumed, was but a temporary set-back in the march of progress and the challenge should, and could, be met by what a Kansas historian called the 'Contriving Brain and Skilful Hand' of the new technologists (Malin 1955). Yet the following case studies from the USA and Australia, while supporting the generalizations made above, will show how technological strategies have had limited success in meeting environmental challenges, and should help to explain why drought is currently, and likely to remain, one of the many potential challenges not only to those two nations, but also to global society well into the twenty-first century.

Drought Mitigation in the United States

Although there had been significant local droughts which had troubled the European settlers before the mid-nineteenth century, it was not until the advancing wave of farmers spread into the drier western half of the United States, specifically the Great Plains, that the impacts of drought became more noticeable. Prior to the farmers, fur hunters, seeking beaver and later bison, had been little troubled by the periodic dry seasons. The ranchers, following on their heels, were affected by the loss of grazing in such dry spells, but as the range itself was free and the claims of the indigenous inhabitants discounted, the response was usually to bear the losses of livestock and move elsewhere to less drought-stricken pastures. The semi-nomadic lifestyle of both hunters and ranchers enabled a shift of location to cope with local droughts. Interestingly, it was a fierce winter and extensive snowfalls on the high western edge of the Great Plains in the late 1880s that is usually seen as the major environmental disaster for the open range ranching system, as starving livestock drifting before the blizzards piled up against the first barbed wire fences of the advancing farmers and froze to death. The farmers were to go on to challenge the ranchers, but drought was waiting in the wings to challenge them both, and the Great Plains were to be centre stage (Wilhite, Svoboda and Hayes 1983, Wilhite 2003).

Drought Relief on the Great Plains from 1850 to 1930

As farmers advanced in increasing numbers after the 1850s, and particularly after the Federal Government's Homestead Act of 1862 allowed for small family farms on 160 acres of land, they were encouraged by the claims of private land developers and the railway companies. The latter had been paid in government land for each mile of rail track laid and were obviously anxious to recoup their investment from the sale of that land. Yet the farmers were also more vulnerable than the earlier 'new settlers', for they became tied to specific locations and did not have the mobility of the native Indians and the later European hunters, and were at least initially forced to grow their own basic foods on site, before they could raise a cash crop. Indeed some of the aspiring farmers, sponsored by the railroad companies, were provided with a few weeks' food supplies to help the initial settlement process. By comparison ranchers tended to have somewhat better financial resources and the bulk of their 'capital' was on the hoof, could be eaten if all else failed, and their livestock's grazing was quickly able to transform the native grasslands into saleable produce.

As a commentator suggested, while settlement of the Great Plains region proceeded rapidly after 1850, 'immigrants arrived with little money, few possessions and scanty knowledge of the climate and other features of the environment. Technological options to cope with the vagaries of climate were also limited' (Wilhite, Svoboda and Hayes 1983: 42). Thus when the grasslands or crops withered and the wells dried up, destitution and famine were not far away.

The initial responses to drought impacts came from the local communities themselves, setting up committees to try to collect and distribute food supplies and channel funds from more prosperous areas, with only irregular and uncertain assistance from government funds. In 1860 in the new territory of Kansas a Territorial Relief Committee raised money and provisions for starving farm families, and in 1874 US President Grant allowed surplus army food and clothing to be distributed to over 100,000 drought victims in Minnesota, the Dakotas, Iowa, Nebraska, Kansas and Colorado. Yet when Texas appealed in 1886 to Congress for money for replacement seed, the then President Cleveland vetoed the proposal, arguing that such use of public funds was 'unconstitutional'! Nonetheless, in 1891 the Nebraska State Government approved $200,000 to purchase food and seed grain for drought-hit farmers. But in 1894 a move in Colorado to follow the Nebraskan example was vetoed by the State Governor on the same grounds as President Cleveland earlier.

There were other attempts to beat droughts, most notably the development of the so-called Dry Farming system of cultivation, which basically tried to save surplus soil moisture from one year to the next by alternate years of fallowing the fields. Developed by a South Dakota farmer, H.C. Campbell, who published his first descriptive pamphlet in 1893 and a series of technical manuals between 1902 and 1916, the dry-farming system required deep ploughing to break up the soil to receive and hold any precipitation, subsurface packing to prevent seepage of the moisture deeper into the soil beyond the reach of the plant roots, and, after precipitation had been received, constant harrowing of the surface soil to maintain a mulch to reduce loss of moisture by evaporation. The intention was to carry out this process on fallowed land for one year and then plant on the same land in the second year, thus in theory providing two seasons' rainfall for the crop. Subsequent research showed that in fact only about a tenth to a quarter of the previous year's rainfall was carried over, but that seemed to have been enough to encourage the believers. Not surprisingly the scheme was strongly supported by the land developers and railway companies and had some scientific backing. But it left the fallowed fields vulnerable to wind erosion, as the frequent harrowing of the topsoil reduced it to a fine dust, which moved with the slightest breeze.

Further distress in the droughts of the 1910s was met by similarly uncoordinated and inequitable official responses and it was not until the 1930s, when the combination of the effects of the Global Economic Depression and a change in political direction with the Democratic Party's control of the US Congress and Senate, that a new philosophy addressed the management of environmental hazards such as drought.

Drought Relief on the Great Plains in the 1930s

Interestingly, the initial response to the droughts of the 1930s by the Republican President Hoover had been to invite the Red Cross to supervise the drought relief efforts, but this required a change in their constitution, as previously they had only been involved in relief programs for disasters seen as 'Acts of God' and drought was not so recognized. Nonetheless the President did approve federal government support for voluntary crop production loans and a [Stock] Feed and Seed Loan, totalling $90 million. These seem to have been the last attempt by central government to enforce a self-help strategy upon the farming community.

As drought returned in the early years of the 1930s the fields fallowed as part of the Dry Farming strategy began to blow away, at about the same time as the desperate farmers were trying to increase production in the face of falling world prices for their produce. Many walked off their land when they could no longer pay their debts, with approximately a third of the rural population leaving between 1930 and 1940. In May 1934 dust from the eroding farmlands fell on the capital, Washington DC, a potent lobbyist that was to introduce a new label 'Dust Bowl' for the plains, and stimulate another, more successful, attempt to meet the drought challenge (Figure 7.1).

The election of the Democratic Party to power in 1934, coupled with the continuing and worsening drought situation, brought a totally new series of strategies to try to cope with the drought impacts as part of the Administration's New Deal. The new President's program was a complex series of emergency food aid and purchase of starving and otherwise worthless livestock, alongside the purchase of failed farms and relocation of the families, together with loans to surviving farmers to buy feed for their livestock and/or seed for the next crop, and public works camps, run by the army, to provide an income for urban workers unemployed as the result of the drought (Table 7.1).

Subsequent additional funding created over 1.2 million acres of shelterbelt tree plantations, to reduce wind speeds across the plains and thus reduce the soil erosion potential of the drought-bared croplands, and funds for water storage construction. Further droughts in the late 1930s continued the pressure to remove cultivation completely from the worst eroded areas and led to the reversion to grassland, e.g. the creation of the Cimarron National Grassland in 1937. The administration of the programs was at local government level, basically county units, but an impressive array of federal government agencies began to be involved in the complexities of

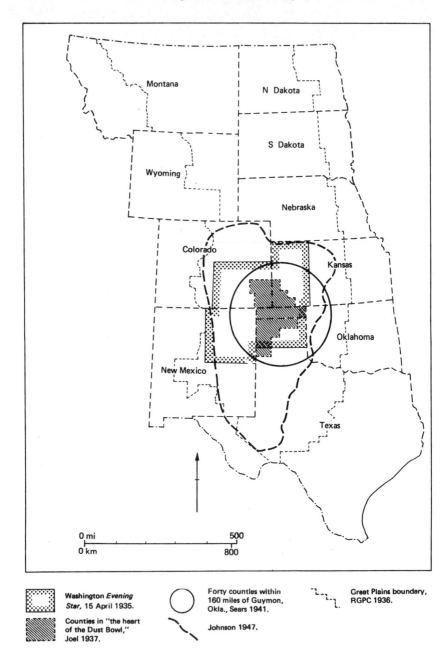

Montana

N Dakota

S Dakota

Wyoming

Nebraska

Colorado

Kansas

Oklahoma

New Mexico

Texas

0 mi   500

0 km   800

Washington *Evening Star*, 15 April 1935.

Counties in "the heart of the Dust Bowl," Joel 1937.

Forty counties within 160 miles of Guymon, Okla., Sears 1941.

Johnson 1947.

Great Plains boundary, RGPC 1936.

Figure 7.1   Defining the 'Dust Bowl' from the Original Definitions to the Reassessments in the 1940s

Source: Map from Heathcote 1980: 36, with specific sources as indicated.

drought relief efforts, and this was to increase and itself become a problem as the years went by.

Table 7.1   President Roosevelt's Drought Relief Program, June 1934

Program	Funds ($M)
Special work program and human relief	125
Livestock purchases in addition to funds already available under the Jones-Connolly Act	75
Shipping, processing and relief distribution of purchased cattle	100
Loans to farmers to finance emergency feed purchases and shipments	100
Emergency acquisition of submarginal farms and assistance in relocating destitute farm families	50
Work camps to afford employment in the drought area for young men principally from cities and towns	50
Purchase of seed for 1935 plantings and for loans to get seeds into farmers hands	25
Total	525

Source: Data from Wilhite, Svoboda and Hayes 1983.

Drought Relief on the Great Plains in the 1950s

Virtually 20 years later drought struck again. Initially, the new President Eisenhower (Republican) was against a similar pattern of programs to those of the Democrats in the 1930s and tried, unsuccessfully, to involve the states in sharing program costs. When they refused he was forced to provide a large and complex pattern of emergency and longer-term strategies to cope with drought impacts, even more comprehensive than those installed by the Democrats in the 1930s (Table 7.2).

Apart from the precedents of food aid and emergency funding for purchase of starving stock, concern for the impacts of drought upon local businesses and industry was evident, and there was research by various government agencies into crop insurance, farm tenures, land price trends and shifts in land use as well as longer term planning for population and water needs. One innovation was the granting of permission for livestock to graze areas previously badly eroded, which had been reserved as 'Soil Bank' lands to allow them to regenerate from the erosion damage of cultivation. Recovery of these environments must have been sufficient to make this a viable bonus. On paper the measures were commendable, but so many government agencies were involved in this enlarged ambit of recognized drought impacts that overlapping responsibilities and bureaucratic protocols seem to have slowed down responses. This was sufficient to give rise to complaints that the drought impacts had disappeared before relief was available!

Part of the grumbling seems to have been quietened by the economic effects of the Korean War (1950–1953), which pushed up agricultural prices. In addition, however, beneath the Plains was discovered the Ogallala Aquifer, providing a vast source of potable water which was beginning to be tapped by deep wells and applied to the fields through diesel-powered centre-pivot rotating irrigation systems. These were developed first in Nebraska, but rapidly diffused over the plains and subsequently the world. Here, it seemed, was the immediate answer to the drought (Illustration 7.1).

Table 7.2   President Eisenhower's Drought Relief Program, 1953–1956

Program	Funds ($M)
Distributed government-owned surplus food free through state welfare offices to needy people in cities, towns and rural areas	100
Distributed government-owned food grains to help farm and ranch families maintain foundation livestock	140
To help purchase hay and other forage to maintain foundation livestock, including dairy cattle	26
To help implement wind erosion control measures	18
Emergency credit and livestock loans	260
Purchased beef and pork products to strengthen distressed livestock prices. Frozen hamburger [meat] was purchased to help stabilize prices of certain grades of cattle	184
Long-term favourable rate loans for small businesses in drought-stricken communities	1
Free grain forwarded to small farm families through state welfare offices to maintain subsistence livestock	na
Special permission in 562 counties in 12 states to graze soil bank reserves	na
Total	729

na = No funding amounts provided.
Source: Data from Wilhite, Svoboda and Hayes 1983.

Drought Relief in the 1970s

But 20 years on drought was again in the headlines. This time, however, it was more widespread, including California and most of the southwest as well as the Great Plains. In fact it was the initiative of all the western states, whose collaboration created the Western Governors' Task Force on Regional Policy Management in 1976, which led the push for federal aid once more. This time the new President Carter (Democrat) was quick to support a broad series of programs which were rapidly passed by Congress (Table 7.3).

As the range of programs illustrates there were the, by now usual, emergency loans to tide farmers, ranchers and small businesses over what was assumed to be a

Illustration 7.1 Centre Pivot Irrigation System
This system has been modified to lower the spray heads and so reduce water loss by evaporation. Photographed in Kansas 1980s.

short and temporary period of low to minimal income. New components were the loans to communities for emergency water supplies and the increasing concern for conservation issues, not only soil erosion, which had been present in the 1930s and 1950s relief programs, but now also wildlife and fish protection against drought impacts.

Quite apart from the President's programs, there were in addition several separate relief programs sponsored by, for example the United States Department of Agriculture (as a traditional player), the Department of Commerce, the Department of the Interior and the Small Business Administration, which together were claimed to provide another $5 billion in drought relief. Again, despite the increased size of federal funding (direct or indirect), there were complaints: that the funds were not allocated in time to be of help, there were inconsistencies in criteria for help between the various agencies, and a lack of coordination between agencies, which could lead to the smarter 'victims' being able to choose the most valuable 'helper'.

At least, however, by now there was a suggestion of the need for a national approach to the problem. Following the end of the drought in 1977, the Comptroller General of the United States recommended a national plan for drought assistance. This was to provide identification and demarcation of the role of the various

Table 7.3   President Carter's Drought Relief Program, March 1977

Program	Funds ($M)
Emergency Loans program of the Farmers Home Administration (5% loans to cover perspective losses of farmers and ranchers)	100
Community program Loans (5% loans [$150M] and grants [$75M] to communities of less than 10,000 population for emergency water supplies	225
Emergency Conservation Measures Program of the Agricultural Stabilization and Conservation Service (Soil conservation cost sharing grants)	100
Federal Crop Insurance Corporation Insurance (Increasing the capital stock)	100
Drought Emergency Program (Bureau of Reclamation) (creation of a Water Bank, protection of fish and wildlife stocks, grants to states, 5% loans for water supply and conservation)	100
Emergency Fund (Bureau of Reclamation) (Emergency irrigation loans)	30
Emergency Power (South West Power Authority, purchase of emergency power supply)	13
Community Emergency Drought Relief Program (5% loans [$150M] and grants [$75M] to communities of over 10,000 population for emergency water supplies (a)	225
Physical Loss and Economic Injury Loans of the Small Business Administration [SBA] (low interest loans for small businesses and individual farmers) (b)	50
Total	844

(a). Only $175M appropriated.
(b). SBA loans funds not used as rates had already declined.
Source: Data from Wilhite, Svoboda and Hayes 1983.

agencies involved to prevent overlap and duplication of their efforts, legislation to define those roles, and 'standby legislation' to allow for more timely response to the drought. Yet as late as 1999 the claim was made that 'to date, none of these recommendations have been acted upon' (US NDPC 1999, 25).

A summary graph of trends in drought adjustment (Figure 7.2) suggested that there were in addition some attempts at water supply protection, improved weather forecasting and further weather modification such as cloud seeding being applied, but they were as yet fairly low levels of application and adjustment. On the farms, summer fallowing, with its erosion risk, was still preceding most cropping in Kansas and Oklahoma and reached 95 per cent of sowings in Colorado. As one commentator observed: 'To me it is striking enough that four decades after the Dust Bowl, we still tolerate serious damaging wind erosion even though we know what causes it and how to prevent it' (Lockeretz 1978: 8).

Thus, despite the evidence by the 1970s that drought was a fairly regular hazard facing the nation, it appeared that the relief response continued to be crisis management rather than based upon careful assessment and improvement from

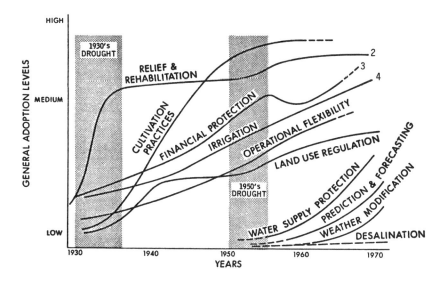

Figure 7.2   Generalized Historical Trends of Drought Adjustments in the USA

Key: 1 = Very rough approximation of relative levels of adoption. 2 = Institutional arrangements for relief and rehabilitation – not actual payments. 3 = Shape of the curve generalized from number of acres insured and amounts of loans in the United States (dip in 1950s reflects lower adoption of insurance at that time). 4 = Based on total irrigated acres in the United States.

Source: Warrick, Trainer, Baker and Brinkmann 1975: 102.

previous experiences. Change was in the wind, but it would take another round of droughts and dust to stimulate it.

The 1988–1989 Drought in the USA

By mid-1988 over half of the conterminus USA (northern Great Plains, Mid West and Southeast) was once more experiencing 'severe to extreme drought conditions', based this time upon evidence from scientific soil moisture indices (Palmer Hydrological Drought Index and the Palmer Drought Severity Index), not rainfall alone. Although not as severe as the worst drought conditions experienced in 1934–1936, the resultant impacts led to financial losses of over $39 billion(US) – 'the most expensive natural disaster ever to affect the nation' (Riebsame, Changnon and Karl 1991: 43). In addition there were the estimated deaths of between 5,000 and 10,000 people from heat stroke – mainly 'poor elderly [people] living in substandard housing without air-conditioning' (Ibid.: 156). What were the details of the impacts and why were they still so costly in monetary terms, but relatively less costly in terms of human lives?

The bulk of the financial costs, some $18 billion, came from lost agricultural production of $15 billion, which varied between crops from a loss of only 9 per cent to a massive 52 per cent, with the main crop (grains) losing 31 per cent of previous years' averages. To this was added the $3 billion cost of claims on crop insurance agencies and emergency federal government drought assistance. Interestingly, not only the obvious field crops were at risk for even the oyster harvest was affected, being only a third that of the previous year. The next largest losses were from increased food prices, some $10 billion, resulting from rises of 1–2 per cent overall. Impacts upon the environment ranked next, with the main losses coming from a series of disastrous forest fires which cost $5 billion in lost timber and fire fighting costs. General Federal Disaster Assistance cost a further $4 billion, while the drastic decline of Mississippi River levels brought the lucrative barge freight services to a halt and forced the redirection of freight to the railways and even to the Great Lakes shipping system. This brought costs of $1 billion to the transport industries involved. Finally, energy costs increased by about $200,000, being the result of losses from reduced hydro-electric production and increased demand for coal-fired electricity for air-conditioning. Less amenable to financial accounting measures were the losses to wildlife from forest fires and the drying up of marshes and other wetlands.

Yet there were beneficiaries from the drought. First were the farmers in non-drought districts who benefited from the increased crop prices brought on by drought-induced scarcity; second were the railways and Great Lakes shipping interests who benefited from the windfall freight traffic diverted from the Mississippi; third were the engineering companies involved in the search for new water supplies; fourth were the power companies who were able to benefit from the increased demand for electricity for air-conditioning; fifth was the commercial aviation industry where the clear skies meant fewer weather related delays and sixth were the construction industries where fewer delays from adverse weather were experienced.

The study concluded that despite the variety and costs of the drought impacts, they were less than expected with Gross National Product only 0.4 per cent below expectations and only 0.3 per cent of the Consumer Price Index rise of 5 per cent being attributable to drought. The farming community had been buffered by government assistance together with some use of crop insurance schemes and by the high prices for what crops were produced. As a result it seemed to have been minimally disadvantaged financially by the drought. Some commercial enterprises had lost money while others had benefited. Urban communities had suffered only marginally from increased food prices and from the periodic bans on outdoor watering of gardens or cars, unless of course they were the elderly poor mentioned above.

Yet the researchers were less than sanguine about the future. They commented that:

> Despite decades of crop breeding, water system development, and other improvements in climate-sensitive technologies, the drought demonstrated that the simple lack of 'normal' rainfall still provokes serious disruptions in agriculture, water supply, transportation, environmental quality, and other areas. It can affect the health and well-being of millions of people and evoke billions of dollars in government aid. (Riebsame, Changnon and Karl 1991: 1)

In fact, they forecast an increasing sensitivity to future droughts:

- as water demand increased faster than available supply with water storage systems already at peak capacity;
- as population aged and became more vulnerable;
- as the natural reservoirs of wetlands were reduced and urban sprawl reduced aquifer recharges;
- and as institutions grappled with the need to anticipate not only the range of historical climatic parameters but forgot to factor in the unpleasant surprises (the events beyond the existing records) which nature can supply.

In such situations, they suggested, even the presence of a similar stockpile of agricultural produce to that which had blunted the impact of this drought might not be of much help. The next decade, however, was to bring some changes.

The Emergence of a National Drought Management Policy

In the 1990s, scientific and political opinions were appearing which pointed out the wealth of evidence of losses but also the wealth of experience of successful mitigation of drought impacts. As an example, the University of Nebraska set up a National Drought Monitoring Center in 1995 with a web site to 'provide users with access to all the information for drought monitoring in a timely and reliable fashion' (Wilhite, Svoboda and Hayes 2005). In fact this predated any routine government monitoring service! But a year later another national drought provided the opportunity to bring this and other expertise together. In response to the 1996 drought a Federal Emergency Management Agency (FEMA) set up a multi-state drought task force to coordinate federal aid and improve drought management. Its report suggested the formulation of a National Drought policy 'based on the philosophy of cooperation with state and local stakeholders', plans to apply past experiences, the creation of a national drought monitoring system for early warning of the threat and institutional structures to assess potential impacts on a national scale, along with regional consultation on specific needs. Finally, it suggested that FEMA, which previously had been focused upon flood, hurricane and storm disasters, should add drought to its repertoire.

FEMA, however, ducked this proposal and passed responsibility for 'drought preparedness and response' to the Unites States Department of Agriculture, on the grounds that agriculture was where most droughts were! The stimulus, however, was sufficient to focus interest, and a National Drought Policy Act was formulated and passed by the US Senate and Congress and became the National Drought Policy Act (Public Law 105–99) in 1998 (Wilhite and Vanyarkho 2000). Part of that initiative was to incorporate the Nebraska monitoring system into a new national system, the 'U.S. Drought Monitor', based on the university's website (http://droughtmonitor.unl.edu/), which by 2002 was receiving more than 5 million hits per year (Wilhite, Svoboda and Hayes 2005: 127).

In 2000 the National Drought Policy Commission, which had been set up by the Act, reported. It acknowledged that there had been no consistent nor comprehensive policy 'driving the federal role to help reduce the impacts of drought'. Any new policy should support, but not supplant, existing efforts to reduce drought impacts, but should as guiding principles:

1. Favour preparedness over insurance, insurance over relief, and incentives over regulation.
2. Set research priorities based on the potential of the research results to reduce drought impacts.
3. Coordinate the delivery of federal services through cooperation and collaboration with non-federal entities (US NDPC 2000: i).

The new policy required 'a shift from the current emphasis on drought relief. It means we must adopt a forward-looking stance to reduce this nation's vulnerability to the impact of drought. Preparedness – especially drought planning, plan implementation, and proactive mitigation – must become the cornerstone of national drought policy' (Ibid.). By 2001 an impressive list of federal research programs related to drought had been created (http://www.fsa.usda.gov/drought/finalreport/fileg/summary_federal_programs_2.htm).

One of the detailed suggestions was that 'For federal action, more rigid triggers [for drought relief programs] such as the 5th percentile drought might be appropriate reflecting truly unusual circumstances' (Ibid.: 5). The regularity of appeals for relief needed to be reduced to identify the real causes for public support. By 2006 that action had occurred with President Bush signing the 'National Integrated Drought Information System Act of 2006', which was to coordinate the multiple state and federal agencies in the flow of drought-relevant information, including forecasting and the sharing of innovative management strategies (Pulwarty, Wilhite, Diodato and Nelson 2007). Time will tell how effective this will be.

Sharing the Experiences

> We are going to have to learn to share, north and south, all of us together. It is the only way we can solve this [drought] problem. (The Governor of California, Edmund G. Brown, 6 January 1977, California 1977: iii)

While the implementation of the new national policy is still in train, what can be said about the collective wisdom available from the US experience on drought mitigation? Several points are worth 'sharing'.

1. *Urban water demand is generally extravagant of the resource and can usually be reduced significantly by official appeals for conservation.* Californian experience in the droughts of 1977 and 1987–1992 showed that such appeals reduced demand by 20 per cent, and some of that saved water could be diverted to other deserving users (agriculture or industry). But there was a catch. Because the urban water was provided by commercial companies, when the demand for water fell so did their income, and as a result after a couple of years of drought the companies proposed to put up their water rates. Not surprisingly, the public was not impressed, believing that these increases were a penalty for conserving water during drought, and became 'incensed by being required to pay more for less', so the Los Angeles Council refused the increases (USACE 1993: 89).

2. *The ownership of water is a vital component in any management strategy.* Ownership of water is intrinsically separate from ownership of land and 'represents an autonomous judicial entity to be bought and sold or left in heritage' (Sternberg 1952: 676). Water law can restrict water management (Table 7.4). Even a brief examination of this table highlights the problems in a drought for irrigators who only have riparian rights versus those who can import water from distant sources owned as prior appropriations, and for governments trying to impose restricted use upon users with riparian rights. In fact, in California where all water is the property of the people of California, riparian rights have been restricted to 'reasonable and beneficial use', which is not usual in common-law usage.

Table 7.4   Riparian and Prior Appropriation Water Rights

Components	Riparian rights	**Prior appropriation rights**
Water ownership	Inherent in the land	Only from prior use or statutory rights
Location of the land to which water rights accrue	Must adjoin stream	Need not adjoin, i.e. can be distant from stream
Use of water	Diversion for 'natural uses' only; no consumptive use	Consumptive; prior rights may claim whole
Amount of water to be used	No limit in theory	Fixed – by prior right Fixed – by licence
Duration of water right	Infinite – does not lapse if not used	Finite – lapses if not used

Source: Heathcote 1983: 270.

In addition, however, California acknowledges three other water rights:

- Pueblo rights of the original Spanish communities before incorporation into the United States, where rights are superior to either riparian or prior appropriation rights and expand with the growth of the community.
- In stream water rights to maintain stream flows to support fish, wild life and recreational uses and prevent diversion for agriculture or urban uses.
- Ground water rights where 'all owners of land over a ground water basin have equal rights and responsibilities to share any controls of use (in drought), [and] Imports [of water] have to be made to users if the basin is overdrawn (USACE 1993: 45).

3. *The successful management of water through the system of Water Banks.* Droughts rarely affect the whole of a nation or large area, so there are usually some locations with surplus water which can be purchased and 'banked' for use in the needy areas. California, where irrigation was supported by a complex system of water transfers and imports, was the scene of initial Water Banks. The 1977 drought saw the US Bureau of Reclamation (whose job was basically to build reservoirs and supply water for irrigation) set up a water bank to buy surplus water and resell to drought affected buyers. In 1988 a Federal Disaster Assistance Act (Aid to Water Transfers) set up the provision for buyers and sellers in the regions, administered by the Bureau. The success of the system rested upon the fact that California was criss-crossed by a network of pipelines, bringing water from Oregon and the north of the state and the River Colorado to the irrigated areas of the Central Valley and the urban agglomerations of San Francisco and Los Angeles. With this infrastructure in place, water could be diverted and delivered relatively simply and quickly. In 1991, in the fifth year of a drought, the Californian State Drought Emergency Water Bank was buying water at $125/acre foot and selling at $175/acre foot (USACE 1993). With the concept of water as a commodity, the market place came into its own.

4. *The importance of Ground Water.* While many of the strategies for coping with drought were aimed at surface water supplies, the value of ground water was highlighted not only as a source of extra supplies, but also as efficient storage for surface surpluses. The 1993 survey of California's drought experiences put as its first confirmed lesson that 'Water in the aquifers continues to be the most effective strategic weapon against drought', and it was claimed to have saved the state in 1977.

5. *Rain making.* Cloud seeding, mainly with silver iodide, to stimulate precipitation had been active since at least the 1950s. The results, however, had been questioned and there were legal problems which had not been resolved, namely: whose was the responsibility for any resultant storm/flood damage; what of the down-wind locations claiming for 'lost' precipitation; and who should decide when to seed?

(Warrick, Trainer, Baker and Brinkmann 1975). Yet in the drought of 1988, 12 different seeding programs were active in the mountain watersheds of California, focusing upon increasing the winter snow pack. An increase in runoff of up to 10 per cent and benefits for irrigators and hydro-electric power companies were claimed (California 1989).

6. *Some management has been successful.* On the other side of the continent the city of New York had suffered serious droughts in 1949–1950, 1961–1966, 1980–1981, 1985 and 1995. The latter had had similar weather patterns to the previous 1980s drought but the impacts had not been as serious, the reason being partly due to the increases in reservoir capacities from the 1950s, combined with increased use of ground water sources from the 1980s, but specifically that water consumption had been restricted and citizens had cooperated to bring down consumption by 10 per cent. In addition agriculture had benefited from the drought by applying more irrigation, and more sunlight had improved the quality of the products, especially fruit (Degaetano 1999)!

7. *Reviewing the Experiences and the Road Ahead.* Looking back from the end of the twentieth century, commentators reviewed drought's impacts:

> The federal government alone spent $3.3 billion during the 1953–1956 drought; $6.5 billion again in the 1976–1977 drought, and about $6 billion in 1988–1989. Generally, there is a costly drought somewhere in the U.S. each year. Moreover, the occurrence of significant water and power shortages and devastating wildfires are, of course, deeply interrelated with drought. Demands for water by a growing population and a dynamic economic system will only intensify. (Truby and Boulas 2001: 14)

By the end of that century, in response, the technologists were hard at work and the national vision had some promise. By 2005, even in Florida, not perhaps the first state to be thought of as drought prone, an impressive array of 'planning and management tools for minimizing the negative effects of drought' had been assembled. They included water conservation and reuse, conjunctive [shared] use, use of marginal resources, desalination, deep ground water extraction, optimization modelling, and decision support systems, together with drought watch alert systems (Alvarez, Rossi, Vagliasindi and Vela 2005). By 2006, while recognized to be a complex process, at least the drought threat was beginning to be tackled on a national scale.

Drought Mitigation in Australia

> A bishop lately arrived in one of these colonies, a very honest man, was requested, during a late drought, to issue a circular prayer for rain. He replied

that an average sufficiency of rain fell every year, and that he declined to petition God to work a miracle until the colonists had done all that lay in themselves to preserve it by constructing reservoirs ... Drought is the worst enemy in Australia, but rain falls sufficient for all necessities, and only asks to be taken care of. (Froude 1886: 111)

Many in the United Kingdom link the thought of Australia with that of drought, and no doubt the rainfall behind the coast belt is deficient. Also periods of terrible dryness have afflicted the continent, and will no doubt afflict it again. It must be remembered, however, that on the coastal belt there is generally quite sufficient rain, and in Queensland even more than sufficient. Also the system of dry-farming is yearly increasing the area that is suitable to the growth of wheat, while the discovery and exploitation of subterranean waters are in many districts a great safeguard against loss of stock by thirst. Furthermore, considerable areas of the various States are now being brought under the influence of skilled and costly irrigation schemes and, in the last extremity, the extension of the railway systems often enables cattle and sheep to be moved from drought-stricken areas to those in which grass and water can still be found. (Final Report of the Dominions Royal Commission 1918: 56–7 [Australia 1918])

Droughts have been a long standing feature of the Australian climate. Described in early Aboriginal myths and legends, they have plagued the invading Europeans from the late eighteenth century onwards. By the late nineteenth century, however, Froude could quote the bishop who urged colonists to make better use of the rainfall, and at the end of the First World War the Report on the future of the new Dominions of the Empire, while admitting that Australia did face a drought problem, was generally optimistic that it would be adequately met. Much was being done to reduce drought impacts through technological innovations and more flexible management. By the end of the twentieth century what was the situation?

A theme report for the State of the Environment report for 2001 covered the main characteristics and effects of past droughts:

The main effect of drought [in Australia] is on agriculture and certain industrial sectors ... including sugar cane, pasture growth and cattle production in tropical Australia and cereals and fruit cultivation in southern Australia. Given the size of Australia, droughts may be ending in one part of a State and at the same time beginning in another part of the same State ... Droughts may last for one season or extend over several.

The cost of drought to government comes mainly through payments, subsidies to agricultural and industrial sectors and family farm restart schemes. However, this is only part of the overall cost to individuals, businesses and communities ... Droughts also result in reduced income for governments from fewer exports and lower tax receipts from primary production. (Manins et al. 2001)

In fact, over the last two centuries, Australia has suffered considerable losses from droughts, with shortfalls in agricultural and pastoral production systems leading to adverse national balance of payments and reduced Gross Domestic Product. Service costs for water provision and communication maintenance were increased as a result of impacts from drought-accelerated soil erosion. Rural settlement was thinned out and locally retreated from economically marginal lands as droughts provided the culminating stress.

There have been benefits, however. Drought-reduced domestic and feral livestock numbers gave the hard-pressed grazing lands a breathing space to make at least a partial recovery. Private transport and government railways benefited from increased freight and livestock movements associated with both drought relief supplies and drought livestock evacuations and restocking. Official drought relief, in addition to the immediate relief (which often channelled funds into areas previously sparsely serviced), provided public works (e.g. improved road construction and maintenance) which were of long-term benefit (Heathcote 1969 and 1988).

Yet, Australian responses to droughts since 1788 have generally seen them as a challenge to the current and future well-being of society, and have shown a preference for technological solutions. Those responses can be classified into three groups: first the perceptions, attitudes and responses of the communities as a whole, second the private strategies of the settlers themselves, and finally the policies of governments.

Community Strategies

In traditional societies, invoking the aid of the gods is a standard response. Invoking God's aid is still practised here in Australia. Although an early Bishop in Australia refused to authorize prayers for rain as we saw above, community prayers for rain are a common response to extreme droughts. Such prayers were made in Queensland in 1902 (*Queensland Times*, 12 October 1987: 6) and in 1983, and a meeting was called in Australia's House of Parliament on 21 October 2007 to pray for rain. Indeed, the Anglican *Australian Prayer Book* has a standard prayer for protection from 'Drought, flood or bushfire'.

> All things look to you, O Lord,
> to give them their food in season:
> Look in mercy on your people,
> and hear our prayer for those whose lives and livelihood
> Are threatened by drought (or flood, or fire).
> In your mercy save both man and beast.
> Guide and bless the labours of your people,
> that we may enoy the fruits of the earth
> And give you thanks with grateful hearts.
> We ask this through our Lord Jesus Christ. Amen

> (Anon 1978: 92)

In the face of drought impacts the offers of help are spontaneous and massive. Australia has never asked for international help, but there have been national Red Cross Appeals and 'Band Aid' concerts to raise funds, the more frequent local mayor's relief fund and appeals for public subscriptions. Bob Geldorf came to play in Goondiwindi in 1993 (*Sunday Mail*, 25 April 1993) and John Farnham's concert in the same year helped raise c.$7 million towards the 'Current Affair' media sponsored FARMHAND appeal (Martin 1994).

For suburban city folk there were a variety of simple self-help strategies such as the use of grey water from the kitchen to save vegetables and if possible flower gardens, and application of mulch to reduce evaporation from the soil. There was even help to create a 'drought garden' by choice of plants and water conservation strategies (Windust 1995). More recently, as a result of the droughts in the early years of the twenty-first century, have come commercial products to improve soil moisture-holding capacity and polymer sprays to reduce transpiration through plant foliage.

Perhaps in Australia we need drought, it has a community function. A recent paper has suggested that drought discourses in Australian culture have a role as a moral drama. In this argument, drought in Australia is seen as a symbolic 'national enemy' against which national energies can be galvanized as part of the national and personal 'character-building qualities of drought'. One reason offered is that because drought is 'recognized as a natural fact rather than correctly perceived as a socially constructed fact' (West and Smith 1996: 99), Australians see drought as an Act of God rather than as it is, a consequence of their own actions in resource management.

A more recent community response, however, has challenged that view. In 2007 a publication by the Victorian Women's Trust (VMT 2007) provided a unique overview of the challenge provided by the droughts of the first years of the twenty-first century – what a later commentator called 'the decade of water, or more precisely, the decade without water, a time when we started to treat water as a valuable and scarce commodity, letting our lawns die for the greater good' (Stewart 2009: 3). The Trust's comprehensive report *Our Water Mark: Australians Making a Difference in Water Reform* provided scientific evidence of drying rainfall trends over the southwest and southeast of Australia during the twentieth century and sharp declines in river flows in the southern portion of the Murray Darling River Basin (Australia's major agricultural area). In addition, the increasing demands for water from expansion of industrial uses (particularly mining), increased commercial tree plantations for chip-board exports, increased allocation of irrigation licences in the Murray Darling River Basin, and the spread of urban areas from the growth of population were highlighted as putting increasing pressures upon surface and groundwater resources. The combination of declining meteorological supplies and the rising demands for water were collectively increasing the risks from droughts.

In fact, by the end of 2006 drought initiated water restrictions were in force in 28 of the 31 cities in Australia of over 50,000 population (Ibid.: 44). The commercialization of public water provision and the unlimited application of water trading and water banks, together with moves towards the creation of a national water market for water used in agriculture by 2014, were viewed with suspicion, as there seemed to be 'no specific democratic reference for this agenda, even though it poses real challenges to the fundamental idea of water as a common good. Instead, we are assured that the market place will effectively regulate future water problems' (Ibid.: 90). The report's proposed '20 principles guiding [future] water reform' provided a carefully constructed blue print for future community and official water management. Whether it will be influential remains to be seen.

Private Strategies

Whether Australians have ever fully accepted drought as endemic to the continent is debateable (Heathcote 1969 and 1988), but there is no doubt that as settlement spread inland down the rainfall gradient, pragmatic and often technologically-based strategies were developed to cope with the increasing incidence of agricultural drought, independent of any official inputs.

For the pastoralists a basic management strategy, particularly in the semi-arid and arid ranges, was first 'opportune use' of the often fleeting livestock feed resources – i.e. the ability to use the resources of feed and water as and when they were available, in order to build up a reserve of profits to survive the droughts when income was negative (Heathcote 1965). Interestingly, the concept has been revived (Westoby, Walker and Noy-Meir 1989) and was still evident in the late 1990s as:

> [A]n attitude prevalent that essentially supports abandonment of management
> of total grazing pressure and instead seeks to run as many livestock as possible
> in order to graze whatever forage is produced before something else eats it.
> (Ludwig et al. 1997: 69)

The environmental impact of such strategies can be imagined!

In part these strategies were a response to the challenges posed by the oscillation between favourable and unfavourable seasons, but they were also aided by improved land transport, including 'beef roads' built into the semi-arid interior to facilitate livestock movement. In effect such roads have often merely improved the official travelling stock routes, originally designated route ways provided for the droving of livestock on the hoof, a technology going back at least to medieval Europe and seasonal livestock movements there. There was one bottom line alternative, however, as one grazier put it to the examining 'Royal Commission into the Condition of the Crown Tenants in the Western Division of New South Wales' in 1901: to 'find a fool and sell to him' (Heathcote 1965: 141).

For the pastoralists with a good sense of geography and the finances to back it, a favoured strategy was an inter-regional spread of properties from summer to winter rainfall areas, thus reducing the risk of total drought loss, and helping to create millionaires such as Sydney Kidman (1857–1935) and James Tyson (1819–1898). The strategy is still visible (Figure 7.3). Technological responses were also favoured: fencing to control access to the range and conserve drought feed by allowing the rotation of grazing areas; excavation of tanks to store surface run-off; deep drilling to tap the Great Artesian Basin from 1887 onwards, and (more recently) drought-resistant breeds of livestock. Finding adequate water, however, did not solve the drought problem. Until the discovery of the underground water supplies, livestock losses were usually from dehydration, but after the exploitation of the underground supplies guaranteed the water supply, livestock tended to die from starvation as all edible feed had been removed from around the watering points. Storage of surplus feed from the good seasons, however, was deemed uneconomic for all except stud livestock (Powell 1963), and starving stock were evacuated to leased grazing lands (as agistment) outside the drought area if available, or sent off to graze the 'long paddock' of the official travelling stock routes if other pastoralists had not beaten them to it, or left to survive as best they could on the range, if those two other options were unavailable.

A recent review of management in various pastoral regions suggested that these traditional strategies to cope with droughts would provide economic benefits provided diversified management strategies were followed. These included reduced stocking rates which enabled 'the biological performance of livestock to be maintained at a more constant level' in drought years; the removal (destocking) of all but breeding stock in the drought; the build-up of stock numbers rapidly when the drought broke; and finally management that recognized that drought was a seasonal risk. In keeping with the technological bent of contemporary management, a computer program 'Drought Plan' was offered as a guide to tailored pastoral management for specific regions (Buxton and Stafford-Smith 1996: 292 and 306).

Some pastoralists of the New South Wales Western Division have already diversified by cropping the beds of the periodically dry lakes for hay or even cereals, thus taking advantage of remnant moisture from the periodic flooding. In the late 1990s some 70 such lakes were being cultivated, with average sizes of some 2792 hectares. Summarizing the land use on these lakes in western New South Wales, a report provided a fascinating insight into the possibilities: 'grazing, irrigated cropping, opportunistic cropping [after rains], frequent dryland cropping [using dry farming methods], fishing, recreation, and water storage' (Seddon and Briggs 1998: 238). You take your season and choose your strategy!

For the farmers, the march out onto the droughty plains has been hard-fought, with human ingenuity stretched to its limits to cope with varying seasonal conditions. The optimal time for seeding was learned by trial and error (Cary 1992); fallowing of land to save moisture from one season to the next began privately in South Australia in the 1880s before the technique received official blessing in 1906, following official contacts with the Campbell Dry Farming system of the USA (Williams 1974). About 10 per cent of the previous season's soil moisture could be carried over locally, but the cost in wind erosion of the finely worked topsoil could be disastrous.

Machinery to harvest drought stunted crops – Ridley's Stripper of 1843 – was developed in South Australia before diffusing rapidly through the wheat lands

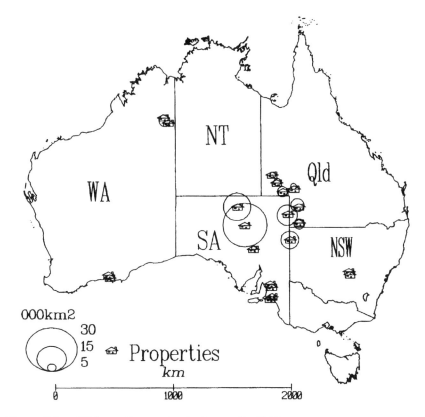

Figure 7.3   Kidman Properties in Australia 1995

Shown are the locations and by proportional symbols the relative sizes of the properties held by the Kidman Company in 1995. A house symbol alone shows properties of less than 5,000 sq.km.

Source: Heathcote 2002: 17.

of Victoria and New South Wales. South Australian farmers were experimenting with drought-resistant wheat varieties before William J. Farrer began experiments. His famous variety 'Federation' (produced during the last years of the 1895–1902 drought) was to push the wheat fields further down the rainfall gradient into the semi-arid lands over the next 25 years.

Water catchments for domestic and stock water were built around bare rock outcrops in Eyre Peninsula (South Australia) in the early 1900s to supplement the wells, bores, dams and tanks (Illustration 7.2). A private strategy for water conservation on farms was developed in New South Wales in the late 1940s. It involved channelling rainfall runoff along the contours – ' key lines' – into multiple farm dams for stock water or irrigation. Despite strong support from the professor

Illustration 7.2   Pildappa Rock and Watershed Channel, Eyre Peninsula, South Australia, 1990s

The rock monoliths shed any rainwater, which is then led off by low walls to underground storages, originally for both domestic and livestock use.

of geography at Sydney University, however, it appears to have been forgotten (Holmes 1960).

On the farms also, opportune use had its advocates and practitioners. The earliest settlers often had to dig their own domestic water storages before the luxury of a piped water supply was provided, and even public rainwater tanks are still evident (Illustration 7.3). Farms in summer rain country are urged to consider 'opportunity cropping': 'When there is not a good amount of subsoil moisture the amount of crop planted should be reduced to avoid risk of costly failure. In better seasons, this trend should reverse itself and result in two crops in 1 year … The farmers who have made money are those who have made the most of the few seasons when it has rained' (Borrell 1994: 21).

When income from the farm proved insufficient, sources of off-farm income were attempted and this strategy, while not solely aimed at drought management, is still very important with 90 per cent of households in the sheep industry in 1979–1980 and 60 per cent of all surveyed properties in 1985–1986 having off-farm incomes (Byrnes 1987: vii, Males, Poulter and Murtough 1987: 6). And this diversification of income sources has much to recommend, if at all possible.

For the urban investor in rural Australia, however, drought seems to have served a useful purpose in that drought-induced losses on rural properties could be offset for taxation purposes against profits made elsewhere. In the vernacular, 'Collins Street' (Melbourne, Victoria) or 'Rundle Street' (Adelaide, South Australia)

Illustration 7.3   Government Built Roadside Rainwater Catchment Tanks, Eyre Peninsula, South Australia
These corrugated iron tanks were for public use.

farmers seem to have successfully written-off rural drought losses against their other business profits.

On Australian farms, however, when the crop starts to wilt there is little to be done if irrigation is impossible, and alternative incomes are not possible. When the crop has died and the soil begins to blow away, there is only emergency tillage (deep ploughing to bring moist subsoil to the surface as protection) to stave off disaster, and that only as long as the subsoil is moist enough to hold the clods together. When that fails, the next step has often been to appeal for official drought relief.

Official Drought Management Strategies

The first official Australian drought relief seems to have originated in South Australia in 1865 when, in response to appeals from pastoralists in the north of the colony, Surveyor General Goyder was sent to delimit the extent of the drought as the basis for subsequent relief. For pastoralists in the drought affected area beyond Goyder's Line this relief took the form of the waiver of their leasehold rents for two years. The relief – essentially an economic subsidy – was justified as the pastoralists on their leases were tenants of the Crown, and as landlord the Crown offered relief which would enable the tenants to survive a natural disaster which they could not control (Heathcote 1981).

Subsequently the scope of drought relief has broadened considerably. In South Australia 39 Acts of Parliament were passed between 1866 and 1990 relating to drought assistance, and this pattern of expanding forms of drought relief was duplicated in all the other colonies, later to becomes states and territories. The variety of official relief measures has included direct assistance in subsidies for in-drought emergency measures, water provision for humans and livestock, feed or evacuation provision for starving livestock, and low interest 'carry-on' loans or cash grants. Post-drought measures have included replacement seed, compensatory purchase and destruction of worthless livestock, public works employment for victims, support for research into drought-resistant animal and plant species, support for irrigation schemes (particularly in the Murray Darling Basin), rain-making experiments, and subsidies for surface and subsurface water storage for both urban and rural needs. South Australia's Marginal Lands Scheme of the 1940s attempted to remove uneconomic farmers from marginal agricultural lands and, by offering their vacated lands to surviving neighbours, encouraged them to diversify into combined crop and livestock activities with, hopefully, less exposure to drought. The problems were nationwide, and the Commonwealth's Rural Reconstruction Commission Reports of 1943–1946 began a long battle by successive governments to stabilize their farming and grazing communities in the face of fluctuating prices and rising production costs exacerbated by the periodic onslaught of droughts. The battle is still on, and the separate story for each of these strategies is to be found in Powell (1989) and Williams (1976).

The story of the 'modern' rain-makers, from Charles Wragge in Queensland with his Vortex Guns to the CSIRO's cloud seeding flights (Figure 7.4), illustrates the complexity of the issues in drought mitigation. Despite the plaintive lament of the researchers when the latter experiments were abandoned in 1981 – that they had the techniques, but not the right clouds (Heathcote 1986b) – there has been some recent revival of interest. A recent paper claimed that while 'Historical dreams that cloud-seeding would drought-proof nations have been abandoned. Modern goals are far more modest, but in the face of global warming even small increases in precipitation in the right place may prove useful' (Luntz 2007: 11). An even more recent paper has claimed that revived Tasmanian cloud seeding over the river catchments of the state's hydro-electricity company had produced some rain: 'A number of independent statistical tests showed a consistent increase of at least 5% in monthly rainfall over the catchment area'. The key seemed to be that the Tasmanian clouds are 'unusually clean and contain large amounts of supercooled liquid ready to trigger freezing [which begins the process of precipitation]' (Luntz 2009: 5). The author did admit, however, that the presence of clouds was a vital requirement, as all the experts have insisted!

For droughts, as with all other 'natural disasters', the states had to bear the initial costs as the Commonwealth had no legal requirement to provide disaster relief. The depression years of the 1930s, however, saw Commonwealth assistance provided under Section 96 of the Constitution which provided for

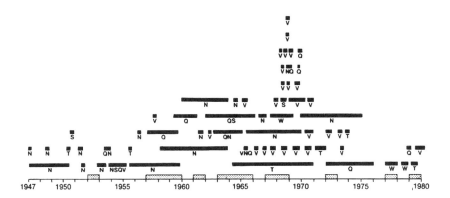

Figure 7.4   Rainmaking Experiments in Australia 1947–1980
Key: Solid bars show periods of official rainmaking experiments undertaken in the states indicated. Dotted bars show major drought periods at continental level. The states are: N = New South Wales; Q = Queensland; S = South Australia; T = Tasmania; V = Victoria; W = Western Australia. Official rainmaking experiments were discontinued by the Commonwealth Scientific and Industrial Organisation in 1981, because of the absence of suitable clouds. However, subsequent experiments have been conducted in Tasmania by the Tasmanian Hydro-Electricity Authority.
Source: Heathcote 1986b.

grants to the states for any purpose. Initially this assistance was to wheat farmers in 1934 burdened by falling prices, droughts and eroding farms, subsequently in 1939 to bushfire victims, and through the 1960s–1980s incorporating drought, bushfires, flood and cyclone impacts (Butler and Doessel 1989, O'Meagher et al. 2000). From 1970 to 1989, with the formation of the Natural Disaster Relief Arrangements, Commonwealth relief came once a threshold of expenditure by each state or territory had been reached. Thereafter the Commonwealth provided initially all, but later only a proportion, of the additional costs (Australia 1996). In effect, drought relief has been for many years the largest component of official disaster relief. From 1962/3 to 1987/8 it absorbed 57.6 per cent of Commonwealth disaster payments (Smith and Callaghan 1988) and from 1983/4 to 1988/9 drought relief formed 47 per cent of all state and territory disaster payments (Joy 1991).

By the 1980s, however, the Commonwealth drought relief programs were under fire. The abandonment of the rain making experiments by the CSIRO in 1981 coincided with a process of revaluation of long-standing policies. The response to the next drought (1982–1983) brought criticism from economists that the relief was inequitable, excessive, and not conducive to efficient rural production. The National Review of Water Resources, published amid the drought in 1983, warned that drought had to be expected and that irrigation was not proof against droughts. In response a National Drought Consultative Committee was set up and a new policy was initiated. Table 7.5 summarizes the changes in South Australian and Commonwealth (national) drought relief policies from before 1866 to 1991.

Australia's New Rural Drought Policy: Drought as an Exceptional Circumstance from the 1990s?

The 1990s saw the introduction in Australia of what was claimed to be a new drought policy. Studies of the actual application and receipt of drought relief in the 1980s had complemented earlier economic criticisms of official drought relief policies and had shown that relief was not equitably distributed. In fact, few producers actually applied for relief (20 per cent of eligible producers in New South Wales in 1982–18983; a similar 20 per cent applied in Western Australia but only 7 per cent were successful), fewer actually got any relief (in South Australia only 10 per cent of eligible producers got relief loans) and in Queensland there was clear evidence of abuse of the system (Smith, Hutchinson and McArthur 1992). There it was shown that over the period 1964/5 to 1988/9 some 36 per cent of the state had been drought declared every 1 in 3 years, and over the period 1984/5 to 1988/9 a total of 65 per cent of relief had gone to 15 per cent of the claimants, with 40 per cent going to 5 per cent of the claimants (Smith, Hutchinson and McArthur 1992: 81–2). In the words of an expert state official who had tried unsuccessfully to reform the system: 'something is wrong when primary producers require government subsidies every couple of years to keep going' (Daly 1994: 2).

Indeed, responsible managers exist and have rarely even considered applications for drought relief. In Queensland in the 1990s, 20–30 per cent of producers in

Table 7.5   South Australian and Australian Drought Relief Policies

Period	South Australian policies	Australian Commonwealth policies
Pre-1866	None	None applicable
1866–1913	Direct economic assistance to: 1) pastoralists 1866 2) farmers 1882,1896,1902, 1904 Research initiatives: Colonial weather service 1881– Tree planting 1882–1888 Irrigation encouraged 1887 onwards.	From 1901–1913 no apparent policy.
1914–1929	Direct economic and material assistance to both pastoralists and farmers 1914, 1919, 1923, 1926, 1927, 1928 and 1929.	No apparent policy.
1930–1964	Direct economic and material assistance to both pastoralists and farmers, but becoming incorporated into broader schemes for rural restructuring as a result of the world depression. Research initiatives: cloud seeding; agricultural. On-farm management assistance as part of debt recovery.	First direct disaster relief (for bushfires) 1939. Thereafter special finance to States for Marginal Lands (rural reconstruction) and subsidises on fertilizers. Initiation of matching States for expenditure for private hardship and restoration of public assets. Cloud seeding began 1947.
1965–1991	Direct economic and material becoming replaced by support as 'carry-on finance', especially after 1977. 1982–1983 a reversion to both direct economic and material assistance as a result of a commonwealth subsidy as part of a pre-election ploy. Thereafter back to post 1977 policy.	1970–1971 change, disaster payments only after State's threshold of payments was passed. Threshold increased 1978–1979 and from 1984–1885 indexed at 0.25% of State's previous 2 year budget. 1981 cloud seeding abandoned. 1989 Drought Policy Review Task Force set up; report May 1990 emphasized need to encourage less dependence on official relief by improved farm/pastoral management strategies.

Source: Compiled from State and Commonwealth legislation, Butler and Doessel 1988, Drought Policy Review Task Force 1990.

drought declared shires did not apply for relief and three-quarters of the successful applicants used less than $5000 per property of subsidies – considered a 'minimal' contribution. Nearly a fifth (14.9 per cent) of users got 65 per cent of the drought relief – these big users were not necessarily big properties or those most in need – greed played its part (Ibid.). In New South Wales over the period 1980–18983 only c.20 per cent of producers in drought declared areas applied for drought relief and only c.10 per cent actually received loans (Smith and Callaghan 1988: 95).

In summary, the major report found that 'In all States the majority of the payments [of drought relief] were to a minority of landholders. The conclusion was that the former system of government drought aid favoured the poorer managers and that climatically marginal areas received proportionately more assistance' (Smith et al. 1992: i.). By then, however, the axe had already fallen.

In April 1989 the Commonwealth removed drought assistance from the list of national Natural Disaster Relief Arrangements (which had been in place since 1970) and set up the Drought Policy Review Task Force to advise on future policies. The subsequent report suggested that a national drought policy should 'encourage primary producers and other segments of rural Australia to adopt self-reliant approaches in managing for climatic variability ... [and] facilitate the maintenance and protection of Australia's agricultural and environmental resource base during periods of climate stress ... [and] the early recovery of agricultural and rural industries, consistent with long-term sustainable levels'. The report specifically stressed that drought was a 'natural, recurring and endemic feature of the Australian environment' and 'the prospect of variable seasonal conditions is a normal commercial risk that must be incorporated into the management of Australian rural enterprises' (Drought Policy Review Task Force 1990, 1: 13–14). Sustainable farm management was expected to be able to withstand drought impacts without official relief efforts.

However, despite its banishment from the ranks of natural disasters, drought relief crept back in. Commonwealth drought support measures over the period 1992–1996 were noted as $21 million in 1993; $42million in 1994; $147 million in 1995; and $211 million in 1996 (O'Meagher, Stafford-Smith and White 2000: 123). Another source gave Commonwealth drought relief of $260 million for 1995–1996 and another $135 million in 1996–1997 (*The Weekend Australian*, 28–29 September 1996). How could this be explained?

The 1996 review of the Natural Disaster Relief Arrangements program noted that although drought had been removed from the list of eligible disasters in 1989, the new arrangements 'also recognize the occurrence of extraordinary drought events for which specific assistance is made available under the Exceptional Circumstances provisions of the Rural Adjustment Scheme ... [and] the Commonwealth has formulated specific exceptional circumstances assistance packages to address severe and sustained drought' (Australia 1996: 19). Entry now appeared to be by the back door, and a recent study has claimed that policy makers seized upon this excuse to continue the relief process because at the time 'there were signs that much of eastern Australia was heading into serious drought conditions' (O'Meagher, Stafford-Smith and White 2000: 120–21).

The identification and monitoring of these exceptional circumstances has proved to be very difficult (Bureau of Rural Sciences 2000, White 1997). Initially, the concern was to establish a time frame and severe droughts (rainfalls in the 5th percentile) occurring once in 20 to 25 years were considered to be eligible. With this as given, however, the main substantiating argument was to be based upon attempts to provide an objective measure of the drought impacts, using rainfall, water supplies, crop and livestock condition, as well as environmental impacts and farm income levels, as indicators of the extreme nature of the drought impacts (White 1996, White and Bordas 1997). As part of this process, a 'national drought alert strategic information system' had been proposed, arising from the apparent dysfunctional relationship between meteorological observations and farmers/ community claims of drought occurrence. The intention was to produce 'land condition alerts' for local government areas based upon the 'best combination of rainfall analysis, seasonal climate forecasts, satellite and terrestrial monitoring, and simulation models of meaningful biological processes' (Brook and Carter 1996: 13). The technique has been actively developed by the Queensland Department of Primary Industry, making use of the Internet 'Long Paddock' site (McKeon 1997), and models of 'Total Standing Herbage Dry Matter' are being tested.

All of these indices, however, were recognized as reflecting the current condition of the land, which reflects in part the management strategies of the farmers or pastoralists; relief based on that condition would reward poorer rather than better managers. A review admitted the difficulties and stressed the learning process still at work, but was hopeful that a combination of objective physical criteria on environmental condition together with information on optimal regional management strategies could provide a basis for equitable drought relief in exceptional circumstances (O'Meagher, Stafford-Smith and White 2000). But by 2005, a critic suggested that the problems had not been solved, concluding that 'there is as yet no objective means of assessing the relative severity and impacts of long-term drought' (Wright 2005: 202).

By late 2009 evidence was accumulating that the concept of exceptional circumstances itself had not produced the key to separating deserving from undeserving applicants for relief. The *Australian* newspaper of 2 November 2009 carried on its front page a photograph of a farmer examining his droughted grain crop in central New South Wales under the headline 'Change or lose assistance, farmers told'. In explanation the report noted that the Commonwealth Minister for Agriculture was 'crafting a shake-up that will shift the focus of drought policy from disaster relief to risk management, ending the spectre of perpetually drought-stricken farmers spending years requiring government assistance'. Support was reported from the President of the National Farmers Federation, who was quoted as suggesting that 'Farmers receiving support should be required to demonstrate a commitment to sustainable and self-sufficient farming' (Franklin 2009: 8)

Interestingly, New Zealand has been paralleling the Australian attempt to evolve an equitable official drought relief policy, having had the same complaints that government drought relief 'had discouraged farmers from carrying out practices that would reduce their vulnerability to extreme climatic events, because the government was there to "bail" farmers out' (Haylock and Ericksen 2000: 108). Their strategy has been to raise the meteorological threshold for 'an event of national significance from a one in twenty year event (5 per cent annual probability of exceedence) to a one in fifty year event (2 per cent annual probability of exceedence)' and to plan the withdrawal of central government support, so shifting the financial burden of drought relief from central to local government and individual resource managers (Ibid.: 111–13).

In the irrigated Australian countryside the American concept of water banks had been accepted as a basis for trading water entitlements by the 1980s, when individuals felt able to share dwindling supplies, especially in the irrigation areas of the Murray Darling River Basin. However, this water trading came under fire when a foolhardy official issue of extra irrigation licences in the latter years of the twentieth century promised water which did not, in fact, exist (Mercer and Marden 2006). This was to cause further problems into the twenty-first century, as we shall see below.

Australia's New Urban Drought Policies

> We are now at a critical stage in Australia's history when it is increasingly clear that past State governments, in particular, have been caught completely off-guard by the rapid onset of climate disruption and have been negligent in planning for future water provision in a country that is becoming progressively hotter and drier, especially in the south. (Mercer 2009)

This criticism came at the end of the period of concern for Australia's climate noted in the Victorian Women's Trust Report. In Australia the standard response to the need for urban water supplies had been surface storage supplemented by ground water where available. Depending upon its intensity, each previous drought brought pressure for additional storages, coupled with searches for further ground water sources. By the late twentieth century, however, the least costly sites for dams had been occupied and there was already a strong lobby opposed to further such storages on environmental grounds.

Responses to increasing drought threats differed between the Commonwealth and the State governments. The Commonwealth proposed the Australian National Water Initiative in 2004 and set up the Australian National Water Commission to be provided with $2 billion dollars over the period 2004–2010 (Pigram 2006). Most of the money ($1.6 billion) was to be spent on encouraging the development and uptake of 'smart' technologies and practices in water use, while the rest was to support surveys and community projects, e.g. water recycling. The States, however, more concerned with the immediate problem of ensuring supplies to their urban population majorities, set about massive programs to construct sea water desalination plants alongside the coastal state capitals to supplement existing surface storages. By 2007 Perth had one plant online and another planned and the remaining capitals were following suit (Table 7.6).

Table 7.6　　Major Desalination Plants Completed or Planned in Australia, 2010

Location	Date completed or planned	Daily output (litres per day)	Output as % of city needs	Cost ($billion)
Adelaide (Port Stanvac)	2011	50BL	Quarter to half?	1.83
Brisbane Gold Coast (Tugun)	2009	125ML	?	1.2
Melbourne (Wonthaggi)	2013?	150ML	One third	3.5
Perth (Kwinana) (Bunningup)	2006 2010?	150ML 50GL	17% 13%	? 0.955
Sydney (Kurnell)	2010?	250ML	15%	2.4

Source: Mercer 2009, Bita 2010.

Desalination, in fact, had been a long-time strategy for the Australian Outback, when homesteads or communities faced inadequate rainfall and used solar power or diesel powered heat to distil brackish bore water. At least 294 such plants were in operation at the time of writing – for communities, mines, and other industrial purposes – while a further 976 were under construction and 925 planned.

Recent research has focused on the potential for recycling of storm water runoff, not least for environmental purposes, but most recently for potential domestic supplies. In 1999 a significant new CSIRO urban water program was launched. This aimed to show that 'urban water, waste water and storm water systems can be provided in ways that are more sustainable, substantially reduce costs, and reduce water availability as a constraint to growth in this country' (Speers 1999: 40). In fact, there is substantial storm water runoff from the extensive urban areas of Australian cities. Estimates of 26.5^2 km for Melbourne, 27.8^2 km for Sydney and 98.9^2 km for Brisbane, suggest an annual runoff of 1285GL would be available for processing (VMT 2007).

The artificial recharge of groundwater by such runoff has been successful in South Australia, where a local urban council, Salisbury, with assistance from the CSIRO and the South Australian Department of Water, Land and Biodiversity Conservation, has created wetlands where storm water is collected, partly purified and pumped into a local aquifer from where it is recycled as 'grey water' through residential areas and public facilities from purple taps (Illustration 7.4). While not on the same scale as the main surface storages, locally such schemes are proliferating and have much promise, although are of necessity dependent upon significant rainfalls (Cribb 2002, Eamus 2006).

A less conventional revisiting of a very old concept, the urban household cistern, has been recently revived as a strategy. Two papers in Australian

Illustration 7.4 Greenfields Wetlands Water Treatment Site, Salisbury, South Australia, 2010

geographical journals suggest that urban catchments (roofs and sealed surfaces) could provide significant amounts of water if collected. For Sydney the estimate was over 1700GL/year (Warner 2009) while for a typical suburb with mixed older and more modern housing, the suggestion was that roof rainwater collection alone could provide 63.4 per cent of all water demand in older suburbs, and more than demand (135 per cent) from more modern structures with larger roof areas (Ghosh and Head 2009). Such a strategy would require the redesign and re-plumbing of housing, but is already being encouraged by official subsidies for new rainwater tanks and installation into existing plumbing systems. The technology exists; all that is needed is the will to adapt and adopt it.

The Question of Insurance

In Australia, one of the first proposals of the newly formed National Drought Consultative Committee, set up in 1985, was for an inquiry into the potential for drought insurance. Earlier, in late 1966, in response to an invitation to prepare a paper on drought impacts in Australia, I had phoned a major insurance company in Adelaide to enquire if drought insurance for crops or pastures were available. There was a pregnant silence at the other end of the telephone line and then the reply came, 'No, we don't – but you could try Lloyds!' Baffled, I was curious as to why drought insurance had not been part of drought management strategies in Australia. After all, the basic rationale for agricultural insurance is clear enough. Individual agriculturalists have a responsibility to protect themselves against injury; insurance would stabilize farm incomes in disaster (drought) years, it would provide welfare support, encourage efficient use of resources, increase the recovery rates of bank loans and help stabilize other rural incomes (Hazell, Pomareda and Valdes 1986: 3).

The further I investigated, however, the more complicated the picture became and I now realize that the full story of drought insurance would justify at least a volume in itself, so all I can do here is to highlight some of the issues.

In the United States private multi-peril crop insurance was apparently offered for the first time in 1899, but only for one year, and similar private schemes were trialled in 1917 and 1920, but the companies lost money and quickly abandoned them. In the light of the agricultural depression of the 1930s and in the spirit of the New Deal, the Federal Government introduced a Federal Crop Insurance Act in 1938, but for wheat only. Other crops were subsequently included, but the scheme was supposedly abandoned in 1981, partly because fewer than 20 per cent of farmers took up the offers, but mainly because the indemnities paid to farmers always exceeded the premiums, so that the scheme was basically a subsidy. In retrospect, it was estimated that an even greater subsidy, probably up to 50 per cent of the premiums charged, would be needed to get a majority of farmers involved. Finally, it was realized that:

> The introduction of a widely adopted crop insurance program, as exemplified by the disaster payments program, appears to encourage crop production in marginal areas and other risk taking in farming. (Gardner and Kramer, 1986: 222)

In other words, it seemed to encourage actions which might be constituted as a 'moral hazard'.

Nonetheless, in 1999 the Agricultural Working Group of the newly formed United States National Drought Policy Commission reported that the Federal Crop Insurance Corporation was still functioning, under the wing of the US Department of Agriculture's Risk Management Agency, which had taken over the job in 1996. Drought was included in the potential risks. There were claims that over 70 crops in over 3000 counties, 'about 75 per cent of the annual U.S. farm production value', were covered. Interestingly, however, the comment was made that 'in order to encourage participation in the program, all crop insurance premiums are subsidized'. In addition, the administrative costs were covered by the Federal Government and thus were not included in the price of the premiums (US NDPC 1999). In other words the concerns of the 1980s had been conveniently forgotten and the whole program was another form of farm subsidy.

Over the border in Canada, prairie farmers had been as badly affected as their neighbours to the south on the Great Plains in the 1930s, but official relief came as food and fuel subsidies, cheap rail freight rates for movement of seed and fodder, and even grasshopper baits, but no national agricultural insurance schemes until 1939. The Prairie Farm Assistance Act of 1939 did allow variable lump sum payments for crop failure if yields were less than five bushels per acre, and if yields were less than four bushels per acre there was provision for payments on up to half the cropped area, and an additional subsidy when wheat prices fell below 80 cents a bushel. Over the period 1939 to 1966 farmers had paid premiums of $C357 million and the federal treasury had paid out $C467 million at 1980 dollar values. This was a ratio of 0.76 of premiums to indemnity, close to the US experience. A further provincial scheme in Saskatchewan begun in 1960 with voluntary crop insurance (which guarded against poor farm management) had better overall results from 1961 to 1979–1980 in the sense that farm premiums provided more funds than were disbursed, but there were years when disbursements were well over premium income. Commenting generally upon the insurance schemes introduced from 1930 to 1980, an official report suggested that they 'have been short-term and reactive in nature' which 'works against the planning of a coordinated drought strategy'. As a result 'these programs have done little to enable those assisted to resist the next drought (and there have been and will continue to be "next droughts") as a result of the direct aid received. Any strategy must be performance oriented' (Saskatchewan 1982: 118). In other words, the relief efforts had been oriented to the immediate problem with no real attempt to anticipate further future reappearance of the problem.

In Australia, where agricultural yield instability is claimed to be among the highest in the world and instability in rural incomes appears to be increasing as export markets become more volatile (Lloyd and Maulden 1986: 157), private crop insurance for hail and fire damage, but not drought, has been available. However, by the mid-1980s the 'cumulative loss ratio' (ratio of premiums paid to indemnities received) was 0.7 (i.e. 70 cents of premium averaged a return of $A1 to the farmer). Remarkably similar to the US and Canadian experience, not surprisingly this was a figure 'which is in excess of levels regarded by the industry as desirable for commercial insurance' (Lloyd and Maulden 1986: 162). The next move was to propose some kind of government insurance scheme.

In response to the call from the Commonwealth's National Drought Consultative Committee for an investigation of the possibilities for insurance, reports were provided by the Industries Assistance Commission (IAC 1986) and the Bureau of Agricultural Economics (BAE 1986). Neither, however, supported the concept of agricultural insurance schemes. They claimed that insurance of crop yields (crop insurance) would prove to be too costly because the large areas affected by droughts and the large number of farmers involved would limit the scope for 'risk pooling' (i.e. spreading the risk to other areas and other hazards), and the insurer would need 'substantial diversification' into other risks and 'reinsurance, or a very high level of reserves. In practice, all three [measures] would probably be necessary' (BAE 1986: 23). Insurance of regional rainfall (rainfall insurance) would be 'cheaper and simpler to provide than crop insurance' (IAC 1986: ix), because local meteorological records could be used. However, the existence of official commonwealth and state drought relief policies removed the perceived need for further insurance and no government assistance for insurance schemes was thought to be justifiable, as it 'would encourage farmers to place greater reliance on insurance to provide in advance against adverse events' and 'may encourage the growing of crops in high risk areas' (IAC, 1986: 38) – the 'moral hazard' problem again.

At the international level global insurance companies are facing more and increasingly costly disasters, whether supposedly natural or human precipitated. Arguments against any form of insurance claim that many in fact are subsidies, which encourage poor resource management and prevent real market forces meeting real needs. 'Yet, most experts agree that even subsidized insurance systems are … preferred to post-disaster aid, and the reinsurance market is not yet prepared to commit sufficient and affordable capital to markets serving the poor' (Linnerooth-Bayer, Mechler and Bals 2008–2009: 11). Added to which is the threat from possible climate change transforming previous estimates of risks. The United Nations Framework on Climate Change has proposed a two tier program for international insurance against disasters. The first tier comprises a pool of insurers to provide a percentage of the losses for high risk nations. A second tier would attempt to assist national insurance schemes facing medium levels of risk by 'capacity building', i.e. technical assistance and possible absorption of some of the costs. Low levels of risk would need to be borne by the national governments and the private insurance sector. The system, however, is as yet untried.

What has been tried, however at a smaller scale, has been a World Bank sponsored drought insurance scheme for groundnut farmers in Malawi begun in 2005. The farmers were allocated credit to grow a specific hybrid crop – groundnuts – which they sold to the National Smallholder Farmers Association of Malawi, and in return they were given partial insurance against drought losses. The proceeds of the crop were paid to the Association which in a good year passed them on to the farmers, after taking out partial repayment of the loan. The repayment and insurance premium represented about 10 per cent of the expected revenue in a non-drought year. In a drought year (defined by rainfall at critical stages in the groundnut growing season recorded at local weather stations) the insurance company paid for at least a portion of the loan. By 2008 over 95 per cent of the loans had been repaid and the scheme has been claimed to be a sustainable strategy to meet drought threats, despite some problems of uneven rainfall data from the scattered weather stations, fluctuating crop prices possibly affecting the size of the insurance premiums and, of course, the threat from climate change and global warming. The reporters, however, were generally hopeful that the scheme provided a viable basic model for local crop insurance in the Developing World (Linnerooth-Bayer, Suarez, Victor and Mechler 2009).

Anticipating the Risks – Would Forecasts Help?

We might ask, however, what would happen if droughts could be forecast? Will the forecasting of drought mitigate its impact? The answer is not as simple as it might appear, and depends first on the accuracy of the forecast, second on the form it takes, and finally on the response of the decision makers, given the commercial context of their decisions.

Most drought forecasts are essentially attempts to forecast the likelihood of effective rainfalls, in other words attempts to anticipate meteorological drought. Such forecasts are obviously relevant to farmers, to city engineers checking their reservoir levels, and to tourist and other commercial enterprises where lack of rain may be either a benefit or a cost. The discovery of the El Niño–Southern oscillation phenomenon in the 1980s and its apparent links to dry periods for 75 per cent of its occurrences for Southeast Africa, northern India, Australia, Indonesia, Brazil and central America was an enormous boost to forecasters' confidence by the end of the twentieth century (Ropelewski and Folland 2000: 27). Despite some fears that it is not infallible, scientists have been encouraged to review a whole host of similar atmospheric linkages across the oceans, not only for the present time but looking back as far as reasonably useful records extend. A recent report listed five ocean and climatic indices:

> Southern Oscillation Index (SOI),
> Multivariate ENSO Index (MEI),
> Pacific/North American Index (PNA),
> North Atlantic Oscillation Index (NAO),
> Pacific Decadal Oscillation Index (PDO),

mainly targeting sea surface temperatures and air pressures, but the list promises to grow.

The framing of the forecasts, however, has proved hazardous. One of Australia's most experienced drought researchers put the problem quite clearly: 'predicting a 30% chance of drought is likely to cause a different response to a forecast of a 70% chance of normal or wet conditions, even though the objective content of the forecast is the same' (Nicholls 1999: 1387). Forecasts were often difficult to interpret, different people were looking for different outcomes from the forecasts, and there were difficulties in translating probabilities of occurrences into purposeful actions. Nonetheless the search for answers continues and the forecasts continue.

But what would happen if the forecasts were wrong? The United States, with its reputation for litigation, has numerous examples of claims for damages from disillusioned resource managers who had believed forecasts and acted upon them. Three examples must suffice. The Bureau of Reclamation provided irrigation water for the Yakima Valley in Washington State, but fearing the effects of a forecast drought over the watershed for its dams, warned farmers that water supplies for the forthcoming season in 1977 would be curtailed. In the end there was no significant reduction in precipitation and normal water supplies were provided. However, in response to the forecast, many managers had modified their seasonal sowings to less valuable and more drought tolerant crops and not surprisingly claimed they were out of pocket as a result (Glantz 1982).

In another example, as part of its national drought mitigation policy the US National Oceanic and Atmospheric Administration issued a forecast of the extension of an existing drought in the mid-western states (Illinois, Indiana, Iowa, Minnesota, Missouri and Nebraska) over the period March to May 2000. However, heavy rains fell in late May and through to July, and the media went to town on this 'failed' forecast, anticipating that major economic hardship would have resulted from managers acting on it. A study showed that most managers in the agriculture and water supply sectors were indeed aware of the forecasts, but while two thirds of the managers did react to them, a third ignored them. In interviews with the responsive managers, 'Most decision makers that took action resulting from the forecast felt that their actions were beneficial' but there were also local reductions in revenues. At the outset there had been disagreements among atmospheric scientists about the validity of the forecasts, and the effect of the failure of the forecast was, not surprisingly, 'a loss of credibility in climate predictions and a reluctance to use them in the future'. The researchers commented that 'credibility is a fragile commodity that is difficult to obtain and is easy to lose', and that in the future, as a precaution, 'Long-term forecasts of drought issued as certain, or interpretable as certain, should not be issued. Levels of uncertainty need to be included' (Changnon 2002: 1042 and 1052).

A more serious error occurred in 1922 when hydrologists based their forecast of the flows of the Colorado River on the previous 18 years of data which had given an annual average of 17.5 million acre-feet. These figures were to be the basis for the international agreement between the USA and Mexico as to the division of the waters. Since then, however, the flows have averaged 11.7 million acre-feet and from 2000–2004 the flow dropped to 9.9 million acre-feet. *The Economist* of 25 January 2003, in an article on 'The western drought: A growing thirst', noted that 'the west squabbles constantly over the difference'.

So it would seem that forecasts are still not as free from error as had been originally hoped. Nonetheless research into atmospheric physics and close study of past records may still provide useful guesstimates of what the future might hold, particularly if the forecast can be both regionally specific and tied to the agricultural calendar. Indeed, one of the most recent studies to hand has claimed that an index of Indian Ocean sea surface temperatures combined with sea surface temperatures associated with El Niño events 'based solely on data for the month of August … provides the best trade-off between model skill and adequate lead time for the Sept–Dec rice planting season' over Java and western Indonesia (D'Arrigo and Wilson 2008: 611). Whether time will substantiate the claims remains to be seen.

We need to remember, however, that for the most part these forecasts have been of projected precipitation and temperature patterns, that is for a meteorological drought. They are not necessarily forecasts of agricultural drought, which reflects more important resource management strategies. As we have seen, agricultural droughts may not reflect meteorological conditions at all, but rather an excessive demand for moisture by ignorant or exploitative resource managers.

Yet another type of 'forecast' concerns the political environment when water transfers are made across international borders. Again this is a topic with the potential for a book, but one limited example must suffice. When Singapore became separated from Malaysia in 1965, the two nations agreed to continue the existing arrangement whereby Malaysian water was supplied by pipeline across the causeway into Singapore. The price of that water was agreed and the arrangement functioned successfully. Since 1998, however, the two nations have been arguing about the future cost to Singapore of that water and an article in *The Economist* in January 2003 on the progress of those discussions brought a response from the Singapore High Commissioner in London. He claimed that the discussions on the price of future supplies beyond 2011 had stalled, with the Malaysians refusing to agree to future agreements until 2059, i.e. two years before the end of the current agreements. As a result Singapore was 'faced with the likelihood of no renewal of supply from Malaysia, … [and] has had to prepare to be totally self-sufficient after 2061', and he was confident that this could be achieved (Letter from Michael Eng Cheng Teo to the Editor *The Economist*, 25 January 2003). Did the discussions suggest the forecast of a political drought? Singapore is now planning to become self-sufficient in water through desalination and wastewater recycling, to end its dependence on Malaysia and the political pressures that sometimes accompanied this dependence (*Straits Times*, 27 April 2010).

Resource managers may or may not accept a forecast, and they could be damned if they took notice and reacted and damned if they ignored it (Pittock 1986)! They face not only the accuracy of the forecast and whether he/she believes it, but also what might happen if it were accurate and he/she did act upon it only to find that the future market was glutted by produce from other resource managers who had also believed and acted upon the forecast? Forecasters attempt to predict only the weather, which might affect production, but not the value of that resultant production on the free market!

A Conclusion – No Quick Fix?

We can conclude, therefore, that technological responses have not eliminated drought from the hazards facing humankind, although they may have enabled many more people to survive its impacts. The technological strategists have not given up however. Plant and crop breeders still attempt to optimize their progeny; engineers continue to improve the efficiency of their water storages, recovery and conduit systems; while atmospheric scientists continue to sharpen their monitoring systems. Meanwhile, politicians continue to try to dodge drought's impacts. Perhaps for them there are some techniques to be learned from both India and Zimbabwe?

In India a climate model, set up in 1980 by the Indian Meteorological Department precisely to warn of potential droughts, appeared to have proved reasonably reliable with its forecasts until 2002, when a drought occurred which was not forecast. Since 'The final product [forecast] is vetted by the prime minister's office because of the monsoon's tremendous impact on the Indian economy', there were 'whispers that the forecast could be manipulated for political reasons', implying that bad news might not have been reported (Bagla 2002: 1267).

President Mugabe, of troubled Zimbabwe, in 2003 had a similar but more blatant technique. In the UK *Sunday Telegraph* of 26 January 2003 a report noted that the 'President [Mugabe] bans forecasts of drought in fear of more riots' and 'The country's Meteorological Office has now been ordered not to reveal its long-range forecasts [of a further two years of low rainfall] before clearing them first with senior presidential aides. They are expected to remove the most negative aspects before authorizing their release'. No news must be good news!

Chapter 8
Drought Reporting and Drought Relief as Moral Hazards?

We make guilty of our disasters the sun, the moon, and stars; as if we were
villains by necessity, fools by heavenly compulsion.

(Edmond in Shakespeare's *King Lear*, Act 1, Scene 2).

Reviewing the twentieth century, Stephen Jay Gould provided in 1999 a
'Commentary: Tragic optimism for a Millenium dawning', in which he contrasted
the material achievements of the century with the moral stagnation which had
paralleled it. He noted Alfred Russell Wallace's similar assessment of 1899.
'Wallace's paradox – the exponential growth of technology matched by the
stagnation of morality – implies only more potential for instability and less
capacity for reasonable prognostication' (Gould 1999: 7). There are some grounds
for seeing the policies for drought mitigation, the actual reporting of drought and
the provision of drought relief itself as being open to question on moral grounds
also.

Drought Occurrence as a Matter of Morality?

As we have seen, the first and obvious example of the occurrence of drought
implying a moral issue is the belief among both traditional and western societies
that a drought was itself some punishment for past human misdeeds, an Act of God
to remind us of the need for moral vigilance.

A more complex argument arose during the Sahel Droughts of the 1970s and
1980s, when political historians suggested that it was the impacts of past colonial
policies of exploitation of resources which exacerbated existing vulnerabilities
to drought. The colonial capitalist economic policies of introducing monetary
economies to raise income from taxes, together with the encouragement of
a switch from traditional food crops to cash crops for export, it was argued,
disrupted traditional drought management strategies by removing food and
livestock as exchange commodities and the basis for traditional kinship ties (Baker
1984, Garcia 1981, Morgan and Solarz 1994, Watts 1984). Before nineteenth and
twentieth century colonialism 'disaster was isolated in space and time. After
colonialism, disaster was symptomatic through space and time … [and] the poorest
classes generally suffer the most' (O'Keefe 1979: 531). By the end of the twentieth
century a global economic review spelled out the results :

African economies are more volatile than most others, because their export
earnings are concentrated on a few primary commodities, and extremes of
weather (droughts and floods) are more severe and have a heavy impact ... the
outlook for the future is more depressing. Levels of education and health are
much worse, population growth is still explosive, problems of political stability
and armed conflict are bigger, and problems of institutional adjustment and
integration in a liberal capitalist world order seem just as great. (Maddison 2001:
161 and 167)

Colonial expansion in the Americas in the nineteenth century saw the creation
of a Frontier Ethic, where apparently abundant natural resources were exploited
until they became exhausted; then the settlers moved on to exploit new lands,
an exploitation which had its own questionable morality. When that movement
involved a shift down a rainfall gradient into drier and more variable rainfalls, as
noted in the previous chapter, drought resurfaced. In North America the advance
of the invading European settlers was basically from east to west, and in the
central plains between the Appalachian Mountains in the east and the Rocky
Mountains in the west, that advance was from well-watered country into the semi-
arid western margins of the Great Plains. Initial reactions labelled the area 'The
Great American Desert' (R.H. Brown 1948) but while this label was discounted
subsequently, droughts certainly troubled the initial settlers and have continued to
cause problems ever since.

In Australia a similar sequence can be identified; initial European settlements
were along generally moist coasts before the march inland headed for the arid
interior. A neat example of the process can be found in the history of settlement
in South Australia. With three-quarters of the modern state receiving less than
200mm of rainfall per year, the initial European occupation in the 1830–1840s
was of the coastal plains and Adelaide Hills where the rainfall was over 500mm
per year. As agriculture was pushed further inland it passed through areas with
between 400 and 500mm in the 1850s–1870s, it was in 300–400mm country in the
1870s–1880s, and by the 1890s the farmers were at the edge of the interior desert
with rainfalls between 200 and 300mm and suffering from increasing frequencies of
droughts (Heathcote 1993). A similar sequence of increasing exposure to declining
rainfalls, but with slightly different dates, could be traced in southern Africa and
in Argentina in South America. In all cases the spread of new settlement was into
more risky rainfall country and was to pay the price over subsequent years.

Reporting the Droughts: Crying Wolf?

Given humanity's tendency to provide aid to fellow sufferers, there would seem
to be evidence that both victims and observers of drought have been known to
exaggerate drought occurrences and impacts for specific ends.

Thus for the evidence from China, where can be found some of the longest human records of weather and climate going back to c.200 BC, the reporting of droughts appeared to provide maximum coverage of impacts in the most accessible areas of the empire. A researcher commented:

> Floods and droughts in these areas would be reported on account of their economic significance, whereas floods and droughts of the same or greater magnitude in other regions might escape notice. The writer is convinced that among all the explanation suggested above … [rainfall, typhoons, terrain, rice cultivation questions] this last is probably the most significant and the most logical. (Shan-yu 1943: 365)

He went on to explain that:

> [F]loods and droughts in important economic areas, such as Chekiang and Kiangsu in the period of the Southern Sung, were reported and conceivably exaggerated in order to affect exemptions from land taxes, while instances of similar or even greater floods and droughts in other less important regions were ignored and consequently escaped attention. (Ibid.: 372)

In addition there was some evidence that reports of such disasters were more than conscientiously reported in order to signal to the central imperial authorities that all was not well with their administration.

The Communist takeover, however, saw something of a reversal in reporting strategies, where drought occurrences were apparently down-played or ignored, partly because of their adverse impact on the success of the national Five Year Plans, but mainly because Marxist Theory had no provision for the recognition of the impacts of natural disasters upon a nation's political economy (Freeberne 1962).

Even international aid agencies have been accused of crying wolf. In a report in *The Guardian* on 21 June 2008 headed 'Ethiopia accuses aid agencies of exaggerating drought', the deputy prime minister of Ethiopia, who is also the minister of agriculture and rural development, suggested that 'some agencies' had been guilty 'of exaggerating the impact of a drought afflicting the country to raise money under false pretenses'. He was quoted as saying 'While we appreciate assistance whenever it is needed, we reject being used as publicity to raise funds under false pretensions'. He was diplomatic enough, however, not to name the offending agencies (Tadesse 2008).

Drought Management as a Moral Hazard?

> There are now two kinds of drought: the real and the rigged. Both can be under way at the same time in the same place. (Sainath 1996: 260)

From moral meteorology or its reporting, is there any evidence of a moral hazard in the management for the mitigation of droughts impacts? The Israeli government faced a specific situation in the 1950s after they took control of the Negev Desert. After initiating government financial compensation to the Bedouin farmers for their crop losses from drought, two researchers found that:

> So *compensation farming* became a most remunerative activity for the Bedouin farmer of the Negev, providing him year after year with a stable income, in good crop years from sale of crops, in bad ones from government compensation. This stability further increased the cultivated area [which increased the risk of drought losses]. (Amiran and Ben-Arieh 1963: 170–71)

After further droughts in 1958 and evidence of unjustifiable claims for compensation, a new policy provided that 'compensation should not be paid to areas which receive so low an average rainfall that no crops should be grown there at all by dry farming methods' (Ibid.: 171). In other words, the government, not the Bedouin, was deciding where crops should be grown, or rather where crops eligible for government compensation should be grown.

Drought in South Africa in 1991–1992 saw the creation of a National Consultative Forum on Drought consisting of representatives from 'churches, civic organizations, liberation movements, unions, businesses, non-governmental organizations, homeland governments, welfare organizations, and the South African governments'. With so many stakeholders it was perhaps too much to expect consistently moral agendas and participants commented that: 'Our experience over the last six months has once again clearly indicated that the present crisis is as much a problem of poverty and institutional inadequacies as it is a drought-induced problem' (Hobson and Short 1993: 4–5).

Suggestions of the mistaken identifications of real droughts have been worldwide. A supposedly serious drought in western Texas in the 1920s was greeted differently by different groups:

> The newspapers outside the state showed a decided tendency to exaggerate the accounts of the drought … But whether the outside newspapers magnified things or not, the [local] real estate agents resented it … It is not strange that those [agents] who persistently insisted upon singing the praises of the country, regardless of whether they were motivated by local patriotism or personal interests, would oppose any movement to discredit their cause. (Holden 1928: 115–16)

By contrast, Texan cattlemen welcomed the drought as helping to remove the threat of 'nesters' (new settlers, who as small homestead farmers were beginning to take up claims on the open range long held at minimal cost by the cattlemen). With fewer resources than the ranchers, the nesters were the more vulnerable to drought impacts and would be the first to abandon their lands.

A study of urban water management in a drought of 1962–1966 in the northeastern USA found that not only had the impacts been less than was officially claimed, but the impacts had been used as justification for expensive dam building for 'drought-proofing'. In fact, more careful management of the municipal water system could have reduced the drought impacts further and reduced the need for expensive new infrastructure. The researchers concluded that 'In short, "drought" need not constitute, as it now does, a convenient natural cloak for hiding past planning failures or garbing for public acclaim plans for building expensive monuments to the "right" to cheap water' (Russell, Arey and Kates 1970: 195).

In Australia by the 1970s the possibility that the reporting of droughts might be manipulated was recognized and at an international symposium on drought a possible reason was provided:

> Drought watch information systems controlled by farmers can become, in time, loudspeakers demanding government relief. While information and participation from local farmers is important to an efficient drought watch system, one must recognize that, as they discover their importance to it, farmers may also learn to alter the emphasis of information given and to dictate values that suit their needs. This apparently happened both in Australia and India. (Sandford 1979: 281)

In November 1966, as part of my own research into the then on-going drought in eastern Australia, I received a letter enclosing a 'Graziers Survey on the Effects of Drought in New England' from the Chairman of the New England District Council of the New South Wales Graziers Association. The letter explained the survey: 'I would point out that this survey was done to try to get an idea of exactly what was happening to the grazing industry on the tablelands at the specific period, with the principal object of making a case for specific drought relief measures, a matter in which, I might add, we were quite successful' (Personal letter from P.A. Wright, 16 November 1966). There is no suggestion of immoral action here, but such special pleading does carry risks.

That meteorological droughts may not coincide in time or space with agricultural droughts, or at least claims of agricultural droughts, has been obvious for some time. In New South Wales the disparity between drought occurrences as defined by soil moisture conditions and as declared at the request of local producers has been noted by Smith and Callaghan (1988). I graphed their data and the disparity increased as the average rainfall decreased (Figure 8.1). The disparity in Queensland resulted in the allegations of rorts and the removal of drought from the natural disaster relief arrangements in 1989, as noted above. Certainly, if one plots meteorological droughts against areas declared to be in drought in Queensland there seem to be anomalies (Figure 8.2).

In the 1980s increasing concerns about claims of drought occurrence and potential misuse of Australian public drought relief funds led to an official enquiry which reported in 1989. In effect each state or territory apparently had its own

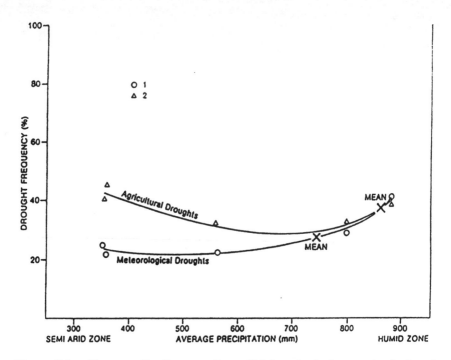

Figure 8.1　　Comparative Cross-sections of Meteorological versus Agricultural Droughts in New South Wales 1976–1982

Key: 1 = Percentage of the period when meteorological drought affected location.
2 = Percentage of the period when agricultural drought affected each location.
The two types of data are shown for five locations, sited on the horizontal axis according to their average annual rainfall, with a trend curve and mean value for both types of drought indicated. Meteorological droughts are months with the Palmer Drought index values below −2. Agricultural droughts are months officially declared as drought months, usually officially verified after a request from primary producers in the location.
Source: Heathcote 1991: 224.

system of drought declarations, and thus 'there are currently eight separate approaches to declaring drought, one in each State and Territory … [and] relevant State and Territory departments confirm that a degree of subjectivity is usually required in determining whether an area should be drought declared' (Drought Policy Review Task Force 1989: 19). Further, there was clear evidence of exaggeration of drought occurrences in at least three of the states:

> In Western Australia, five shires in agricultural areas have been drought declared for ten out of the last eighteen years. In Queensland, some shires have been either partially or completely drought declared for 70% of the time since 1964, i.e. almost seventeen of the last twenty-four years. In the Western Division of

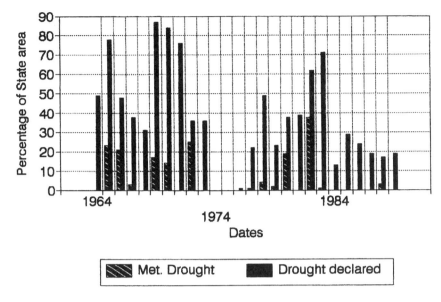

Figure 8.2　Queensland Meteorological Droughts versus Drought Declared Areas 1964–1989
Meteorological droughts are annual rainfalls in the first decile (lowest 10% of records), while Drought Declared Areas are local government areas officially proclaimed as drought affected in the specific years.
Source: Heathcote 2002: 25.

> New South Wales, some districts have been drought declared for three months or longer in twenty years out of the last thirty. (Drought Policy Review Task Force 1989: 20)

Such frequent occurrences belied the real definition of droughts as unexpected or unusual occurrences for which society cannot be prepared, and led to reconsideration of official drought relief policies as we have seen.

One of the criticisms of twentieth century drought mitigation policies in the USA was the multiplication of official agencies involved. For the droughts in the 1970s:

> Some sixteen agencies administered as many as forty separate drought relief programs, often with overlapping functions and geographical jurisdictions. No one definition or objective climatological measure of drought was employed by the agencies as the 'trigger' for initiating these programs. (Rosenberg and Easterling 1987: 2)

Again, with so many interested parties involved there was considerable scope for possible relief applicants to play the field for the best deal.

Given such evidence, we must recognize that there may be more than scientific interest in the claims of drought occurrence, but are there also moral issues in the provision of drought relief measures?

The Justification for Drought Relief

A good case can be made for phasing out all forms of drought relief. The basic argument is that droughts are normal for most of Australia, not aberrations; if pastoralism is not viable under this 'normal' climatic regime, it should not be publicly subsidized. (Cocks 1992: 196)

Justification for the awarding of public funds for drought relief seems to rest upon three 'traditional' premises:

1. *Humanitarian Grounds: the implicit liability of all humans to come to the aid of fellow creatures in distress, whether from natural or human-derived hazards.* This is the thinking behind the Biblical 'Good Samaritan' story and is implicit in the medical Hippocratic Oath.

2. *Social Welfare Grounds: the implicit liability of the rich to help the poor, particularly if the impoverishment is the result of forces beyond human control.* The assistance here is generally privately sourced and usually identified in such examples as 'The Lord Mayor's Fund' or 'Live Aid Charity Concerts' types of fund raising activities.

3. *Political Grounds: the implicit liability of political organizations or governments to support their constituents and protect them from harm to either life or livelihood from either natural or human-derived hazards.* Among the duties of such governments are the protection of the lives and property of the citizens from the impacts of natural hazards or the effects of human intrusions through terrorism or warfare. That support is assumed to attempt to return the victims to some semblance of their previous conditions, and to protect national territory and infrastructures.

In addition, however, the latter quarter of the twentieth century saw two other premises added:

4. *Animal Welfare Grounds: where drought has led to death or significant injury or trauma to domestic or wild animals, relief in the form of feed or water or, in extremity, slaughter to avoid further suffering, might be justified.*

5. *Environmental Grounds: where drought has led to or threatens significant changes in environmental conditions, e.g. massive wind-blown soil erosion threatening future national productivity, or desiccation of wetlands or surface water flows needed for rare or endangered vegetation species, subsidies for environmental reconstruction might be justified.*

Assuming these premises have some validity, the question arises as to how drought relief itself could be claimed to incur a moral hazard, in other words be questionable on ethical grounds? The first problem to be faced is whether the relief is equitable. Jointly editing a collection of papers on *Hazards: Technology and Fairness* in 1986, R.W. Kates suggested that the papers reflected the third generation of ethics in environmental management, the first two generations being characterized by environmental absolutism (the Earth Era of the late 1960s and early 1970s) and the weighing of economic costs and benefits of environmental management (beginning in the mid-1970s). The third generation was 'concerned not only with the overall balance of benefit and harm but with their distribution to specific groups or individuals, with the fairness of the process as well as the outcome' being considered (Kates and Weinberg 1986: 253).

The second problem is the basic one of moral hazard, where individuals expose themselves or their property to loss or injury in the expectation that any such loss or injury will be compensated by others, usually private or government aid agencies. The problem is not limited to drought, as the 2007 floods in the United Kingdom demonstrated. An editorial in *The Daily Telegraph* of 26 July spelled out the issue quite clearly:

> In a country with a thriving civil society, there is a limit to what the state should be expected to do [as disaster relief]. First, its emergency response should be complemented by the activities of voluntary bodies ... The second limit on the state's involvement in disaster relief concerns those who have not insured their properties. People have a choice as to whether to protect their possessions against disaster, and with that choice goes responsibility for the consequences of their decision. The presumption should be that taxpayers' money should not go towards indemnifying those whose failure to insure ... is knowingly to run a severe financial risk. Accepting responsibility for one's actions is, after all, part and parcel of a free society.

The third problem faced by all those involved in providing disaster relief services is how far should the 'helpers' interfere in the actions of the 'victims' (actual or potential)? We have already seen how, in frustration at the cost of continuing drought relief subsidies for Bedouin 'compensation farming' in the Negev, the Israeli authorities told the Bedouin when their crops would or would not be insured. Such a simple strategy overlooks the complexity of reality.

Do elected officials have the right, or even the responsibility, for instance, to prevent what in their judgement is terrible public policy (e.g. preventing heavy industrial growth in high-hazard areas) when the majority of their constituents disagree with them? Can we arrive at fair and just public decisions about hazard management in a context of tremendously unequal political and economic power among interested and affected parties (e.g. will the power and resources of the business community result in local development patterns that favour maximization of profits over public safety?) (Beatley 1988: 2)

This is the dilemma for civil servants as public officials, but a similar problem is faced by any private individuals considering helping out neighbours or other 'victims'. Such questions have troubled philosophers for many years. John Stuart Mill said that individuals' actions should only be interfered with if their actions impinged adversely upon others. Jeremy Bentham fell back upon a utilitarian argument that the decision needed to be based upon an assessment of the social costs and benefits of the action and that citizens had 'social rights' which had to be protected by the State. To the above social rights, however, need to be added the 'rights of the environment' as identified by 'Green' philosophers; not only the rights to life of plants and animals but also the rights of rivers and lakes. Morality in environmental management in such a context becomes very complex indeed.

Drought Relief and the Moral Hazard

Given the above as background, how could the provision of drought relief be considered to be liable to create a moral hazard? Several possible scenarios can be suggested:

1. *Provision of Drought Relief to Non-victims.* In this case, because drought relief is often made available to all resource managers in specific administrative districts, unless there is some kind of assets test there may be a strong suspicion that the better managers, who do not need the relief, can nonetheless apply for and be granted benefits.

2. *Provision of Drought Relief to Excessive Risk-takers guilty of 'Indecent exposure'.* Here relief is offered to and accepted by resource managers who deliberately expose their activities to likely failure in the belief that they will nonetheless benefit from any economic loss thereby. The Bedouin use of 'compensation farming' in Israel has been noted above and there have been many examples around the world since then. An American commentator quoted from an editorial in *The New York Times* on 26 August 1988 on American farmers' repetitive appeals for Federal Drought Relief. The editorial suggested that 'Perhaps the only way to persuade individual farmers to take charge of their fate will be to deny them special relief next time disaster strikes'. In the commentator's view: 'The editorial may have signalled

exasperation with federal farm support during endlessly cycling crises and the end of nation's loyalty to the myth of the independent farmer. Ultimately farmers consumed $20 billion to protect 9 million acres during the 1988 drought' (Opie 1994: 281–2).

3. *Provision of Drought Relief to Resource Exploiters.* Here relief is offered to resource managers who have so exploited the natural resources of their locations that their management systems have collapsed; they have destroyed the natural resources themselves, for example by soil exhaustion and erosion or the exhaustive use of available water resources above or below the ground. The motives for such management will be many and varied, ranging from the pressures of population numbers and the need to provide food for starving families, to plain avarice for profits. Both ends of the spectrum raise moral hazards, but the first case might be condoned provided relief does not become automatic. Summarizing his experience in Africa, a researcher suggested that 'The aim of drought assistance must be to aid the unlucky without subsidizing bad management ... Drought must first and foremost be considered as an unusual event in which its victims are unlucky as opposed to the bad managers. Otherwise, drought can become a chronic ailment in agriculture' (Dyer 2000: 226). Another commentator on drought management in South Africa suggested that 'periods of drought declaration were excessive (up to 70% of the time over a 30-year period for some districts) and were more related to overstocking than to climate' (Smith 1993: 293). Greed for profits should not be condoned by drought relief! 19The current philosophical climate on management of the environment would certainly classify as a moral hazard any support to multinational companies (as opposed to starving peasants) claiming drought relief on failed agricultural or other exploitative environmental enterprises.

4. *Provision of Drought Relief merely to maintain the Status Quo Ante.* Most of the history of drought relief has demonstrated that the relief was provided on the assumption that it was needed to meet a temporary and unexpected hazard, in other words it was given to provide the means to maintain the situation as it was before the drought struck, i.e 'carry on support'. Yet history shows that drought reoccurs with monotonous regularity; it is only in the last 30 years that drought relief operations around the world have attempted to encourage changes in resource management in order to prevent the reoccurrence of droughts.

The unwise provision of continued 'carry on support' relief was noted in the Depression years of the 1920s–1930s in Australia. A drought in South Australia in 1914 had been met with State funded drought relief. A report in 1917 on the administration of the State's Drought Relief Act of 1914 contained the pious hope 'that the lesson of the drought will be remembered by the farmers, and that in future every farm will have its haystack and strawstack as a standby in the event of the State being again visited by a serious drought' (South Australian Parliamentary Papers 1917, No. 10: 40). Yet in 1931, following further droughts and official relief, the State's Agricultural Settlement Commission Report (set up to examine the whole future of farming in the State) complained (and in capital letters for emphasis!):

> It should be borne in mind ... THAT NO FORM OF CROP INSURANCE CAN MAKE AN INDUSTRY PAY IF OTHER FACTORS, e.g. THE QUALITY OF THE LAND, THE RAINFALL, AND THE METHOD OF FARMING ARE NOT FAVOURABLE. Moreover, there is an undoubted risk of careless and indifferent farming if farmers are led to believe that the final returns are covered by insurance. The granting of drought relief has already created the feeling that the Government will always provide funds when crops fail. (South Australian Parliamentary Papers 1931, No.71: 42)

Despite the capitalization of the message, the situation does not seem to have changed there or elsewhere in Australia subsequently and by 1985 an internationally recognized agricultural economist reminded officialdom of what should be the basic assumptions behind drought relief.

> Policy analysts must assume that farmers are aware of fluctuating seasonal conditions and their effects, that farmers take a long-term view when making investment decisions, and that they have contingency arrangements for coping with shortfalls in income and cash in the event of significant climatic variations. In fact, most Australian farmers do cope successfully with droughts. Those that do not should not be farming. (Dillon 1985: 316)

In effect, by the 1980s the process of, and justification for, drought relief was being reconsidered in both Australia and in the United States.

In Canada also there were concerns that insurance was an expensive option and open to misuse in droughty areas. On the Canadian prairies droughts affecting wheat or other cereal crops were reported in 26 of the 51 years from 1929–1980, and in part of the Palliser Triangle (the semi-arid prairies) they were reported every year. 'Saskatchewan Crop Insurance records show that drought related indemnities were paid in almost all of their 23 crop risk areas in every year during 1972 to 1980' (O'Brien 1988: 241). Methinks they did protest too much.

As Australia was reconsidering the justification for any drought relief in the 1990s, in the USA a 'Bipartisan Task Force on Disasters' provided a remarkably similar report on the official role in any disaster relief:

> If homeowners mistakenly believe that the Federal Government will rebuild their homes after a natural disaster, they have less incentive to buy all-hazard insurance for their homes. If state and local governments believe that the Federal Government will meet their needs in every disaster, they have less incentive to spend scarce state and local resources on disaster preparedness, mitigation, response and recovery. This not only raises the costs of disasters to federal taxpayers, but also to our society as a whole, as people are encouraged to take

risks they think they will not have to pay for. (US Congress 1994, Executive Summary)

Official largesse could also discourage other sources of financial support. Reviewing the official and private lending situation in the USA in the 1930s, it was suggested that 'state relief legislation that altered debt contracts to the detriment of lenders substantially reduced the supply of loans and had different effects on different types of private lenders. This past experience suggests that, although such measures as moratoria on farm foreclosures and the recently enacted Chapter 12 bankruptcy law may provide relief for some farmers, they may also result in substantial reductions in the supply of agricultural credit' (Rucker 1990: 24).

5. *Provision of Drought Relief for Political Purposes: Conducting the Gravy Train.* Here drought relief is offered usually as 'carry on support' as recompense for past political support even though the impact of drought is debatable. In the late 1980s a conscientious civil servant 'blew the whistle' on resource managers in Queensland who were rorting the established system of 'carry on support' drought relief. His concerns were contained in his book *Wet as a Shag. Dry as a Bone* (Daly 1994.) Summarizing his experience as an officer in the State's Drought Secretariat, he claimed that:

> In the past, a large number of primary producers, their accountants and other advisors worked at maximizing subsidy payments. Properties got drought-declared when not really droughted and subsidy payments were made for normal management practices. There was a lot of double dipping.

> In many cases, subsidy assistance contributed to an escalation of assistance provided over time. The assistance cushioned the [effects] of adverse seasons on overstocked marginally viable properties, discouraged the adoption of more appropriate stocking rates and postponed rural adjustment. In the past, a lot of properties got drought-declared because of overstocking rather than a severe rainfall deficiency. They received assistance to do more of what they had always done. (Daly 1994: 107)

The assistance came from a Liberal and National Party government whose traditional support had come from the rural sector and which traditionally had been very liberal with public funds for drought relief. Introducing his book he noted that:

> Undoubtedly, the frequency of official drought declarations in large areas of the state had encouraged people to expect droughts to occur often … In Queensland, where large areas had been drought-declared every couple of years, there were obviously serious misconceptions about 'normal' seasonal conditions and the amount of cropping or stocking that land can be expected to carry. (Ibid.: viii)

A leaked critical memo from him to rural stock inspectors led to media exposure of his concerns and his removal from the Drought Secretariat. However, the political exposure helped to lead eventually to the revision of national drought policy in Australia. Daly was eventually exonerated and his removal from his post noted as 'unjustified'. His book remains one of the clearest sources of whistle-blowing on the political context of public disaster management.

6. *Provision of Drought Relief for Political Purposes: Drought as the Scapegoat.* Perhaps the most blatant international evidence of the use of drought impacts as a convenient smokescreen for political incompetence or corruption came from the Sahel region in the 1970s. For the first time, with the help of television, live graphic images of the apparent devastation of the droughts were provided with images of starving children amid the carcasses of dead livestock and withered crops. Local political leaders were quick to use the international public's concerns for the apparent impacts of drought in the rural areas of their West African countries to ask for and receive large amounts of financial support from the United Nations and international aid agencies. There were logistical problems and aid did not always reach the victims. Inadequate and inefficient transport and too few distribution points did not help, but there is no doubt that considerable amounts of the aid were siphoned off, disappearing into government coffers, politicians' pockets and possibly even Swiss banks (Glantz 1976).

7. *Provision of Drought Relief in spite of the existence of the Hydro-Illogical Cycle.* The final example of the danger of a moral hazard in drought relief efforts comes from the unwillingness of either the public or its civil servants to admit that drought is a regular threat. This unwillingness was first identified and labelled as the Hydro-Illogical Cycle by D.A. Wilhite (Wilhite 1993b: 93) and is illustrated in Figure 8.3. He described the cycle as the process by which the threat from what is considered to be a natural disaster is not recognized until the disaster occurs. At that time private, community and official concerns peak and a crisis management strategy is implemented, aimed at containing the worst aspects of the disaster and compensating the victims as much as possible. Once the initial impacts have passed, however, the interest of all parties soon wanes and the experience of the disaster tends to be forgotten as attention switches to other problems. Further reoccurrences of similar disasters result in the same cycle of crisis management with heightened concerns, followed by rapidly declining interest. There appears to be little attempt to learn that crisis management does not address the inherent complex nature of the drought hazard. A neat example of the existence of the cycle comes from the suspension in 2004 of the production of two official drought related leaflets by the Queensland Government, *A Survey of Seasonal Conditions* and *Regional Crop Outlook: Wheat*, due to lack of demand and funding.

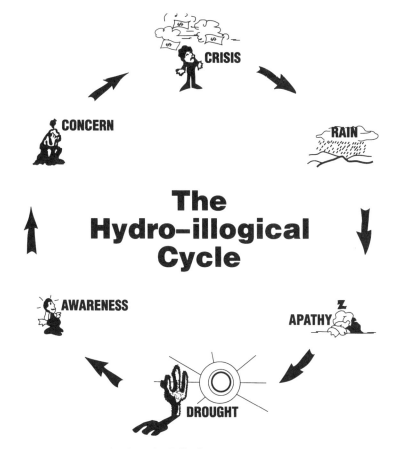

Figure 8.3   The Hydro-Illogical Cycle
Source: Wilhite 1993b: 93.

Explaining the Hydro-Illogical Cycle

The question might be asked, why does this illogical cycle keep reoccurring and recreating the conditions in which the process of drought relief raises questions of morality? Various explanations have been offered:

1. *Drought is regarded as a random, non-normal event, which cannot be predicted and therefore mitigation of its impacts cannot be planned for.* This is the result of:

> The 'variability inherent in drought. This contributes to the view that drought is a random, rare event – by definition one for which there can be no planning'. (Easterling 1987: 1)

2. *Lack of prior knowledge of drought impacts and response strategies.* This may result from:
- a) The lack of information available (i.e. stored in the communal memory) 'in tracking the development of drought and its impacts';
- b) The lack of knowledge on the social and economic effects of drought;
- c) The lack of knowledge of the mechanisms for coping with drought;
- d) The above deficits may result from a turnover of population, resulting in the presence of newcomers, with no prior experience, when the next drought occurs. (Ibid.: 2)

3. *Because the impacts of drought are so slow of onset and usually so spatially extensive, communities find it difficult to devise a comprehensive approach to such a large scale threat.*

4. *Because drought is usually first experienced in a rural context, societies dominated by an urbanised population, somewhat buffered against drought impacts, tend to play down the significance of the event and are unwilling to contribute resources to mitigate the impacts.*

5. *Because the spatial location of droughts varies.* This may result from:
- a) Colonization of new lands which brings new settlers into country more prone to drier climatic stresses;
- b) The introduction of more moisture-demanding crops or livestock management systems into an old established settlement area resulting in new vulnerability to droughts.

6. *Because the temporal pattern of droughts has changed.* This may result in:
- a) Changes in the duration of droughts requiring a change in management or mitigation strategies;
- b) Changes in the frequency of drought occurrence which, if less frequent, may result in loss of memory of past management strategies, or if more frequent may result in the failure of traditional mitigation systems (e.g. reserves of fodder or money) based upon infrequent drought stresses.

7. *Because recent technological innovations had supposedly reduced the perceived threats from drought, or in some way 'drought-proofed' the society.*
- a) Irrigation is the best example here as it has been traditionally seen as the answer to the threat from drought. In fact there is a large body of evidence to suggest that irrigation does not drought-proof a society; rather a series of factors may make that society more vulnerable. These are:
- i) Presence of irrigation may encourage a more intensive land use, which will be more vulnerable than the previous less intensive uses should the water supply fail. The best example is the Green Revolution which encouraged planting of more productive rice varieties but which required irrigation for

optimal production. In drought in such areas there would be no production whereas traditional crops would have produced something;

    ii) The intensive land use is unlikely to have surpluses available as drought relief supplies to other areas, because there is the need to maximize land use for immediate profits rather than for future disaster relief.

b) The widespread use of fallowing on the Great Plains is claimed to have resulted in rises in water tables and reduction of croppable areas (Blakeburn 1993: 19)

8. *Because people/communities can handle only a limited number of worries and if drought is not present, or imminent, other concerns are more pressing.* The worry bead concept comes from the closing address to a conference on Natural Hazards in Australia by Robert Kates:

> I will leave you [with] … a vision of the world carrying a set of worry beads … they are common in Mediterranean lands. They are carried, rubbed, twirled, in a comforting manner. Now imagine a bead for each of our worries and all having a fixed number of worry beads … even the society has a limit to the number of things it can worry about … and the question remains when and where shall it [society] rub its worry beads. (Kates 1979: 520)

9. *Because people are optimists and expect stressful situations to improve, and drought, after all, so far in history, has disappeared, although it tends to reappear!*

10. *Because of the impersistence of memory, or not learning from experience.* As George Bernard Shaw is supposed to have said: 'We learn from experience that we do not learn from experience'. What is the evidence?

A Convenient Memory Loss?

Claims that farmers do not learn, and have not learned, from their past experiences of drought impacts and associated problems are many. In 1967, in the midst of a severe drought in eastern Australia, a Bankers' Conference was told:

> It was said that many of the New England graziers had had no previous experience of drought and made panic decisions in 1965 [to sell stock] … However the view was expressed that farmers located in areas which experience drought no less frequently than at 10 or 15 year intervals tend, with the passage of time, to forget the skills acquired from a previous drought. (Reserve Bank Australia 1967: B7–4)

A decade later a timely warning against complacency based upon lack of on-going drought stress was voiced by an agricultural economist:

In the 1970s there has not been a drought of the intensity or duration of that of, say, the mid-1960s. Although the Commonwealth-State system of negotiation has in the past few years handled a variety of localised disasters, perhaps the range of measures relating to drought should now be matched against the current state of the rural sector? With reduced agricultural maintenance and investment rates in the past decade ... the pastoral resource base has been run down. This may predispose increased sensitivity to future drought. (Anderson 1979: 28)

This forecast was borne out in the subsequent severe drought of 1982–1983.

Experience of drought management has been no better in the USA. In 1977, in the midst of an extensive drought, an experienced researcher noted that the objectives that would improve drought mitigation strategies were known and that indeed much knowledge of such strategies existed, but research to achieve those objectives and coordinate that knowledge was not being undertaken and, if the past was any guide, the research would not take place. (Kates 1977: 2).

Ten years later, the earlier complaint had not been resolved. At an International Symposium on Drought, held in Nebraska in autumn 1986, another researcher noted the paradox that while drought is 'a normal feature of climate':

> Yet, as droughts come and go, left behind are the visible scars of human suffering along with the usual debates over the efficacy of ad hoc relief efforts and, at best, inadequate or incomplete plans for dealing with future droughts. With the first rains comes a new sense of security, relief efforts are dismantled, plans for the next drought forgotten, and society resumes its so-called harmony with climate until the rains fail and the cycle begins anew. (Easterling 1987: 1)

As a result crisis management remained the basic response to drought threats.

On the American Great Plains, an area long troubled by extensive and destructive droughts, similar complaints were being made in the late 1990s, although more than memory loss was identified as responsible. Paul Blakeburn, Director of the US State Department's Office of Ecology, Health and Conservation sounded several warnings:

> In the farming areas of the great plains we are currently witnessing at least four disturbing phenomena [which are increasing the vulnerability to droughts]:
> i) As farming equipment has gotten larger, many farmers have stopped using soil conservation measures introduced after the dust bowl. Shelter belts are being cut down and strip farming is becoming less common.
> ii) In some areas, ... methods have been so successful at conserving water [that] we are now experiencing a phenomenon we call "saline seep" [where] salt laden ground water is flowing into areas where the water table surfaces and destroying the ability of these areas to grow crops.

iii) In some of the great plains where sprinkler irrigation systems were introduced after the Second World War, the farmers are in effect mining the groundwater – the consumption rate exceeds the recharge rate.

iv) In some areas, farmers are plowing land that probably should not be plowed in order to be eligible for certain government programs.

He went on to observe that:

i) In some areas, capital intensive projects were constructed that have not been maintained or, in the alternative, have only been maintained with constant infusion of funds from the government.

ii) In some areas, the composition of the vegetation is significantly changing. (Blakeburn 1993: 19)

One feels that he was being diplomatically circumspect in his ubiquitous use of 'In some areas'. Pointing the finger may not have been politically acceptable?

In conclusion, there does seem to be some support for Gould's concern for the Millenium, noted at the beginning of the chapter, if society's attempts to apply drought relief are any guide. But this may be just the complex context of humanity trying to provide relief for fellow victims of what have been traditionally seen as Acts of God. What is the evidence when we widen the enquiry to the arts, literature, and the sciences?

Chapter 9
Drought in World Literature, Art, Philosophy and Community

So far we have examined the nature and impacts of drought from an academic point of view using the publications of researchers, officials and, where possible, resource managers themselves (from farmers through to commercial and government planners). But it is not surprising that the recurring phenomenon of drought, with such immediate and potentially dangerous impacts upon people, their property, their livelihoods and environments, should have a resonance upon human cultures and find its own image in literature, art, and communities' philosophies and attitudes to life. We might now ask:

- What form does that image take? Is it always threatening and destructive of all life forms or does it provide a challenge to stimulate humanity's 'Contriving Brain and Skilful Hand' (Malin 1955)?
- How is drought perceived as compared with other 'natural hazards'; has its relevance and significance suddenly increased in the wake of the tide of concern for global warming?
- Is it again the imminent destructive force threatening all life on earth, or do the technologists still think that they can fix it?
- Can societies survive drought, or perversely perhaps, do they need it as a kind of moral imperative?

To identify changing attitudes to drought and the management of its threats we will look back at some of the evidence already provided in this book, but also at the evidence from general literature, graphic art, and general philosophical writings. This will be very much a personal assessment and flawed as a result, but hopefully it might identify some issues of interest

The Perception of Drought in Literature

Literature is defined as both travel-writing/documentaries and fiction. It is a vast and varied Aladdin's cave of perceptions and experience, but we must recognize that it omits much of human experiences which were, and still are, never put to paper, and of which only a portion may ever be 'locked up' in oral histories. Introducing a series of articles on the deserts in literature for the *Arid Lands Newsletter*, its editor commented on the non-western examples provided, suggesting that 'writing may not be the best way to probe the ways in which indigenous cultures see their inter-connections with the deserts they live in, particularly if the cultures in question come from a largely oral tradition' (Waser 2001).

Yet the written word has usually been the umbilical cord by which societies have maintained their links with the past, rationalized their traditions and projected their visions of the world around them, real or imagined. The daily routines, the attachments to places, the intrusions of the natural cycles of the seasons or the disruptions of natural disasters, and the structures by which societies govern themselves and try to protect themselves against threats from fellow humans or nature itself, comprise a treasure house of riches. This would justify at least a volume to itself, but all that is offered here is a brief overview of what that treasure house has to offer on human appreciation of drought and its impacts upon human societies, actual and potential.

As an example, as part of the scientific investigations into the extensive drought in west Africa from 1968–1973, researchers noted for the Toureg and Somali peoples that drought was portrayed in their literature 'as a short term or cyclical phenomena, connected with the short-comings of human societies. Religious leaders are expected in times of such disasters to intercede with God'. The poetry of the pastoral Muslim Fulani people, while praising God, was also used as 'a psychological support in times of adversity. References to water, and its over-riding importance to society, occur constantly'. Within the Muslim farming communities in The Gambia, 'If Islamic religious leaders fail in their intercessions [with God], then there is a tendency for the women in particular (as also among the Somali) to return to traditional pre-Islamic customs designed to encourage rain' (Dalby and Harrison Church 1973: 18–19).

Literature in which drought appears might be hypothesized to include at least three situations where drought is seen to be an important characteristic of location described in the literature. First, might be those regions where drought was a frequent and, indeed, perhaps the dominant, characteristic of the region. Here the droughts etched the regional character; it was the phenomenon which completely dictated life and livelihood, the landscape and life style depicted the dominance of drought. Second, might be the accounts of places where, or times when, periodic droughts appear to challenge the livelihood or even the survival of societies, and where the literature shows the attempts to meet those challenges. This literature might provide an insight into a specific culture's understanding of the role of humans in the global environment, especially the culture's attitude to the management of environmental resources, and particularly its beliefs as to the scope of that management and the details of the successes or failures in coping with the drought threat. Finally, in the realms of science fiction can be found examples of literature which portray drought as the ultimate global natural catastrophe, destined to destroy civilizations and even the global ecosystem itself. Some examples of each can be suggested.

Drought as the Distinctive Characteristic of a Region

The regional novel traditionally has been accepted as the medium by which a society, its members and its immediate environment are portrayed most vividly. In one view it provides:

> A conscientious presentation of phenomena as they really happen in ordinary everyday life on a clearly defined spot of real earth, a firm rejection of the vague, the high-flown and the sentimental, an equally firm contact with the real: these are the marks of the regional novel, which occupies in fiction the place of the Dutch school of painters in art. (Bentley 1941: 45)

Several examples exist of the portrayal of a region regularly devastated by drought, and one where drought is established as the major characteristic of the environment and the human life within it, and I have chosen three to indicate the scope of the fictionalized image, although the last comes closer to the documentary medium. First is the classic *The Grapes of Wrath* by John Steinbeck (1989), describing the droughts which enfolded the 'Dust Bowl' on the Great Plains of the USA in the 1930s. There is no doubt as to the real impact of those droughts as crops and livestock died and fields blew away. Steinbeck provided a family saga to parallel the dust storms, as commodity prices collapse in the economic depression, properties are seized, and farmers ultimately flee west to the supposed haven of California, where irrigated crops survive and need hands to harvest them.

Second is John Updike's *The Coup* (1978), set in the Kingdom of Kush in the Sahel of west Africa, where the 1970s droughts are in full swing and international aid is hovering. The ruler President Colonel Ellelou had deposed the previous monarch and now rules by combining Marxist Socialism with the theocratic populism of Islam. With a background in the French Army in Vietnam in 1954 and despite an American university education, he is opposed to any foreign aid, especially American, and gets his people to burn the breakfast cereals, spam and milk products which arrive as that aid. Despite feeling responsible for the miseries of the drought, which he witnessed as he traveled incognito through the country, he sees 'the humanitarian catastrophe … [as] the human condition'. Himself deposed by the next coup, whose leaders hope to profit from the foreign food aid from the West and military aid from the USSR, he is exiled to Nice with one of his four wives. His moral dilemma, influenced as he is by East and West, a strong conscience and strict Moslem beliefs, has led to his downfall and inability to come to terms with his main problem, the drought. The novel is a political commentary upon the complications of drought impacts and responses in a part of Africa where drought is a traditional visitor and foreign aid often inappropriate.

The third, perhaps the most striking example, is the Brazilian writer Euclides da Cunha's *Os Sertões* (1901) (translated as *Rebellion in the Backlands*), which was set in northeast Brazil and was one of the stimuli for Arons's book to be noted below. The book described the crusade of a Catholic priest, Counselor Antônio

Conselheiro as he became known, to create a religious community in the semi-arid back country of northeastern Brazil. Finally settled in a small town, Canudos, they appeared to be successfully establishing a communal subsistence lifestyle before they were hit by drought in 1893. The drought sparked the customary social unrest and the authorities thought the community was the core of a rebellion. In reply, a series of military expeditions were sent to crush it. The military campaign took four years, involving four separate military columns and the rebellion was only crushed by the last in 1897. This last campaign involved virtually the whole of the Brazilian Army and resulted in the slaughter of the inhabitants of Canudos, the levelling of the town, and a final death toll of over 15,000 military and civilians.

The retreat of the group to Canudos, and the approach of the invading military expeditions was through semi-arid vegetation, the *caatinga*, the thorny scrub covering the dry and hostile uplands. The narrative of the peasants' attempts to cultivate amid the scrub, and the barrier it posed to the invading forces, created an image which has fired the imaginations of all who have read the book. As a student I found it breathtakingly vivid, and the conflict between the sophisticated military machine and the primitive guerilla warfare of the peasants leading to the eventual bloody massacre has the dimensions of a Greek tragedy.

Da Cunha had served in the Brazilian army as an engineer, but resigned in the early 1890s. He visited the ruined settlement after the final battle and was appalled by what he saw and by what he had found out about the tragic events. The result was his masterpiece, which continues to provide the most impressive commentary upon the relationship between humanity and nature in a semi-arid environment up to the present day. In Arons's words:

> The epic story represents the first time that an author defined the region as a metaphysical space of war and peace where humanity and nature were in constant combat, reflecting not just the state of humanity in nature but the inevitable conflict between civilization and barbarism. (Arons 2004: 33)

Da Cunha had described in detail the flora and fauna of the region as playing a significant role in creating the independent but also somewhat fatalistic outlook of the peasants. At the same time the region comprised a formidable terrain which took costly toll in men and equipment of any invaders, no matter how sophisticated their armaments. For Da Cunha the region's drought was worse than the war itself. 'Unlike the heaven that is Canudos, drought is the hell of the desert' for peasant and soldier alike. His narrative 'etched into the Brazilian imagination a landscape of memories, illusions, and realities. The war that took place was a brutal one; it was also a metaphor for the interactions of humans and their natural environment' (Ibid.: 33). And his narrative was to provide the template for most subsequent descriptions of the northeast and has been plagiarized subsequently on several occasions by journalists and novelists. Da Cunha had created the archetypical drought-debilitated environment.

Drought Seen as a Manageable Challenge to Society

Here, while drought is depicted as a natural disaster, its impacts are seen as challenging the livelihoods of the community or main family/characters, but not necessarily seen as automatically life-threatening. The documentaries or novels show how the challenges are faced and sometimes, but not always, successfully met. Such literature will show how the characteristics of the society itself may make it more vulnerable to drought, and may also, alternatively, show how some attempts to cope with drought may be successful, while others can be shown to be flawed and to exacerbate its impacts.

Two Field Reports on Drought Impacts and its Mitigation: Northeast Brazil and Rural India

Two books, N.G. Arons *Waiting for Rain: The Politics and Poetry of Drought in Northeast Brazil*, 2004, and P. Sainath *Everybody Loves a Good Drought: Stories from India's Poorest Villages*, 1996, condense within their pages the complex relationships of humanity with drought and with humanity's attempts to cope with its impacts. I will try to précis their common findings because they have fundamental issues which need to be recognized in our perceptions of the phenomena. This will not do justice to the eloquence of the authors, but it may encourage readers to sample the originals for themselves.

Arons, who graduated as a lawyer from New York University, spent time with the United Nations High Commissioner for Refugees and civil liberties organizations. Inspired by N. Scheper-Hughes's *Death without Weeping: The Violence of Everyday Life in Brazil*, he obtained a Fullbright Fellowship to investigate the 'culture of drought' in Brazil. His field experiences from the two years he spent there were the basis for his book. Sainath trained as a journalist and joined the *United News of India* in 1980, but left editing positions in Mumbai in 1993 to take up a *Times of India* Fellowship to research rural poverty in India. His 84 newspaper field reports, which form the basis of his book, brought international awards for journalism, and the book was rated by *The Australian* newspaper as worthy of comparison to Engels's *Condition of the English Working Class*, but written with more humour.

Here then were two field-men who each brought their experiences from two great nations to the common challenge of drought. Despite contrasts in the detail of culture and religion, the overall message from both is of the grinding poverty of the peasant farmers in the face of periodic droughts, whose impacts are exacerbated by the socio-economic context of peasant life and the corruption and incompetence of official drought relief efforts.

There is no doubt of the significance of drought events for both nations over a roughly comparable period from the late eighteenth century onwards (Tables 9.1 and 9.2). In both areas, there was loss of life, rural distress, local famines and out-migration. The story did not begin in the eighteenth century, for it had been going on for centuries beforehand, but that it should still be going on in the twentieth century was disturbing. Why was this so? The answers seem to rest partly in the socio-economic context of rural life in both nations, together with the human tendency to exploit public monies for private purposes, particularly where political advantage can also be gained by the situation.

Table 9.1   Droughts in Northeast Brazil from 1790–1999

Source	Dates
Jesuit Archives	1790–1793, 1824–1826
Arons	1826, 1877, 1889, 1893, 1915, 1932, 1936, 1947, 1952, 1957–1958, 1979–1983, 1992–1993, 1999

Source: Data from Arons 2004.

Table 9.2   Droughts in India from 1790–2000

Dates
1792, 1804, 1812–1813, 1833–1834, 1838–1839, 1848–1849, 1850–1851, 1853–1854, 1868–1869, 1877, 1891, 1896, 1899, 1905, 1911, 1915, 1918, 1920, 1941, 1951, 1965, 1966, 1972, 1974, 1979, 1982, 1987, 1988, 1999, 2000

Source: Data from Rao, Giarolla, Kayano and Franchito 2006: 297.

The Socio-Economic Context of the Drought Victims

The 'Drought Polygon' of northeast Brazil, as first defined officially in 1936 and later revised finally in 1957, by the 1990s had a population of 18 million within its 1 million square kilometres (at 18 per sq.km). Here were contained two-thirds of the nation's poor, with an infant mortality rate of 75 per thousand compared with Brazil's national figure of 58 per thousand. Rainfall averaged 700mm per year but ranged from 250 to 1000mm within the area and the summer peak was notoriously variable. Also included here was the site of the *Rebellion in the Backlands* mentioned above.

The Indian evidence came from five states where the drought impacts were worst and the reports came from the two poorest districts in each state. The ten chosen districts totalled 101,769 square kilometres, with a population of 14.5 million (at 142 per sq.km). Rainfall varied from 350mm to 1100mm and again was notoriously variable.

In both cases a poor peasantry struggled to provide sufficient nutrition, even in the good years, on their small landholdings or from poorly paid employment on large estates, often in debt to money lenders or the local landlords, so that the periodic failure of crops and pastures in a drought usually meant more debts or dispossession, and eventual flight as out-migration. Over time the poor's access to water supplies appears to have been colonized by landlords and entrepreneurs, the story being brought up to date by the development of deep well bores. Sainath traced the story in one of his districts:

> In Kovilpatti, many things happened once the borewells appeared. The first effect was the drying up of the traditional dug-wells of the small farmers [from the draw down of the water table by the new deeper wells]. The next step was for the bore well owners to sell water to those [small] farmers, or to deny it to them at critical moments, ravaging their crop and often entrapping them in debt as a consequence. The third step was to demand a part of their produce. The fourth was to dictate what they would produce. The fifth was to move from exploiting their dependence on water to acquiring their lands. (Sainath 1996: 278)

The result could be drought-affected peasants even at times of average rainfalls. As Sainath suggested, 'That happens when available water resources are colonized by the powerful' (Ibid.: 257). Similar situations were found in Brazil. Here, the government agencies set up to solve the drought problems created engineering solutions by 'drilling deep water wells, and building a large number of dams' both for rainfall storage and to control river flows. Despite the infrastructure benefits 'these actions proved ineffective in reversing the situation of poverty and social convulsion worsened by the occurrence of droughts, since the roots for the misery in the region lie in factors other than climate' (Nobre and Cavalcanti 2000: 77). Only some basic land reform to restore some kind of equality of resources and land ownership would seem to address this privatization of communal resources in both these areas.

The Exploitation of the Public Purse

In India one of Sainath's most surprising discoveries was of drought relief regarded as 'the third crop'. A confidant told him in Bihar in the drought of 1993: 'We have a good drought going here … The Big People are making much money out of it. And the Block Development Officer (BDO) [district government administrator] has gone to harvest the Third Crop … Drought relief … the money that comes in as relief makes the powerful richer than they were. It's quite good business. We like a good drought here' (Sainath 1996: 1). As the result of his investigations Sainath concluded that 'Drought is, beyond question, among the more serious problems India faces. Drought relief, almost equally beyond question, is rural India's biggest growth industry. Often there is little relation between the two' (Ibid.: 255).

Yet drought victims seldom get the full benefit of this third crop. It usually goes to the contractors and sub-contractors who build the roads, canals or reservoirs as the major parts of the official drought relief when it appears. A local contractor bragged: 'In this year's drought, all I did was sub-contract one small dam. I bought a scooter. If there's a drought next year, I shall buy a new jeep' (Ibid.: 17).

The Politics of Drought and Drought Relief

In Brazil Arons likened drought to a 'slow-motion war ... the forgotten natural disaster, the one no one wants to watch. And as with the unnatural disaster that is war, political issues invariably lie behind the madness (Arons 2004: 22). There was no doubt in either case that the declaration of drought itself was subject to political manipulation, and that the allocation of relief was conducted through a political sieve.

Sainath noted the rapid increase in number of drought-declared areas from 1990 to 1996 (Table 9.3). The areas come under a central government scheme, 'but bringing blocks into the DPAP [Drought Prone Areas Program] is now a purely political decision. The central application for DPAP may be nominal. But once a block is under DPAP, a phalanx of other schemes follows bringing in huge sums of money ... employment assurance scheme ... anti-desertification projects, drinking water missions ... Well, <u>some</u> people do benefit' (Sainath 1996: 256). At least those people who are in the construction business and the bureaucrats do.

Table 9.3   Number of Drought-prone Areas as Noted in the 'Drought Prone Areas Programme' (DPAP) in Various Indian States

State	No. of areas in 1990	No. of areas in 1996
Maharashta	c.90	147
Bihar	55[a]	122
Madhya Pradesh	c.60	c.135
Total of 3 states	205	394

Note: a. The 1990s showed an increase of one from the 1980s, the one being the home of the new union minister.
Source: Data from Sainath 1996: 256.

The process of application for drought relief is itself open to manipulation. In India and Brazil there seems to be plenty of evidence. Sainath suggested a drought scam spiral which I have freely adapted as Figure 9.1.

In Brazil Arons suggested that:

> The money lent by the [World] bank is often channeled into personal pockets, a phenomenon referred to as the 'politics of drought'. By perpetuating the myth that drought has been the primary cause of rural poverty, politicians not only have freed themselves from blame, but have also obtained World Bank funds

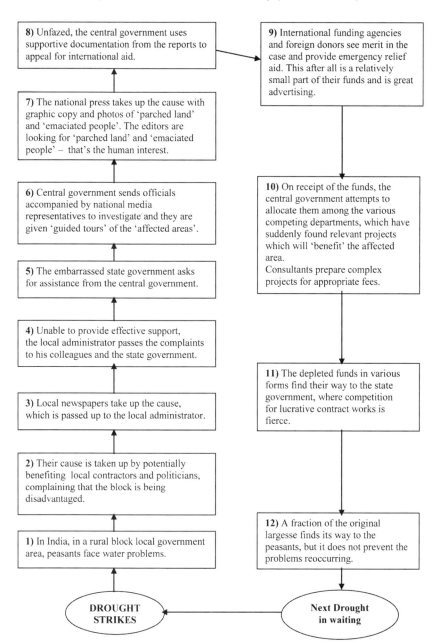

8) Unfazed, the central government uses supportive documentation from the reports to appeal for international aid.

9) International funding agencies and foreign donors see merit in the case and provide emergency relief aid. This after all is a relatively small part of their funds and is great advertising.

7) The national press takes up the cause with graphic copy and photos of 'parched land' and 'emaciated people'. The editors are looking for 'parched land' and 'emaciated people' – that's the human interest.

6) Central government sends officials accompanied by national media representatives to investigate and they are given 'guided tours' of the 'affected areas'.

10) On receipt of the funds, the central government attempts to allocate them among the various competing departments, which have suddenly found relevant projects which will 'benefit' the affected area.
Consultants prepare complex projects for appropriate fees.

5) The embarrassed state government asks for assistance from the central government.

4) Unable to provide effective support, the local administrator passes the complaints to his colleagues and the state government.

11) The depleted funds in various forms find their way to the state government, where competition for lucrative contract works is fierce.

3) Local newspapers take up the cause, which is passed up to the local administrator.

2) Their cause is taken up by potentially benefiting local contractors and politicians, complaining that the block is being disadvantaged.

12) A fraction of the original largesse finds its way to the peasants, but it does not prevent the problems reoccurring.

1) In India, in a rural block local government area, peasants face water problems.

DROUGHT STRIKES

Next Drought in waiting

Figure 9.1   The Hydro Politically Logical Drought Scam Spiral

Source: Captions for the stages of the spiral are from Sainath 1996: 257–60. With acknowledgement of the basic concept from Wilhite's 'Hydro Illogical Cycle' (Figure 8.3 above).

and loans intended for water-related projects. The individual politicians spend
the money as they choose, while the debt piles up on the state. (Arons 2004: 105)

Certainly there seems to have been generally no reluctance in declaring areas
drought affected and in a survey of *Drought, Policy and Politics in India*, the
suggestion was made that in parliament the MPs 'have not been concerned with
questioning the Government for its failure in drought proofing as much as with
attempting to procure as large a volume of relief funds for their constituents as
possible' (Mathur and Jayal 1993: 119).

But there may also be a political motive behind not declaring an area drought-
prone or drought affected. In India, in the 1993 drought, only four of the district
of Pudukkottai's blocks were drought declared, but in fact the whole area was
affected. The Tamil Nadu government, however, had not declared the whole
district drought affected because 'That would mean waiving land revenues and
other dues. It feels this would be a "bad example", one that could prove expensive
for the state. Demands ... have left both centre and state unmoved. Apparently, it
costs too much' (Sainath 1996: 279). In this case government deliberately turned
a blind eye.

Summarizing the Field Reports

In both cases drought seems to have been viewed officially as a natural disaster, a
specific, but not necessarily surprising, event in time and place. In India 'drought
is still mostly viewed as a natural calamity, despite considerable evidence to the
contrary' (Sainath 1996: 266). There seems to be no recognition that drought's
appearance may not only be linked to the shortage of the seasonal rainfall, but
may also reflect the vulnerability of the people in that place to that shortage. In
northeast Brazil, drought is seen as a characteristic problem. In a foreword to
Arons's book, Nancy Scheper-Hughes suggested that 'Peasants, politicians, and
professors alike have invoked the drought to explain everything from the romantic
melancholy that haunts the music to the penitential versions of folk Catholicism, to
the bipolar politics of patron-client dependencies interrupted by heroic banditism
and failed messianic movements, to the predictable, periodic famines and their
inevitable body counts' (Arons 2004: xiv). Drought here has a necessary certainty
as a descriptor of the region.

Seen as a specific event, it is not surprising then that official drought relief is
aimed at providing just a replacement for that immediate moisture shortage to try
to restore the prior situation. That relief is usually most conveniently provided in
the form of engineering projects for water storages, canals, well-boring equipment,
or irrigation systems. There seems to be no recognition that not only physical
engineering technology should be applied, but that also there may need to be some
social engineering as well; some need to address the inequalities of access to land
and to water itself, and for less crippling credit systems.

As a so-called natural disaster, drought has been made the scapegoat for socio-economic inequalities, particularly 'feudal' land tenure systems, and made a medium by which public money can be siphoned into private pockets. This latter has been made easier by the reliance upon minor as well major engineering projects, in which the contractors have been able to profit without supervision. And yet as the years go by, these technological 'fixes' do not solve the problem and more are needed in the next drought, and the contractors and consultants reap the 'third crop' again.

Yet not all evidences of drought are officially acceptable. In Tamil Nadu as we have seen, drought was not recognized in Pudukkottai because recognition would cost too much in lost revenue. In northeast Brazil local poetry illustrates attitudes to life and particularly life in the droughts, but these attitudes may not be acceptable to officialdom. These poems, as pamphlets privately printed or transcribed, and known as *literature de cordel* 'literature on a string' or *folheto*, provided personal views of the impacts of the droughts and the social injustices which made those impacts more horrendous. They were combined into a play *Morte e vida savernba,* translated as 'Life and death severely', which was performed in Rio de Janeiro in December 1965. However, a film version was officially prohibited from screening overseas because it showed depressing images of the country. None the less it was shown in Paris and encouraged other filmmakers to produce similar epics (Pereira's *Barren Lives*, Guerra's *The Guns*, and Rocha's *Black God, White Devil).* 'Rocha's film depicts the massacre at Canudos in the context of drought, suffering, salvation, and the spiritual forces that choose to make the sertão the site of their holy war' (Arons 2004: 49).

But these are documentary accounts of the 'here and now'; what if criminal minds realize that water is becoming a valuable commodity whose value will increase in proportion as its accessible global volume decreases? One novel must stand for several. Varda Burstyn's *Water Inc.* (2005) creates a future situation for the USA where years of drought have revalued the humble commodity. There, an 'agrichemical magnate and corporate visionary' organizes a consortium to plan to pipe in water from northern Quebec. Corrupt politicians in USA and Canada cover the proposal, but appalled conservationists recruit like-minded journalists, lawyers, and politicians to expose and defeat the scheme. The novel's blurb comments, with some justification, that the book is 'an exquisitely timed political thriller … a tale of greed, heroism, clashing loyalties, love, and mortal risk'.

Drought as the Ultimate Global Catastrophe

Here the scale of the narrative moves up to the world, with drought not merely a continental problem. Here drought is depicted as slowly enveloping and ultimately destroying all civilizations and removing all humans from the scene. Two novels are reasonably typical: C.E. Maine's *Thirst* (1977, first published in 1958), and J.G. Ballard's *Drought* (2001, first published in 1965). Maine's book first appeared entitled *The Tide Went Out*, but was revised as *Thirst! The Searing Novel of the Ultimate Drought* for the 1977 edition. This was probably done to take advantage of the global coverage of the West African (Sahel) droughts of 1968–1973. His hero, a Fleet Street reporter, catalogues a disastrous series of earthquakes caused by the structural stresses of the Great Powers' underground atomic tests. These earthquakes successively drain the Pacific and Atlantic Oceans, whose waters react with the magma underground to create more earthquakes. As the earth's atmospheric moisture is reduced, droughts become global, societies disintegrate and government elites retreat to the Arctic icecaps where the main potential source of surface water remains. The hero remains in London as his family is evacuated to the Arctic but, on the last flight north out of London, his plane is shot down as it approaches the Arctic refuges.

Ballard's *Drought* hero is a doctor, estranged from his wife, and taking refuge on his houseboat. A global drought, however, caused by the formation of a molecular film on the sea surface which prevents any evaporation of water into the atmosphere and so prevents any rainfall, is draining all the lakes and rivers. The molecular film is the result of the combination of industrial and radio-active wastes and sewage dumped in the sea over the previous 50 years. Abandoning his stranded houseboat, the hero joins refugees officially encouraged to make for the coast, where desalination plants will supposedly keep them alive. The water supplies, however, are inadequate and society disintegrates as supplies diminish. He retreats inland from the salt-encrusted coastal dunes where primitive settlements are slowly starving, to search for rumours of water and is finally seen heading inland in desperation but with little hope, just as rain starts to fall.

Drought and the Artists

How have artists, photographers and painters portrayed drought? Whether or not the title of the image includes the word 'drought', there seems to be a strong possibility that somewhere in that image a pattern of what seems to be cracked dried mud forms an essential component. The image carries the basic message that here was water, either in a river, lake or reservoir or, at least, as a puddle, and that it has been desiccated as the result of drought.

The United Kingdom's Meteorological Office Hadley Centre used a very simple image of cracked mud to illustrate the contrast with flooding of an urban area, entitled 'Some regions of the world will become more prone to flooding, while others will experience water shortages', on the first page of its *Climate Change, Rivers and Rainfall* publication in 2005. *Time* magazine had used a similar image to illustrate its 'Bone Dry' note in 2002 with the caption 'As floods ravage parts of Europe and Asia, drought is the problem in the US, Africa and Australia. Johnny McClendon surveys a dry Hempstead Lake on Long Island, New York' (*Time*, 26 August 2002: 13).

One frequent and common theme of images of drought's impacts – the sun-baked expanse of cracked mud crowned by an animal skull or skeleton – has been used both as a propaganda document for soil conservation policies and as a factual statement of dead livestock. I noted both uses as part of an advertisement for water pipe and irrigation many years ago (Heathcote 1969). In Australia the theme has been used in an advertisement for Caterpillar earthmoving machines to 'Green' [de-drought?] Australia (*The Bulletin*, 6 June 1970: 8–9) and to launch a public appeal for funds for drought relief (Martin 1994). But perhaps the most striking example was provided by a South Australian newspaper during the El Niño event of 1997, when a cracked mud outline of Australia carried the blanched head and part of the backbone of a steer (Illustration 9.1). The image has a graphic intensity which reminds us of the limited success of human engagements with nature.

At the outset let me say that I could not hope to do justice to the multiple graphic depictions of drought, so I will confine myself to my limited knowledge of Australian art, but before I do that let me share the comment of one UK artist who exhibited in Ottery St Mary, a small town in Devon, in 2008. With fellow artists of the Otter Vale Art Group she had debated whether pictures which could

Illustration 9.1   El Niño and the Archetype Drought Image

be disturbing, and which certainly would not be commercial, should be displayed. Responding to the current (2008) news out of Africa, she offered two unfinished pictures (Illustrations 9.2a and 9.2b), as she said 'I felt deeply angry that a continent of so much mineral and agricultural wealth should be ruined by conflict and corrupt misuse of resources leaving millions dispossessed and starving. Long periods of drought add to their suffering ... You could say that these are depictions of hope and despair, couldn't both be fit subjects for us as artists?' (Moya Lobb, August 2008). Acknowledging that she might be out of the usual 'comfort zone', she wished nonetheless to make her statement.

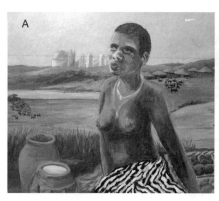

Illustration 9.2   Out of Africa, But Out of Our 'Comfort Zone'?

Two visions of news coming out of Africa by a British artist, Moya Lobb

    A. 'The first picture depicts a well-nourished young woman with an abundance of fruit, grain and milk for sale. Herds of animals drink at the water's edge, while children play. In the background is a suggestion of the modern cities which also bring wealth and progress. The two ways of life here coexist.

    B. The second painting is a harrowing image of a skeletal mother holding her dead baby while another child holds out an empty bowl for food which she cannot provide. The lake has dried up and the animals are dying. You could say that these are depictions of hope and despair; couldn't both be fit subjects for us artists?' Moya Lobb, 2008.

The Australian Images

Aboriginal art, whether as rock peckings or paintings, or paintings on bark or more recently on canvas, have shown a consistent recognition of water resources in the arid interior of the continent. Translated now into standardized products for a tourist market, they nonetheless maintain the traditional components seen by early European explorers. One major continental source is Tindale's 1974 summary of his extensive field studies, mentioned in Chapter 6, and one example must suffice for many. Illustration 9.3 shows a 'Geographical drawing by Katabulka of the Ngaddjara tribe in the Warburton Ranges of Western Australia', who met Tindale at a traditional campsite in August 1935.

Tindale explains the drawing as geographical but partly symbolized:

> Across the top of it Tjurtirango ... the rainbow lies, and between it are two concentric spirals representing Kalkakutjara, the 'heavenly breasts' ... which give rain that flows into ... waters. These are the balance of the concentric spirals. Five darker ones possess mythical ... carpet snakes and therefore are considered never failing; the others are temporary waters. Down the middle runs a stream bed, dry except during rain. On it are marked three waters, of which the top one is Warupuju [which was the camp site at the time of meeting]. Zigzag lines from water to water are the tracks or native roads of men wandering in search of food. The original was done in red and black by Katabulka in May 1939. (Tindale 1974: 68)

In 2007, the drawing was used to compare a contemporary Aborigine's personal map of ancestral country using basically the same technique of red, yellow and black dots (Moggridge 2007). In both, not surprisingly, the dominant features are the water points.

Like many European invasions of the 'New Worlds' in the eighteenth and nineteenth centuries, British colonists landed and took control of the continent of Australia from the coast. Not surprisingly, the initial artists were too busy depicting the new 'picturesque' landscapes of the coastal ranges and the new examples of the 'Victorian cities', established on trade with the mother country and an injection of gold mining, to be much concerned with the harsher interior of the continent and its droughts. There were, however, isolated exceptions, such as William Strutt's 'Martyrs of the Road' (1851), showing dead horses beside an inland track.

The first bloom of a 'national genre' in Australian art is usually associated with the 'Heidelberg School' of Impressionist painters in the 1880s and 1890s, originally based on the outskirts of Melbourne. While they focused on romantic views of mountain ranges and woodland glades, panoramic expanses of cultivated countryside and coastal views of leisure activities, some took up the drought theme in the 1890s. 'The Breakaway' (1891), by Tom Roberts, depicted an arid landscape and a stockman vainly trying to stop a rush of drought-crazed sheep to a remaining waterhole. 'The North Wind' (1891), by Frederick McCubbin, showed a man leading a horse and cart through swirling red dust. His wife rides in the cart with their meagre possessions, sheltering their small child. Have they been driven from their small-holding by drought? The 'Spirit of the Drought' (c1895), by Arthur Streeton, depicted a nude woman standing above the skeleton of a steer against a background of dead trees and hovering hawks. These paintings, however, were greatly outnumbered by views of successful rural settlements set in wooded ranges and grassy valleys.

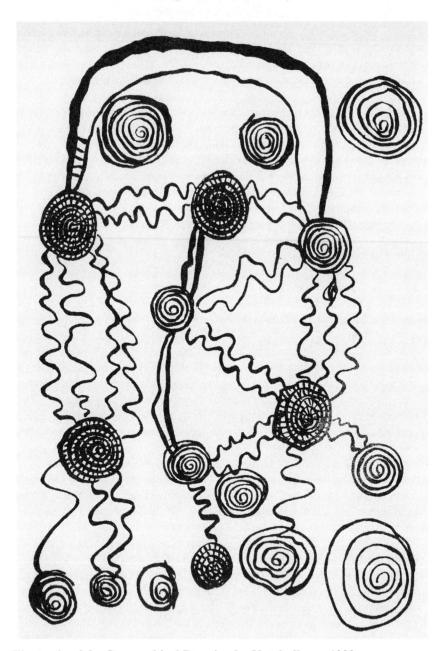

Illustration 9.3   Geographical Drawing by Katabulka, c. 1939

Katabulka was from the Ngadadjara tribe in the Warburton Ranges of Western Australia. 'Tjurtirango the rainbow yields water to storage wells, pools, and sand soaks symbolized by concentric spirals. Tracks made by men join the various waters.'

Source: Tindale 1974: 68.

By the early twentieth century settlement had spread inland to the edge of the deserts and the threat from droughts had become more evident. Yet at the British Association for the Advancement of Science meeting in Australia in 1914, a patriotic Commonwealth Meteorologist claimed that 'the preconceived notion that Australia is the particularly drought-stricken and precarious area of the earth's surface' had been emphatically contradicted by the climate of the preceding decade. He admitted that seasonally dry periods were liable to be experienced, but argued they were not to 'be regarded as drought, and an evil, but rather as Nature's wise provision for resting the soil' (Heathcote 1994: 263–4). But as he spoke, having carefully omitted mention of the drought of the first three years of the century, another serious drought was unfolding over the southeast of Australia. This would reduce the largest river, the Murray, to a trickle easily bestrode by a government minister. In a dramatic frontispiece to a seminal work by Joseph Powell on *Water, Land and Community in Victoria 1834–1988,* Sir Ronald East, later long-serving chairman of the [Victorian] Stock Routes and Water Supply Commission, is photographed astride 'a drought-humbled River Murray' near Nyah, Victoria, in March 1923 (Powell 1989).

One of the first artists to focus upon landscapes distorted by drought was Hans Heysen (1877–1968), who made his early name painting bucolic pastoral scenes amid the magnificent blue gums of the Adelaide Hills and Barossa Valley in South Australia. His lasting contribution, however, was to result from a visit to the Flinders Ranges in the semi-arid north of the state in the 1920s, when a combination of several years of drought and overgrazing had revealed 'the bones of nature laid bare', ready to be translated into stark rugged terrains dominated by bare slopes of reds, browns and yellow tones. Here the colours and a scoured landscape shouted out the characteristics of a drought. Several large paintings resulted over the following years and further visits, and one, 'In the Flinders–Far North' (1951) was to hang in the Australian Embassy in Paris for several years as a celebration of the fiftieth anniversary of the creation of the Federation of Australia. His art in fact foreshadowed a 'red-shift' in the depiction of the Australian landscape, and was featured in a special publication *Arid Arcadia. Art of the Flinders Ranges* (Bunbury 2002).

> Since the early twentieth century, artists have been highly selective in which parts of the [Flinders] ranges they chose to paint; they have censored the land, first foreshadowing and then reinforcing a red shift in Australian landscape imagery. Australia is now an ochre-hearted country. (Bunbury 2002: 10)

Droughts had been significant in creating that 'ochre-hearted country'.

By the 1940s, that country was blowing away and again drifting in the streets of the state capitals. As a result two established artists were commissioned to see for themselves the impacts of drought and provide evidence for metropolitan newspapers. The first was George Russell Drysdale (1912–1981), who was engaged by *The Sydney Morning Herald* in November 1944 to accompany a reporter covering the drought then devastating New South Wales. His line drawings of landscapes, dead crops, wizened stockmen and dying livestock were published on 16, 18 and 19 December 1944. In 1945 he worked up several of his earlier sketches and perhaps his most evocative painting, 'Angry Harrison's Store', appeared in 1950, depicting the proprietor leaning on the verandah of his derelict store in a country town where a stray dog rummages in a crumpled newspaper on the otherwise deserted dirt main street, while in the background isolated trees and a low building with water tank and windmill are starkly silhouetted against a red-brown arid sky.

The second artist was Sidney Nolan (1917–1992), who was invited 'to make a series of black-and-white drawings for publication of the worst drought in recorded history' in the Northern Territory. The invitation from the Brisbane *The Courier Mail* in June 1952 was itself flattering: 'In our opinion, no artist could apply himself better to such a project than your good self and feel that your drawings would help greatly in bringing home to the public of Australia the tragedy and waste of this disaster' (Smith 2003: 95). Nolan was able to use the local mail service aircraft to provide an additional aerial view of the devastated countryside, in addition to interviews with stockmen, Aborigines, and local business folk.

The drawings appeared in the newspaper on 23 and 30 August 1953 and were preoccupied with death. Nolan himself saw the droughted landscapes as visions of the end of civilization, and a caption to a crayon sketch of a metal bedstead and ribs of a ruined shed, remnants of an abandoned property, read 'All that remains is the mirror of their names' (Smith 2003: 128). Subsequent exhibitions of paintings from the trip were held in Melbourne in June and Sydney in July 1953. In each case 'the long main gallery was hung with 27 paintings, the majority being carcase subjects, with an occasional still life, "Desert thorn" … or "Old diggings". They were apparently hung away from the wall to give the impression of great distances and light affected by winds'. None of the paintings sold, but Nolan was not surprised, quoting Picasso that they were 'not done to decorate apartments' (Smith 2003: 99–100).

Drawings from Nolan's trip were exhibited in the 1954 Venice Biennale, along with John Heyer's film, 'Back of Beyond', of the mail run along the Birdsville Track through the Simpson Desert, which showed the isolation and stresses of remote cattle station life, exacerbated by droughts. Although rewarded with the first prize, both were criticized as being 'too harrowing' by exiled Australians in London (Smith 2003: 102). The 'outrage' at the depressing images, particularly on film, paralleled the Brazilian concerns over films on the droughty northeast noted earlier, and echoed similar Australian concerns about the national image portrayed in an earlier 1920 film, 'The Breaking of the Drought', directed by Franklyn Barrett. That film was based on a stage play of the same name which first appeared in 1902, in the year when the 1895–1902 drought in eastern Australia finally ended. But the film was actually shot in 1919 when again drought was devastating New South Wales. It told of the fortunes of a pastoral family forced off the land by the bank when they can no longer service their mortgage, and who retreated to the city. There 'their savings are frittered and gambled away by an errant son'. Their fortunes and their return to the property are conveniently brought about by a neighbour, in love with the daughter. This happy event 'coincides with a bushfire, the son's reunion with his father and the breaking of the drought' (Gaind 1995: 14). Nonetheless the film was deemed offensive and, after a question was asked in the New South Wales parliament, it was banned from overseas distribution (Heathcote 1987: 7).

Both Drysdale and Nolan provided views of the Australian outback which became standard images of inland Australia, to be elaborated upon by subsequent artists such as Fred Williams (1927–1982). In literature these images were expressed in the classic poem by Dorothy Mackellar (1885–1968), 'My Country', where the second verse is:

> I love a sunburnt country,
> A land of sweeping plains,
> Of rugged mountain ranges,
> Of droughts and flooding rains;
> I love her far horizons,
> I love her jewel-sea,
> Her beauty and her terror –
> The wide brown land for me!

Interestingly, the poem was specifically illustrated in the book *A Sunburnt Country* (Beavan 1978), where a combination of the full poem and accompanying paintings was used to communicate 'the flood and fire and famine, the beauty and the terror of the Australian outback' (Beavan 1978: 7). In the twenty-first century Australian artists have revisited the theme, as the implications of a series of droughts in eastern Australia have combined with past and recent flawed official water management strategies to raise questions as to the future viability of agriculture in the nation's breadbasket – the Murray Darling Basin (Neylon 2009).

Drought, the Community, Despair and Humour

Facing the slow death of a landscape starved of water, the human ability to cope with stress is seriously tested. At one end of the spectrum is the mind numbing realization that one's fate is in the hands of the gods. Writing of his Australian drought experiences in the early 1890s, one commentator philosophized that:

> After a good spell of drought, endured on a diet of mutton, bread, jam, and stewed Bohea [possibly *Boerhavia diffusa*], one's indifference to life becomes all but complete. There is nothing wild or hysterical about it. It is merely a profound and passionless headlessness of danger and death … [and] The slow,

pitiless, everlasting horror of a drought means ten years added to the life of the squatter or manager. (Adams 1893: 148–9 and 151)

For an efficient and reasonably well-off wheat farmer in the semi-arid Mallee of South Australia in the 1960s his 'Wheat Country Drought' poem spoke of resignation once the onset of the drought was recognized: 'You know your country and you know your crop; You shrug; You can't fight the wind'. He has some fodder saved from last year 'and the banker still smiles and you can always sell a few sheep to take the pressure off. You won't really suffer, it is just the futility of a wasted year and the frustration of a dying crop. And instead of a fertile greenness, a dead-stalk rattling. Men looking at the sky all day and cursing it before bed' (Napper 1967). For an experienced mid-west American farmer, but who was also an official in charge of the United States Department of Agriculture's drought relief operations in 1988, the experience of day after day of no rain caused him to reflect that 'There isn't anything more disheartening. You're totally helpless. You just watch your crops wither away' (Quoted in Opie 1994: 281).

But for the wives there are other concerns. The daily round of house-keeping becomes basically a strategy of survival. In the 1920–1930s droughts in the newly settled Mallee of southeastern Australia, the wife of a government 'expert' distributing advice on new scientific dry-farming methods from an official travelling rail exhibition, dreams of 'baths with heady oils and perfumes'. And this at the time when the local newspaper is relating how a local housewife is 'saving water by boiling potatoes in beer or using wool fat to clean her infants'. But relations between the wife and her husband have 'dried up' and eventually when the scientific advice proves ineffective and the drought continues, he flees into the army as the Second World War looms (Tiffany 2005).

In the face of despair, however, humanity's ability to laugh, often at itself, is one of our most agreeable characteristics. Not surprisingly, therefore, in the face of the threats from most of the natural hazards facing human life on earth, one human reaction is to see the funny side of the potential disaster. In effect, humour seems to be a standard mode of coping with many risky and potentially life-threatening situations, society's way of reconciling itself to extreme adversity. It is not surprising therefore to find evidence of humour being applied to many of the situations created by extreme drought events. As one researcher pointed out, 'all scholars of humor know that laughter comes not from comfort and ease but from pain and suffering' (Welsch 1978: 4).

The psychologists tell us that humans use humour to cope with life's stresses, a way of creeping out from the depths of depression in the face of a potential disaster. Some hospitals contain 'laughter rooms' as part of health-restoring therapies (Cousins 1979). For Freud humour was a means to escape or avoid potential adversity or stress; for R. Moody it enabled us to laugh at but remain involved with the situation; while for Willibald Ruch and his collaborative authors humour has a broad spectrum of roles (actual and potential) in the relationships both within society and between society and the environment around us (Ruch 1998). What might be the links with drought?

Looking at the evidence from round the world, there seem to be at least five slightly different ways in which drought and its impacts have been, and can be, translated into humourous situations to help cope with its impacts. My guesstimates are set out below.

1. *Humour as a safety valve*. First is the way in which humour acts as a safety valve to relieve emotional stresses and at the same time provide some small measure of hope. Summing up a booklet of North American rural humour, entitled *The Summer it Rained*, the author, who had drawn upon his experiences as a farmer in Nebraska, commented:

> Here's me, sorrowed by the dried farm ponds, listening to the jokes about drouth – jokes that are not meant to bring forth laughter but give a communal ground for the sufferers, jokes that blur the pain and sharpen the hope. Here's me, on the Great American desert, with no intention of leaving it. Because I love the Plains. (Welsch 1978: 12)

One of his jokes told of the farmer who claimed that 'it had been so dry the summer before that he had seen two cottonwood trees fighting over a dog, and another farmer had told the census taker that his son, a strapping young man, had been born "the summer it rained"' (Ibid.: 3). On field work in the semi-arid Mallee region of South Australia in the 1970s I came across virtually the same story about the fight over the dog, the only difference being that the trees were now the local mallee (*Eucalyptu*s species).

Cartoons are often both humourous and poignant – the farmer picking up the mail on the edge of his farm which is littered with the bones of his dead cattle, exclaims to his wife; 'It's the water rates!' (*The Advertiser*, 13 June 1990, Atchison cartoon). Then there's the character with feet set in a concrete block as he sits in the dried up dam: 'Remember this Spud ... if you ever need to cross the Mafia ... do it in a drought ...' (*The Advertiser*, 1 March 2003, 'Beyond the Black Stump').

2. *Humour in the absurd*. A second type would be where the situation is absurd or nonsensical as an offset to the tragic qualities of reality. Examples have a Biblical context and Noah's Ark features in several. A cartoon during the 1972 drought in South Australia showed a disconsolate Noah sitting beside his completed Ark on which was a sign 'Going Cheap' (*The Advertiser*, 24 June 1972, Atchison cartoon). A traveller asked a Nebraska farmer if it had been dry recently? He was asked in turn:

> 'Son, do you know your Bible?'
> 'Well, a little,' I answered uncertainly.
> 'Remember Noah and the Ark?'

'Yes'.
'When it rained forty days and forty nights?'
'Right!'
'Well, that summer we had half an inch'

(Welsch 1978: 4)

Other stories reverse the norms and transform the accepted conditions. Like the farmer in the drought hit by a drop of rain, at which he fell down in a faint and could only be revived when showered with a bucket of dust (Welsch 1978: 5 and *The Advertiser*, 12 May 1987, Atchison cartoon). Not only the farmers, but also the local wildlife had difficulties coping with the dust from eroding droughted paddocks:

> Dust came to be known during pioneer days as Oklahoma rain or Kansas snow, and the observant farmer could tell when the dust storms were coming by listening for the sneezes of the rattlesnakes ...

> Early settlers told of catching Platte River catfish wearing goggles – to keep the sand out of their eyes ...

> In the 1880s there was such a thick dust storm that the prairie dogs in the state thought they had been buried. They dug up through the dust storm to get out and for three hours after the dust had settled it rained prairie dogs. (Welsch 1978: 4 and 8)

And in Australia, 'Back of Bourke', the crows fly backwards to keep the sand out of their eyes (Wannan 1957: 9).

3. *Humour highlighting the No-Hopers*. A third type pokes fun at the pessimists and claims that they always look on the black side. Perhaps the best example is the Australian poem, 'Said Hanrahan', by P.J. Hartigan 'John O'Brien'. It is too long to reproduce in full, but the message can be condensed at the risk of losing the quality of the original poem. The opening scene is outside a local Catholic Church just before Mass as the congregation is gathering. Hanrahan takes the opportunity to declare that 'We'll all be rooned ... before the year is out' – by the current drought. His listeners generally agree and debate how much rain would save them, at which he declares 'If we don't get three inches, man, or four to break this drought, we'll all be rooned, before the year is out'. Eventually the drought breaks with heavy rains, at which Hanrahan declares 'We'all be rooned if this rain doesn't stop'. But the rains stop and the crops flourish until the grass is knee-deep. But before Mass as usual Hanrahan thunders 'There'll be bush-fires for sure, me man, there will without a doubt; We'll all be rooned ... before the year is out' (1965: 141–3).

4. *'Drought Country'*. A fourth type of humour seems to present a wry image of the country, possibly as a bonding mechanism for the community. Describing the droughty grazing range in southwest USA used by the Navaho from 1870 to 1975, researchers noted the locals referred to it as '60–40 country': 'This is a range where a cow must have a mouth sixty feet wide and move at forty miles-per-hour to be able to find enough to eat' (Richmond and Baron 1989: 219). In Nebraska it was reported that 'frogs had grown up to be seven years old without learning how to swim'. Elsewhere crop yields were so low in the drought that a farmer claimed that 'for lunch we just et four acres of corn' (Welsch 1978: 2 and 3).

In a 'Special Drought Edition' for September 1994 of *The Country Web: A Newsletter for Rural Women and their Families* (Orange, New South Wales, Australia), a poem by Barbara C. Hore 'The Farmer's Strife' graced the front page. It tells how the farmer's wife drove the farm truck behind her husband on the tractor across the partly drought-dried lake where bird life has concentrated around the receding waters. Distracted by the beauty of the birds, 'I lost the bloody track' and the truck was bogged. In the process of being pulled out she accidentally pulls on the truck's brake, which doesn't help the process. She goes on:

> We talk no more of swans and birds, and trucks that go astray.
> Said he, 'You need a holiday where roads are hard and straight,
> I'll even help you pack your bag', so subtle was my mate.
> I know a place where he could put his tractor, truck and chain,
> The words I fear I'd have to use are termed as being profane.
> As we reflect on years of drought, such episodes as this,
> Test mind and body to extremes, and threaten married bliss.
> But somehow we survive it all – togetherness of course,
> 'Cause in a drought there's neither time nor money for divorce.

5. *The Pun.* Finally, perhaps, the dreadful pun. 'During the recent drought the sign by the roadside at one of Melbourne's greenest golf courses explained: "We are using dam water". A few hundred yards further on, a wry citizen had written near his dying lawn: "We aren't using any damn water"' (*The Australian*, 30 July 1968, 'Water Resources Survey': 3).

Conclusion

In the light of the questions posed at the outset of this chapter, what tentative findings might be put forward? Drought was seen in various forms and as carrying various messages, but the standard graphic image is the patch of cracked earth, originally mud, denoting the desiccation of a previous water point. The message is immediate and clear, life-giving water has been sucked away and life is at risk. Additional components of the image may be shrivelled crops and bleached animal bones, but they are not essential to the main message, which is that death is hovering.

Drought has been traditionally regarded as a signal of the gods' displeasure, to be mollified by prayers and the intercession of the priests. And prayers are still offered in Australia.

The association of drought with death, actual or impending, is for the Third World small farmer not only a potential threat to his crops and livestock, but also a threat to his community, his family and himself. For the western farmer it is rather the potential for economic distress and the psychological frustration of lost effort, lost investment and lost time, and the feeling of helplessness in the face of nature. For some authors, indeed, drought is seen as the ultimate catastrophe, the ultimate payment for human mismanagement of the global resources

Some regions of the world are recognized to be drought prone, while some are regarded as drought dominated. In both the impacts of drought are suggested as intensified by human mismanagement of the local resources. Unequal access to land and water are put forward as precursors to an increased vulnerability to meteorological droughts, a vulnerability not significantly reduced by technological attempts to improve basic water supplies, but a vulnerability offering unscrupulous entrepreneurs potential commercial benefits from possible monopoly of the water.

For the artists, droughted landscapes, with their striking colours and destroyed plants and animals, often containing the remnants of unsuccessful human endeavours, have both a tragic appeal and an aesthetic grandeur. In part the aesthetics are the colours and shapes of aridity, but the implied human failures in their contests with nature are also a powerful attraction. Yet, amid the distress and disappointment humour remains a solace and strength.

Drought is not compared with other natural hazards nor, because of the timing of the sources, is it considered in the context of global warming. But from the national political viewpoint it seems that drought is not welcomed, since its acknowledgement is seen as an admission of official impotence in mitigating the impacts of this natural hazard. The fact that drought might represent other than natural causes is to be suppressed. This final suggestion leads into the question as to how science has treated drought and how scientists have coped with such a dangerous phenomenon.

Chapter 10
The Scientists and Drought

The prospect of domination of the nation's scholars by federal employment, project allocations, and the power of money is ever present – and is gravely to be regarded. Yet, holding scientific research and discovery in respect, as we should, we must always be alert to the equal and opposite danger that public policy could itself become a captive of a scientific-technological elite.

(President Dwight D. Eisenhower, Farewell Address, 17 January 1961, quoted in Michaels 2004, 'Epigraph': vii).

Indignant Surprise?

So far this book has tried to show how an extensive scientific knowledge of the phenomenon of drought has been created over the years through community knowledge and the work of academics, business and government reports and inquiries. However, that knowledge has not yet led to a complete understanding of the phenomenon and, as a result, each new drought still tends to be greeted with what can only be described as indignant surprise (Heathcote 1969). To try to explain this inadequacy, we need to ask the question whether the study of drought has itself influenced the nature of scientific enquiry and whether from that study have come questions for the basic tenets of scientific enquiry in general and specifically for the mitigation of drought impacts upon society.

The Role of Drought in the Evolution of Scientific Thought

The explanations for that indignant surprise, I believe, lies in the evolution of scientific knowledge itself, part of the historical growth of interest in, and knowledge of, the global environment. This growth seems to reflect several developments in human thought, and in his book *The Condition of Man* Lewis Mumford (1963) set out a series of stages by which the earth knowledge of the ancients was converted to the scientific knowledge of the contemporary world of the 1940s. And those stages, I believe, help to explain why scientists are still surprised by the nature of the global environment, not least by the vagaries of the onset of droughts.

Mumford recognized a first stage of expansion of knowledge as that of the 'Dawn of Naturalism', drawing upon the daily experiences of craftsmen, where 'Prayer would work only if one added to its efficacy by intelligently manipulating the environment' (Mumford 1963: 139). The growth of empirical knowledge in the seventeenth, eighteenth and nineteenth centuries partly encouraged that development. Once the power of the Church declined, metaphysics had to give way to the facts of this world rather than concern for the fancies of the next, and the complementary relationships between science and magic which had predated the growth of empirical thought were broken.

In this explosion of empirical knowledge about the global environment the relevance to the well-being of human societies of the earth's geology, soils, plants and animals, and weather and climate, was recognized. It was in the context of the weather and climate that concerns about drought – its origins and particularly its impacts, directly upon plants and animals, and indirectly upon human subsistence – began to appear. From being just another example of the wilful challenges of the Gods, drought was recognized in Western thought to be part of the oscillations of natural weather and climates.

Increasing evidence of the role of drought came particularly from what Mumford suggested as the second stage of the expansion of human knowledge, namely from information stemming from the New World – the environments previously beyond the ken of Western Europe which were being rapidly explored and their resources exploited from the end of the fifteenth century onwards. Those environments contained many more areas subject to aridity and frequent droughts than the 'heartlands' of the European explorers, and as a result the adverse impacts of periodic droughts became more commonly known and the subject of enquiry. In addition those areas contained extensive ruins of past civilizations, and European explorers began to surmise, and archaeologists to demonstrate, that several appeared to have declined, at least in part, because of extensive droughts.

That flowering of knowledge encouraged what Mumford suggested was a new direction for scientific enquiry:

> The concentration upon exploring and settling this world … gave Western man a new task to fulfil and a new drama to enact. People turned from their inner life, which was disordered and confused … they centered their attention upon the outer world and turned impulses that might have been suicidal into acts of aggression and mastery, perpetuated against nature and nature's children. (Mumford 1944: 231–2)

This conquest of foreign lands did not just appear with Columbus's voyages, of course, but it was now accompanied by a new spirit of scientific enquiry into the characteristics and resources of those new lands. This was not merely colonialism, but a new scientific colonialism where the knowledge to recognize, find, and create new resources fuelled nationalistic politics from the sixteenth century onwards. These three goals of the emerging Industrial Revolution in Europe involved scientists in the political process of nation building, and its accompanying colonial exploration and exploitation of new lands and peoples through the Mercantile System of international trade between home nation and the colonial appendages with their new resources.

The new lands, however, held not only resources but hazards, and scientists had to recognize both. Given this expanding knowledge, they began to see part of their role as being to protect society against the mischiefs of inclement weather and climate, including the 'natural hazards' of fire, flood or drought. The tools for that protection were coming from the successes of the Industrial Revolution, a process described by Siegfried Gideon as *Mechanisation takes Command* (1948). In effect this process reflected the rise of technological innovations and the flowering of engineering. Against the threat from droughts were arrayed the developing technologies for the manipulation of water resources, and an increasing capacity for their storage, diversion, and recovery from underground sources through new drilling technology. Later were to come attempts to improve the efficiency of their use by plants and animals and to curb human wastefulness.

A third stage in the search for knowledge created what Mumford called the Scientists' Dilemma. He suggested that a split had developed between the discoverers and what was discovered, between the search and the findings:

> Science opened up the external world and bade it welcome; but it shut out the self [of the scientists]; it enlarged the horizon but contracted the center ... The separation of positive science from nominative science, of instruments from ends, [and] of causal knowledge from final knowledge tended to encourage the pursuit of the first and to belittle and devitalize the concern for the second. (Mumford 1963: 243)

The ability to act to create new knowledge became dominant over any feeling of responsibility for the implications and likely results of that knowledge. In a sense Mumford was anticipating the moral dilemmas outlined in Jungk's *Brighter than a Thousand Suns* (1958) and C.P. Snows's *The New Men* (1959). Mumford put it more bluntly:

> To encourage a mature technique for controlling the external world and enlarging all of man's physical powers, whilst permitting man himself to remain at infantile level, was to place dynamite in the hands of children. (Ibid.: 243–4)

This distancing of scientists from responsibility for their findings was to become a serious concern over the years, and one example must serve for many at this stage. Thus the ability to make rain by seeding clouds with silver iodide has had some success, but the legal question as to who owns the clouds and has the rights to their incipient moisture has not yet been resolved.

The problem of scientific responsibility became even more contentious as Mumford's fourth stage of scientific thought emerged. This was described as the Romantic Fallacy, the romanticising gloss placed upon Nature as a separate entity from the civilized environment. The industrialization of the European environment was paralleled by an attempt to retreat into a pre-industrialized natural world of Rousseau's 'Noble Savage', who was supposedly at one with the natural world.

This cult of nature was plainly, a compensation for a society that was fast becoming too cultivated to remain healthy, too neatly ordered to leave any play to the free imagination, too fully rationalized to use the full force of the id. (Ibid.: 280)

Out of this Romantic Fallacy came the concept of the holiday away from home, the back to nature movement and tourism in all its ramifications. But periodic droughts and natural disasters from floods, storms and earthquakes continued to challenge any suggestion of nature as a benign haven from the perils of the civilized and industrialized world.

Mumford's book was published initially in 1944, and within seven years an international scientific programme was to appear to combine much of the scientific thought noted above into a study of those parts of the global environment where drought was the dominant characteristic – the arid lands. The UNESCO Arid Zone Research Programme was set up in 1951. Its original aims were the collection of data on the global arid lands, the standardization of data-collection units to allow international comparisons, study of different land use systems and the planning of future resource uses in the arid lands. Arising out of the program came a series of publications – The Arid Zone Research Series – which laid the foundations for the scientific global study of lands dominated by intermittent or continuing droughts (Heathcote 1983, Appendix 1). The program ran until 1971, when it became part of the United Nations Environmental Programme with a more specific focus upon combating the effects of desertification, and at the time of writing was still running although on reduced funding. In effect, some scientific research since the 1950s has been concerned directly or indirectly with the phenomenon of drought. For the scientists involved, however, the research has proved to be capable of creating a love/hate relationship with the subject.

The Scientists' Love/Hate relationship with Drought

The main danger to any disaster organisation is the absence of disasters. (Shields 1979: 447)

The study of drought has immediate attractions for scientists. Meteorological drought is recognized to be a regular component of global weather patterns, posing questions of definition, the recognition of characteristics in space and time, and, hopefully, the scope for prediction. Once identified it needs to be explained, rationalized, and its role in global climate and weather patterns established. Historically droughts have been shown to impact on societies, and their continuing impacts constitute challenges to be explained by the experts. And because those challenges have proven so disastrous, scientists have risen to attempt to protect society, their work defending against nature providing their efforts with a moral and political kudos which helps to provide sympathetic research funding for their efforts.

Yet scientific study of drought has proved to be irritatingly frustrating. An immediate and continuing problem has been how to provide an all-embracing and absolute definition. As we saw previously definitions have been many and varied, reflecting relative values rather than steadfast certainties. In addition there is a very blurred boundary between drought and aridity, as the UNESCO Arid Zone Research Programme demonstrated, with the boundaries of the arid zone fluctuating with the effects of seasonal droughts. Further, as we have shown that at least four types of drought are recognized, research cannot be limited to the realms of the natural sciences but must venture into the even more complex fields of the social sciences. Thus there is a case for examining the way in which the scientific study of drought has influenced the broader scientific study of the global environment.

The Scientific Study of Drought and its Contribution to Environmental Knowledge

Why should the study of droughts have more than immediate relevance to the wider study of the global environment? We might try to answer this by posing a series of questions:

1. Has the study of drought led to significant insights into the nature of global ecology, and thus raised understanding of linkages and relationships between the elements?
2. Have the threats from drought required significant innovations and scientific developments to try to protect society from those threats?
3. Have the threats from drought influenced the ebb and flow of land settlement over the globe, and have the influences been always negative as might be expected?
4. Has the study of drought led to any serious challenges to existing theories in general science?
5. Has the study of drought raised significant issues in research in the social sciences?
6. Finally, have the experiences of trying to cope with drought's challenges educated scientists in the complexities of nature?

The following sections attempt to answer these questions.

Drought's Role in Global Ecology

As noted in Chapter 5, Köppen's *Klimat der Erde* (1931) linked global climate and vegetation and in the process recognized a distinctive type of vegetation species – the Xerophytes – which were specifically adapted to drought stresses. This basic classification has stood the test of time, was used by Palmer in his drought indices, and is still considered a useful analytical tool. Greater western contact with the droughty regions of the Americas and Africa brought attempts to understand in particular the processes by which moisture was lost into the atmosphere in the semi-arid lands. Research in the USA on the links between evaporation off surfaces and transpiration from the leaves of plants paralleled Palmer's efforts and led to the recognition of the process of what was termed evapo-transpiration, which provided a viable estimate of moisture actually available for plant growth (Thornthwaite 1973).

Interest in the processes by which plants and all life forms coped with the debilitating effects of extreme temperatures and dehydration brought classification of levels of drought evasion. Schmidt-Nielsen (1964) and Cloudsley-Thompson (1965) provided evidence of the processes in insects, animals and mankind. Paralleling this scientific inquiry into processes were the experimentalists trying to find plants whose drought resistance traits could be incorporated into new varieties of bread grains, and animals which could cope with the drought-prone semi-arid ranges. Breeding for drought-resistance, however, had its inherent problems: was the aim to breed to survive or to breed to give maximum produce? As one author put it: 'Should scientists be aiming at maximizing yields with [associated] risks of increasing variability, or [aim] at yield stability, with possible penalties in lower yield potential?' (Oram 1985). It has not yet been proven that you could breed successfully for both.

Protecting from Drought: The Role of the Foresters

Research into the relationships between vegetation and climate, specifically air temperatures and precipitation, brought the realization of the role of vegetation as a buffer to extreme conditions. The nineteenth century spread of farms in the 'New Worlds' continued the process of global deforestation documented in Michael Williams's *Deforesting the Earth* (2003). Official concern initially was a fear for future supplies of timber, but with some concern for the effect on rainfall. The results were policies attempting to encourage replanting of trees. George Perkins Marsh's *Man and Nature* (1864) proved an international stimulus to what was already concerning officialdom in Australia and the United States, where clearance of woodlands and forests for agriculture seemed to some observers to increase the threat from droughts. The destruction of perennial vegetation has always had a recognized role in the spread of desertification, and specific droughts could bring calls for re-vegetation of the bared soils.

In Australia droughts in the early 1860s saw the Government Botanist of the colony of Victoria, Ferdinand von Mueller, advocating large scale tree planting not only to replace the destruction of native timbers but to 'increase … rain, in the retention of humidity, and the mitigation of burning winds' (Bonyhady 2000: 175). In the drier colony of South Australia the same droughts brought concern for future timber supplies and the creation of the post of Conservator of Forests in 1875. Official timber plantations had begun in the 1870s, while 200,000 government-grown seedlings were sent out free for settlers to plant in 1881. By the time the scheme was abandoned in 1924, over 11 million trees had been distributed to over 63,000 settlers (Lewis 1975). Few trees survived in the drier areas but the sites in the higher rainfall areas were later to become the core of the government's forestry plantations. But these were designed for timber production, not climate amelioration.

On the Great Plains of the USA parallel concerns led to the creation of 'Arbor Days', where citizens, later school children, went out to plant seedlings for future forests. Initiated in Nebraska on 8 April 1872, when 3 million seedlings were planted, and created a public holiday in 1885, the rationale was a combination of timber and climate amelioration (Dick 1975). Further droughts in 1910–1911 encouraged concerned scientists to push for tree shelter belts. By 1941 they covered significant areas of the plains and some of the earlier plantings helped reduce the impacts of the Dust Bowl conditions of the 1930s (Williams 1990).

It was not droughts alone, of course, which stimulated concerns to revegetate the earth, but their timing which made them more significant. In the mid and late nineteenth century they occurred in the midst of the rapid occupation of 'new lands' in the Americas and Australia, which had brought hopes of rapid economic improvements for the new settlers, only to be dashed by the crop and livestock losses of the droughts. In the World Depression of the 1930s droughts exacerbated the global economic depression by again destroying crops and killing livestock which might have brought some economic return to bankrupt settlers.

Protecting from Drought: The New Rain Makers and the Weather Forecasters

Drought and the Rain Makers

From the rain drums of Africa, through the human sacrifices of the Mayans, to the aircraft seeding the clouds over North America and, most recently, to the Chinese gunners bombarding the clouds at the Beijing Olympics in 2008, the rain makers fighting drought have been hard at work. As we have seen in Chapter 7, the results have been contested. A review was provided by W. Harris in 1971 but a survey in 1992 was not optimistic, concluding that 'Cloud seeding is often used in desperation as a last resort to combat water shortages and it is unlikely that this will continue, but the method is still not fully proven [hence] … it will continue to be a "chancy" technology' (Agnew and Anderson 1992: 196). A further overview in 1997, checking the various methods for cloud seeding, concluded that 'They have not demonstrated ... that rainfall can be increased over fixed ground target areas consistently' despite some innovatory techniques. The run down of official research funds in the United States from the peaks of $19 million per year of the 1970s to the half a million of 1997 was seen as an indicator of loss of confidence, and Israel's abandonment of funding for weather modification that year seemed to confirm the uncertainties (Cotton 1997). Thus scientists have not yet provided drought-proofing for humanity through weather modification, but they are still hopeful.

There were other mechanisms involved in rain making, however, which briefly offered hopes for the pioneers. One of these was the belief that 'rain follows the plough', the argument being that ploughing up of natural grasslands released moisture into the air which then returned as rainfall. Kutzleb's thesis (1968) provides the detail of the rise and fall of the theory on the Great Plains of the USA over the period from the initial claims of the 1870s to the high point of the debates in the scientific literature in the 1880s. By the 1890s, however, the hypothesis had been discounted as not proven despite the support of some scientists and the enthusiastic support of the railway companies, anxious to sell their semi-arid land which had been the government's reward for the railway construction! With the failure of that theory in the face of continuing droughts on the plains, the interest of the scientists, promoters and politicians switched to the potentials for rainmaking, which was to continue well into the next century as we have seen. Elsewhere the rejection of the idea was more related to the adverse effects of the plough-up of the semi-arid areas of the world and the resultant initiation of desertification with, in effect, the acceleration of drought impacts rather than their reduction. This was noted for desertification fears in Chapter 5, and the relationship labelled rather that of 'drought follows the plough' (Glantz 1994).

Drought and the Weather Forecasters

Droughts in fact stimulated meteorologists not only to try to provide explanations, but also encouraged them to investigate apparent links in drought occurrences around the globe, and the possibility of forecasts for drought occurrence. As one researcher put it:

> The distribution of drought and its ill effects has geographical, ethnic, occupational, sectorial, social class, age, and sex dimensions, all of which must be taken into account. Concentration should focus on the most important and urgent, not merely on the most interesting and noisy. (Hinchley 1979: 281)

Among the more 'interesting and noisy' fields for drought research has been that of drought forecasting.

Back in the 1960s, however, claiming to be a drought forecaster was not popular, as a Texan researcher noted: 'In general people just don't want to be known as "drought forecasters", but some don't object to being called "long-range climate" or "long-range precipitation" forecasters. There are not many of them either. Drought forecasting as such seems to be taboo' (Carr 1966: 7). In effect such forecasts would be seen as an admission of scientific impotence in the face of what was considered to be a natural phenomenon. He went on, however, to identify areas where the scientists were more comfortable: 'On the other hand, classifying and defining the intensity of the droughts of the past, or working out formulas for determining the severity of the droughts of the future seems to be acceptable. Also, it seems to be acceptable to analyse the causes of past droughts and to postulate about the causes of future droughts' (Ibid.). Carr, as a Texan, was commenting at a time before sea-surface temperatures and atmospheric pressure contrast across oceans emboldened the drought forecasters.

Yet general seasonal forecasts, which included warnings of possible droughts, go back to at least the early 1920s in Australia, with the work of Inigo Jones (1872–1954) (ADB 1983: 515), later Lennox Walker on sunspot cycles, and more recently Leon Morandy (*The Weekend Australian*, 5–6 November 1994), and Roger Stone (*The Courier Mail*, 26 June 1999) in Queensland. In the USA, with the issue of the Department of Agriculture's 1941 *Yearbook of Agriculture* titled 'Climate and Man', the scientific interest in weather and its forecasting was clear enough. Although it was admitted that 'the science of meteorology does not yet have a universally accepted, coherent picture of the mechanics of the general circulation of the atmosphere', yet there were tests of five day weather forecasts in train and 'for agriculture, flood control, water supply, and many other interests, [such] quantitative forecasts are of tremendous potential value' (Rossby 1941: 600 and 646).

As part of the growing scientific interest in possible climatic change in the 1970s, however, drought forecasts received significant stimulus from the association of worldwide droughts with the El Niño–Southern Oscillation (ENSO) event of 1982–1983. Walker's studies of the failure of the Indian monsoons and the links with sea surface temperatures in the Indian Ocean led to Australian scientists discovering the links with the Pacific as the ENSO phenomenon, which were spectacularly demonstrated in the 1982–1983 droughts (Glantz, Katz and Krantz 1987, Oguntoyinbo 1986). Indeed the claim has been made that 'Australia is the only continent on Earth where the overwhelming influence on climate is a non-annual climatic change … [the] "El Niño Southern Oscillation" or ENSO for short' (Flannery 1994: 81). In fact, the Australian Commonwealth Bureau of Meteorology bases its seasonal forecast upon the ENSO process, which is claimed to be influential in national crop production patterns (Nicholls 1985, Whetton 1989) and seems to have been at least partly responsible for the devastation of the pastoral rangelands (White et al. 2004). Thus since the 1960s forecasters have become more confident, but mistakes are still made and getting a forecast wrong is still embarrassingly possible, even when based on the supposedly tried evidence of mechanisms such as ENSO or the North Atlantic Oscillation. Part of the difficulty of course is that droughts may be claimed to exist without any evidence of shortage of rainfalls. Scientists, in effect, have to deal not only with natural elements but also with the effects of human activities – the natural system confronted by the human management system. The rules for the two systems are not the same and, indeed, reports may be falsified.

Protecting from Drought: The Role of the Irrigators

'More crop per drop'.
        The mission statement of the International Irrigation Management Institute

Perhaps the most obvious human environmental manipulation stimulated by drought is the process of irrigation, the supplementation of inadequate local rainfall by application of water diverted from other areas of surplus. Long associated with the rise of human civilizations, Karl Wittfogel saw the complex organization needed to build and maintain diversionary canals and aqueducts, and the skills to time and regulate the application of the waters, as creating specialist 'Hydraulic Societies' which came to dominate the histories of China, India and the valleys fringing the deserts of southwest Asia (Wittfogel 1957). While the details of his claims have been questioned, there is no doubt that the empires which were created based upon irrigated agriculture required complex bureaucracies, encouraged concepts of time and the recognition of seasonal conditions, and were associated with centralized political controls.

Apart from the many historic, contemporary and planned irrigation schemes around the world, there have been several which were planned but never initiated, despite zealous lobbying in the face of hard facts. In Australia there have been plans to even modify the climate of the central deserts, by gravity fed canals from the sea into the Lake Eyre Basin to create an inland sea and so supposedly ameliorate the regional climate. This was first proposed in 1883 after the inland railway surveyors found that the basin was in fact below sea level, was revived after the 1901–1902 droughts, but scotched by the low gradient of the canal (1.5 inches per mile), high cost of construction, enormous loss of water by evaporation, and potential filling of the lake by evaporated salts from the sea water. A reinvestigation in 1937 concluded 'Even if it were practicable, the flooding of Lake Eyre might do as little good to Central Australia as the Dead Sea in Asia does to its barren basin' (Australia 1937: 925).

Further Australian schemes have focused upon diverting supposedly surplus moisture from the coast inland for agricultural purposes. One of the earliest was the brain child of the successful designer and engineer of the Sydney Harbour Bridge (Bradfield) who planned to divert the rivers from the coastal highlands of northern Queensland inland to supplement the drainage systems already flowing into the Lake Eyre Basin. Despite a negative expert report by the CSIRO stressing costs, and the loss of flows from seepage and evaporation along the way (Australia 1945), the idea has periodically resurfaced. It was again officially rejected in 1982 by the Queensland Government (Queensland Government 1984: 110), but bounced back in 1989, and again in 1998 when *The Weekend Australian* (18–19 July: 48) reported 'Inland Plan flows along: A new partnership of farmers, irrigators, local and State politicians is asking the federal Government to engineer drought-proofing', with rivers not only in Queensland but also in Western Australia and the Northern Territory to be tapped. In 2007 the Premier of Queensland urged the national Government to again re-examine the Bradfield Scheme, but without success (ABC NewsOnline, 19 February 2007).

There is also no doubt that the long history of irrigation, despite the equally long history of the associated problems and failures, has been based upon its attractions as a resource management system. These have been itemized as: maximizing returns (on capital invested, on water used, on cultivated area, and on value of agricultural production); increasing food production, household income, and employment; redistributing income; satisfying political ideals and establishing social security; adding to national economic efficiency, and creating foreign exchange funds; helping to settle nomadic folk as agriculturalists; stabilizing and modernizing agricultural systems; and [supposedly] providing drought protection. After listing the attractions researchers went on to suggest long held views that 'It is clear that irrigation is expected to alleviate world food shortages through increasing agricultural production and that arid lands have scope for development' (Agnew and Anderson 1992: 174).

Alleviating drought in the arid and semi-arid lands, then, would seem to be a long standing and still attractive challenge and pertinent goal for human endeavours. As we have seen already, however, the solutions through irrigation may not be long lasting. On the American Great Plains a commentator in the 1990s suggested that the 1950s 'detour into irrigation may simply prove to be a precursor to the return of dry land agriculture, and perhaps of Dust Bowl conditions in the next century' (Riney-Kehrburg 1993: 150). By 2000, in fact, as he forecast, the ground waters supporting the initial irrigators had already been exhausted in parts of the Great Plains.

Drought and the Fortunes of Land Settlement

One of the themes which, hopefully, has emerged throughout this book is the constant role of droughts in the history of human societies. Often destructive of places and peoples, the scars of past drought disasters can be seen in the ruins of past empires in the sands of Central Asia and the tropical forests of Central America. Desertification is the obvious process and while there is much documentation from the past, there is plenty of evidence that the process is still at work in Central Asia, China, India and Africa, and the threat also hangs over the Americas and Australia. Yet drought can bring benefits, however briefly, to those elements of society less disadvantaged and so able to profit in troubled times. Whether the benefits come from careful husbanding of scarce resources, farsighted spatial spread of drought risks, successful anticipation of drought impacts, or even official drought relief itself, all may play a part in individual survival under the threat of drought. The problem is that such successes are rarely given the recognition they deserve.

Drought and the Confounding of Theories: Querying Drought as a Natural Hazard

One area where the experience of research into drought brought some questioning of accepted knowledge was in the definition of the phenomenon and associated impacts. One problem which has caused some confusion among scientists is the extent to which drought can be identified as a 'natural hazard'. In the United States in the 1970s 'natural hazards' were defined as unforeseen natural events causing loss of life and significant property damage (White 1974). As a new field of scientific inquiry, the study of the origins, impacts and possible defence against such hazards initially focused on the threat of flooding, but rapidly came to include storms, hurricanes and droughts, and drew in an international body of researchers (Burton, Kates and White 1978). Although not explicitly stated, the initial philosophical context was one in which the objects of study, the natural hazards, were by implication separate phenomena with causes independent of human actions, i.e. Acts of God.

The results of the investigations were published in ongoing issues of *Natural Hazards Research*, but it became increasingly obvious that the hazards were only partly natural. A significant proportion of their creation lay in human mismanagement of the environmental resources which exposed society to latent dangers. Building houses or factories on flood plains or on crumbling coastal shores, or trying to grow crops on the edge of deserts were obvious examples, where nature was not wholly guilty. In fact, the 'Acts of God' had help from the 'parishioners' and any serious investigation of their causes needed to take account of this.

Not only that, but the scientific inventions and technical advances of the Industrial Revolution themselves could create disasters from human activities alone. These were the so-called Technological Hazards (Cutter 1993, Zeigler, Johnson and Brunn 1983), of which nuclear accidents such as the Chernobyl disaster were obvious examples. The technology by which the challenges of nature were to be met could itself provide hazards to test society. As one geographer put the problem:

> Once the human contribution [to environmental processes] has become more than trifling, the evolution of the new landscape is no longer natural, and catastrophes lose their status as simple acts of God. If there are floods or droughts, hurricanes or calms, hot weather or cold: all might be our efforts. Science ... [by ignoring the human inputs] ... will have to wear the blame. (Webber 1994: 131)

This recognition that natural scientists must look beyond the narrow confines of their disciplines, if they were to understand the realities of what was happening and why it was happening in the global environment, has been evident in the conservation movements and most recently in the focus upon claims for Climate Change and Global Warming (Glantz 2003). Recognition of the need to seek beyond discipline boundaries had its own problems, however. A senior scientist castigated his colleagues in the early days of the debates:

> Shrill exchanges on whether climatic fluctuations or social organization is 'responsible' for the suffering of peoples and landscapes in drought zones have obscured the complicated interrelations that characterize such situations. Case study chronologies and consequences have been transferred indiscriminately around the globe, with little regard for the special circumstances of place or the stage of historical development. Studies of long-term climatic impacts have swung between approaches assuming that no adaptation is too great for societies or ecosystems to make, and equally unrealistic analyses that simply impose possible future climates on today's animal, crop, and human distributions and tally the resulting disruptions. (Clark 1985: 6)

The criticisms here were not that his colleagues had merely a blinkered view of the world, but that the view was not even good science.

In the same year a review of an attempt to forecast the trend of climate change to the end of the century by interviewing 'scientific experts' was criticized as focusing merely upon common ground opinions. The review did not show the wide range of actual opinions, and thereby gave a false impression of a consensus supposedly believing that 'climate in the next twenty years would be similar to that of the recent past', whereas in reality opinions were much more varied (Stewart and Glantz 1985: 175). So how have scientists coped with these problems in their examination of the role of drought in climate change ?

Drought and Climate Change: The Threat to Accessible Global Water Supplies?

> The greatest risk in thinking about the future of climate-sensitive systems is to assume that the climate of the last century will be the climate we will face in the next. (Barnett et al. 2004: 8)

Getting to Know Climate's Ups and Downs

Among many problems, the opening years of the twenty-first century have been concerned with the attempt to understand the nature of what appears to be evidence of a global warming of the earth and the associated possible changes to the familiar climate. With scientific interest in climate increasing in the early decades of the twentieth century, the World Meteorological Organization decided in 1935 upon 30 years as the minimum period of data needed to provide sufficient information to identify a 'normal' set of climatic conditions. The first such period was created for 1901–1930; the second 1931–1960; the third 1961–1990; and the fourth 1991–2020. Yet the average data for each of these first three thirty-year periods has varied considerably, and the differences have had to be defined in order for logical interpretations of any trends to be attempted. And all this in the context of some claims of special pleading, as we shall see.

Defining the Terminology

There is much confusion, not least among the scientists, as to the meanings of terms such as climatic change, climatic fluctuations, climatic variability and climatic noise. In 1982 a researcher proposed guidelines:

> 1) Climatic noise may be defined as that part of the variance of climate attributable to short term weather changes ...

> 2) Climatic variability is ... the manner of variation of the climatic parameters within the typical averaging period [expressed as standard deviations].

> 3) Climatic change is said to occur when the differences between successive averaging periods exceed what noise can account for, i.e. when a distinct signal exists that is visible above the noise ... [S]hort-term changes lasting only a few decades [are termed] as climatic fluctuations, especially if conditions then return to the earlier state. Many authorities reserve the term change for longer-term variation (such as the Little Ice Age, or the mid-Holocene desiccation of much of the sub-tropical world). (Hare 1982)

Given that drought manifests itself over time and space as part of the characteristics of climate, we can expect that any apparent variations in that climate will affect the nature and incidence of drought. The difficulty, however, is that there is considerable scientific as well as popular debate as to the evidence, characteristics and causes of any climate change.

Debating the Evidence on Climate Change and Significance for Drought

Given that drought is recognized to be part of the climate process, we can expect that any evidence of, or concern for, climate change will raise the question of whether such changes will affect the incidence of drought. Scientific concern for climate change is currently very high and has been increasing since the 1970s, stimulated by the extensive droughts in west Africa. By the mid-1970s scientific reports from the International Federation of Institutes of Advanced Study (Bonn, Germany) had suggested that climate change would affect world food supplies and had already been responsible for the increased drought frequency in southern Asia and the Sahel (Gibbs 1979).

The Eighth Congress of the World Meteorological Organization (WMO) considered the inauguration of a World Climate Program since:

> It has been realized that in spite of our increasing technology (and in some respects because of it) the world with its ever-increasing population and demand for food and other resources is now more vulnerable to variations or changes in climate than ever before. [Using the example of the Sahel drought of 1968–1973, the argument went on]… The margin of safety in food production is shrinking and changes or variations in climate can generate international problems on an unprecedented scale. (WMO 1979: 42)

The program proposed increasing international climate data collection and analysis of any evidence of variations and their potential effects upon human activities. At the same time it was recognized that human activities might have already had significant effects on climate, and part of the new research was to investigate any such evidence. Climate itself was to be seen as a resource to be used wisely and to be protected by the community of nations.

Alongside the scientific activities were parallel extravagant popular accounts in the media of impending climatic disasters. Two books appeared in 1975: H.H. Wilcox provided *Hothouse Earth*, which claimed that the increasing release of energy from industry and the expanding human population would lead to a global warming effect, while N. Calder's *The Weather Machine* forecast a contrasting ice age resulting from reduced solar insulation because of blanketing air pollution, with increasing snow blizzards. The latter interpretation was taken up by the Australian media in 1976, with the *People* newspaper on 23 September trumpeting that the 'Ice Age is coming' and *The Australian* newspaper in December headlined 'Millions will die as Ice Age nears'.

Debates about the scientific data continued through the 1980s and 1990s, with the WMO, the United Nations Environmental Programme and, more specifically, the Intergovernmental Panel on Climatic Change (IPCC – set up in 1988) providing direction and reports from the early 1990s. A second World Climate Conference in 1990 created The World Climate Impacts and Response Strategies Program (WCIRP), in recognition of 'the need to adapt to climate events and mitigate human-induced climate change'. With the stimulus of new funds, reports began to document a definite warming in global temperatures since the beginning of the twentieth century. The evidence came from melting glaciers and retreating mountain snowlines, and was accompanied by forecasts of overall increases in global temperatures of between 1.4 and 5.8°C, apparently tied to increasing air pollution. Even Arctic air temperatures were reported in 2003 to have increased by half a degree Celsius over the last 30 years, when 'temperatures were at their highest level in 400 years' (Sturm, Perovich and Serreze 2003: 44). In southern Chile in 1992, I was personally impressed by the evidence of recent glacial retreat.

What seems to be the contemporary scientific evidence on climate change and its relevance for human activities, and in our specific context the relevance to the risk of droughts? The general conclusion is that global warming is occurring, and that it is a result of the trapping of increasing amounts of solar radiation in the atmosphere by the rising concentration of carbon dioxide and other greenhouse gases resulting from human activities. The global warming trend seems to have been borne out by studies of glaciers collected by the World Glacier Monitoring Service. The 30 reference glaciers with a continuous record of observations since 1976:

> [S]how an accelerated thinning, with mean annual ice losses of 0.14 m w.e. [meters of water equivalent] (1976–85), 0.25 m w.e. (1986–95) and 0.58m w.e. (1996–2005), which gives a total average ice thickness reduction of about 10m w.e. ... [This is] a dramatic ice loss compared with the global average ice thickness which is estimated ... to be between 100m ... and about 180m. (Zemp, Hoelzle and Haeberli 2009: 101 and 106)

Glaciers are regarded as sensitive recorders of global climate change.

For China, a study covering the period from 1880 to 1998 found that the highest temperatures since 1880 had been in 1998, and there had been warm periods from the 1920s to 1940s associated with major droughts and in the late 1980s also with droughts, but less intensive. Dust storms had peaked in the 1950s but had declined subsequently, while the annual periods of reduced flow along the lower reaches of the Yellow River extended from the 1970s to 1990s, suggesting possible reduced rainfall effects and human diversions. Both floods along the Yangtse River and 1990s droughts in northern China were claimed to have caused enormous losses (Qian and Zhu 2001). A further study claimed evidence of a northerly shift of the subtropical climate zone of 3.7 degrees of latitude along the 116th meridian, i.e. a regional warming effect (Ye, Wenjie and Yundi 2003). Bearing in mind China's explosive industrialization and the associated industrial air pollution in the last decade, a study of the 'solar dimming' effect of aerosols in the atmosphere has predicted that the effects may be wider: 'Simulations suggest that if current trends in emissions continue, it is possible that the South Asian continent may experience a doubling of the drought frequency in future decades' (Lohmann and Wild 2005: 22).

For Europe overall temperatures increased by 0.8°C during the twentieth century, with the last decade being the warmest on record, while precipitation increased by 10–40 per cent in northern Europe but decreased by up to 20 per cent in southern Europe, giving rise to fears of increased summer drought risk and wildfires (Maracchi, Sirotenko and Bindi 2005). Low flows on the Elbe River from the Czech Republic to the German port of Hamburg over the last two decades brought commercial barge traffic to a standstill for between four and six months in 1991–1992 and again in 2000 and 2003, and declining rainfalls from global warming are threatening to negate official plans to upgrade the navigation potential of this important shipping route (Paterson 2006). Using integrated global water models (basically trends in river flows) a further study suggested that northeastern Europe would probably have increased levels of flooding by 2070, whereas southeastern Europe would have increased occurrences of droughts. Interestingly, the study admitted that the eastern droughts might be the result of increasing water demand, not lower rainfalls (Lehner et al. 2006).

For Australia, a national official report in 2001 showed that mean surface air temperatures had increased by 0.76°C since 1910. Most increases occurred since the 1950s. Minimum daily temperatures increased slightly, leading to 'greater heat stress on humans, livestock, ecosystems, agriculture and building materials, increased frequency of bushfire, higher energy demand for air-conditioning and increased demand on water supply' (Manins et al. 2001: 65). Further increases in temperature with decreases in rainfall in the south and increases in the north were predicted as a result of global warming. Rises have been confirmed recently in the official report for 2009, which claimed a mean annual temperature rise of 0.9°C over the 1961–1990 average, and that each decade since the 1940s has been warmer than the preceding decade (*The Australian,* 6 January 2010: 4). Not surprisingly, earlier reports had forecast an increased risk of meteorological droughts (Chambers and Griffiths 2008, and McMahon 2008).

What is the relevance of the current thinking on trends in global warning to forecasts of future droughts? Using the Intergovernmental Panel on Climate Change reports of 2001 and 2007, the evidence from some 20–30 different climate models is set out in Table 10.1. In 2007 the claim was made that:

> More intense and longer droughts have been observed over wider areas since the 1970s, particularly in the tropics and subtropics. Increased drying linked with higher temperatures and decreased precipitation has contributed to changes in drought. Changes in sea surface temperatures, wind patterns and decreased snowpack and snow cover have also been linked to droughts. (IPCC 2007a: 8)

Looking well ahead, the IPCC forecasts of drought resulting from decreases of between 10 and 20 per cent in precipitation for 2090–2099 compared with 1980–1999 showed there would be droughts in the winters in the northern hemispheres in southwest USA/Mexico, the southern Mediterranean and northern Africa. For the southern hemisphere winters droughts were likely in northeast Brazil, central Chile, southwest Africa and southwest Australia. Certainly droughts in these areas are already well-known and seem likely to continue, if these winter moisture shortfalls follow on from the usual summer dry spells. Such a long range forecast, however, seems highly uncertain and I would seriously question its relevance for any planning purposes.

Table 10.1   Basic Summaries of the IPCC Reports for 2001 and 2007

Findings	2001 Report	2007 Report
Increase in global mean temperature	About 0.6°C (over twentieth century)	About 0.75°C (1906–2005)
Temperatures in the lower atmosphere have risen	About 0.1°C per decade (1960–2000)	About 0.2°C per decade (1960–2005)
Snow cover and ice have decreased	Yes	Yes
Precipitation has changed, increasing in some areas and decreasing in others	Yes	Yes
Most of the global warming over the last 50 years	Seems to be attributable to human activities	Very high confidence … effect of human activities since 1750 has been of warming
Further human activities will result in increased temperatures and sea level rises in the twenty-first century	Yes	Yes
There is likely to be an increase in various extreme climatic events	Yes	Yes

Source: IPCC 2001, IPCC 2007b.

New Arguments for the Twenty-First Century?

While the existence of global warming is accepted by all but a very few, divisions remain over whether the causes of warming are natural or human-induced, or varying combinations of the two, as well as over whether warming will increase the incidence and severity of droughts. There are differences of opinion and several alternative interpretations of the available evidence. For example, volcanic eruptions (such as Mount Pinatubo in 1991) were shown to have 'significantly cooled regional climates', while direct solar energy had influenced at least half of the temperature gains (Hengeveld and Edwards 2000). Future catastrophic changes as a result of global warming were denied, with no 'increased risk of natural disasters caused by increased frequency and severity of climate extremes'; rather the result would be 'a more robust biosphere with more forest, crops and ground cover for more animals and people' (de Freitas 2002). For droughts specifically, 'not only does global warming not produce more frequent and severe droughts, it does just the opposite. In fact, evidence from North America had shown that past droughts (in the last millennium) were much more severe in duration and impact than any encountered in the last 100 years' [of global warming] (Ibid.: 36).

    A somewhat similar forecast has been made for Australia, based upon findings that evaporation from standard Bureau of Meteorology pans has been declining over the last 30–50 years, despite the fact that Australian and global temperatures had been rising. Thus the traditional understanding, that evaporation from open water rose as the air temperature rose, has had to be reconsidered and as a result 'a robust prediction is for global precipitation to increase annually by 17mm for every 1°C of warming (Pockley 2009). On the face of this claim, the future incidence of drought might be expected to be reduced?

Nonetheless, in 2002 southeastern Australia suffered a severe drought with associated wildfires which burned out part of the national capital, Canberra, destroying 530 homes, and causing four deaths and over \$A250,000 billion damage (Lavorel 2003, Wright 2005). At that stage in the debate, while the majority of scientists expected global warming to bring more extreme weather conditions, including more intensive droughts, others suggested that the warmer temperature's boost to plant life would offset the adverse effects of the increased droughts. As one experienced researcher put it, for humans at least, global warming was preferable to global cooling (Ruddiman 2005).

The debates, however, continued. While admitting the beneficial effects upon some plants from the global warming-driven increase in carbon dioxide CO_2, especially on wheat and rice crops, a separate study claimed that these benefits would decline as the warming increased and a temperature increase of more than 2.5°C 'is likely to reverse the [current] trend of falling real food prices … [which] would greatly stress food security in many developing countries' (Easterling and Apps 2005: 183). This division between the contrasting forecasts of the impacts of global warming/climate change on agriculture in the developed versus the developing world was illustrated in the same year. For Mali, declining crop yields and livestock weights, together with drought induced land degradation, were forecast to bring the level of hunger from currently 34 per cent of the population up to 64–72 per cent by 2040 (Butt et al. 2005). In contrast, for the USA a 2.5°C rise by 2050 was forecast to show a relatively modest net economic loss to the national society of 'a few tenths of a percent when compared with GDP. But substantial local changes – both positive and negative – in economic well-being' might be possible (Edmonds and Rosenberg 2005: 159).

On the other hand, for China similarly contrasting views can be found within the same country. Liu and colleagues in 2004 found that 'In summary, all of China would benefit from climate change in most scenarios' (Liu, Li, Fischer and Sun 2004: 125), whereas Fu and his colleagues in the next paper in the same scientific journal claimed 'If the … (IPCC) projections of continued warming in the region [China] during the twenty-first century are correct, the present results suggest that the trend towards reduced runoff is likely to lead to exacerbated problems for agriculture, industry, urban communities, and the overall regional environment' (Fu, Chen, Liu and Shepard 2004: 149). No wonder policy makers might be perplexed.

Over the last few years the arguments about global warming and its causes have intensified. There are disagreements over the projected rate of warming, over the role of human activities in this warming, and over its effects on rainfall.

Consequently, uncertainites remain over the projections about future droughts. These debates have been extensively reported in both the scientific literature and the press, and have become somewhat politicized. A book-length review of the arguments over global warming up to 2008 by Morgan and McCrystal concluded that the biggest uncertainty was over how future climates will respond to increased concentrations of greenhouse gases. However, the authors (who were non-scientists using two panels of scientists supporting opposing views) noted that scientists on both sides of the debate had seemed to close their minds to alternatives (Morgan and McCrystal 2009: 246).

There have even been claims of evidence being manipulated to support particular opinions on the question of climate change, a perverted form of 'crying wolf'. A climatologist, Patrick Michaels, provided damning criticisms of scientific band-waggoning in his book *Meltdown: The Predictable Distortion of Global Warming by Scientists, Politicians and the Media* (2004). I chose his quote from President Eisenhower to head this chapter as I think Eisenhower's comments are still very pertinent. The two dangers he mentioned, first that science would become driven by government policies and funding, and second that 'public policy could itself become a captive of a scientific-technological elite', are potential problems. Michaels claimed that 'When it comes to climate change, there's a culture of distortion out there. But it shouldn't surprise you. Its development was logical, predictable, and inevitable' (Ibid.: 6). Basically, he suggested that there were a bevy of stakeholders, scientists, business folk and politicians, who could potentially benefit from scaremongering.

Fred Pearce, a believer in global warming, hinted at this phenomenon in *The Guardian* in June 2008. Having recently attended a conference of climatologists he was impressed by their admission that many of their models of future El Niño events or even future ice ages were less than optimal. However, he had to add that 'This sudden humility was not unconnected with their end-of-conference call for the world to spend a billion dollars on a global centre for climate modelling. A "Manhattan project for the 21st century", as someone put it' (Pearce 2008: 33). President Eisenhower might have claimed to rest his case.

Morgan and McCrystal would reply that both sides were guilty of scaremongering for the benefit of their own interests. They wrote:

> The polarising effect of the climate change question is partly due to its tendency to appeal to forces massed along the borders of existing political territories, such as between laissez-faire (that is, dog-eat-dog) 'free marketeers' on the one hand and interventionist, anti-consumerist Greenies on the other. Each side accused the other of using climate change as a Trojan horse for their own, ulterior agenda (Morgan and McCrystal 2008: 246)

Uncertainties and suspicions of politicization and self-interest continue to confuse the messages from science. All that can be concluded is that, at present, our ability to be certain about what science has to say about future droughts remains unclear, and much work remains to be done.

Drought and the Social Scientists

> Mike Brennan [pastoralist] watched the flooding Macintyre River engulf his yard and encircle his Goodiwindi house yesterday and declared: 'You beauty!'. Mr Brennan, 50, said he felt 'bloody good' because 'the five year drought is over'. (A Queensland newspaper report, quoted in West and Smith 1997)

The interest of social scientists in drought has not only been concerned with the social dimensions of its impacts, but also with the nature of its role in societies and the social implications of attempts to limit its damages. Under the current economic theory of globalization, traditional common resources, water included, are being commodified, commercialized and privatized at an alarming rate and in so far as the definition of drought includes denial of access to water, such a globalization process has relevance for any attempts to mitigate drought impacts. The omission of water from both the original United Nations Charter and the Universal Declaration of Human Rights has allowed private monopolies of water to be created, which have effectively created droughts through the pricing of that water beyond the reach of sections of society. Evidence of the existence of such a process has already been shown for Brazil and India. At the international level the claim has been made that 'Water is a sacred component of the commons [global resources]; it belongs to our common humanity, the Earth and all living species'. Under such a banner an international pressure group 'Friends of the Right to Water' is agitating for communal rights to be restored above private ownership (Barlow 2006).

Australia has been seen by sociologists as a 'harsh and drought afflicted landscape' in which the Australian national character has been forged by droughts – seen as disasters that were 'outside human control'. In combating those droughts Australian society appeared to need to 'reaffirm social morality and solidarity in the face of an unexpected and unprecedented challenge from nature' (West and Smith 1997: 205). While the challenge can hardly be seen as 'unexpected', given the history of droughts in Australia, nonetheless it has been the 'dominant force' whose influence and impacts might be muted by the more 'friendly' hazards of floods and cyclones, as might appear from the newspaper report quoted above!

Ironically, despite the frequency of drought occurrence in Australia, the media continue to use the 'language of war and disaster, and the imagery of suffering' in their coverage. Until this biased view is modified and the survivors and successful drought managers are brought into focus 'it will be difficult to conduct a fully informed, rational debate about drought policy in this country' (Wahlquist 2003: 85). Certainly, public policy towards drought in Australia has varied over the years, with each new drought apparently needing modifications (often the watering down) of an innovative policy initiated in 1989 (Butterill 2003: 65)

Natural Science and Social Science in the Management of Droughts

> Climate-society-environment interplay is too important to be left to the climatologist, or for that matter, to any single set of discipline-focused researchers. (Glantz 2003: xvii)

The recognition of complexity as a basic characteristic of any understanding of environmental conditions has arisen from the coming together of natural and social science researchers, who were faced by challenges which overlapped their supposedly separate jurisdictions. Those complexities might show up failures in comprehension or might force revisions which lead to new insights. Part of the problem lay in the way the term 'complex' was itself used by the scientific community. Before making his own challenging statement in *Climate Affairs* which I have quoted at the head of this section, Michael Glantz had shown that 'complex' had been used variously to suggest that understanding of a phenomenon could not be complete, that to achieve understanding might take much time and money, that the information achieved might be flawed, and that comprehension of the reality might be beyond the ability of the reader (i.e. the public) (Glantz 2001). Apart from this possibility, however, critics of natural scientists have claimed that 'While it is the case that natural scientists have a language and the means for governing scientific processes that are often lacking to social scientists, the latter possess a notion of global ethical responsibility which is not always found in the former' (Bianchi 1994: 12). This long-recognized division has had to be met in the consideration of the phenomenon of drought as part of the environmental condition. Those divisions might show up failures or omissions in comprehension or might force revisions which lead to new insights.

That these issues were still a problem has been shown by criticisms levelled at apparent inconsistencies and omissions between the various *International Panel on Climate Change Working Group Reports*, which up to now had been seen as the authoritative source of scientific consensus on the Climate Change–Global warming issue. Criticisms have come from both natural and social scientists and were in addition to the issues raised in the previous section. The initial questions came in a paper to the journal *Global Environmental Change* (Ha-Duong, Swart and Bernstein 2007) which, interestingly, was slightly revised but basically reprinted in the flagship natural science climate journal *Climatic Change* (Swart, Bernstein, Ha-Duong and Peterson 2009). Although there had been debates about the findings among natural scientists, the new criticisms focused upon the weaknesses in the communication of these findings. The critics claimed that while the reports had focused upon providing a consensus of scientific opinions, they had not adequately recognized nor reported that there were divisions of opinions and some uncertainties in the strength of the data and arguments presented.

Although not mentioned in the later paper, a further article in *Global Environmental Change* provided evidence of how 'quality' newspapers in the United Kingdom (*The Guardian, The Independent, The Daily Telegraph* and *The Times*) had covered climate change over the period 1997/98 to 2006/07 (Doulton and Brown 2009). Five 'stances' on climate change were identified, from beliefs that climate change would bring benefits through to a belief that overcoming climate change would benefit the world's poor. While the authors claimed that the newspapers did not reflect the uncertainties in scientific understanding of climate change, the evidence suggested that they were reflecting varied interpretations of the scientific evidence available to them and were presumably passing on those varied interpretations to their reading publics.

Another problem (as Glantz had noted above) lay in the vocabulary used. The specific problem in this case was the potential variation in understanding and application of the word 'uncertainty'. Recognizing that the international scope of the research brought together scientists from many disciplines, the critics noted as background that 'More than a century of philosophical discussion on uncertainty has shown that there are divisions which cannot be resolved' – in other words uncertainty meant different things to different people and the reports had forgotten this. Despite official guidance for the report writers on 'the meanings of words used to describe probabilities' there was no recognition that the readers' interpretations of those words might not coincide. As the editorial introducing the first paper above suggested: 'uncertainty is a multi-dimensional concept that is omnipresent in our society ... Uncertainties about climate change not only shape international, national and local climate policy, but they also influence perceptions of, and responses to, climate change at the level of individuals, communities and businesses' (Dessai, O'Brien and Hulme 2007). By overlooking this problem the reports had given a false sense of certainty to their findings (and this may have been behind the extravagant treatment of the findings by the media).

A further criticism was that the reports highlighting the threats of climate change facing global communities did not recognize that similar threats had occurred in the past, and that historically human societies had managed to adapt and survive. By implication this omission appeared to reinforce the 'view of indigenous peoples as passive and helpless', which 'is not new, with roles going back to colonialism and recurring in contemporary discussions of development, conservation, indigenous rights, and indigenous knowledge' (Salick and Ross 2009: 137).

To be fair, however, the targeted reports were intended to be merely reporting on the contemporary research, 'to summarize the literature', which was predominantly in the natural sciences, and the search for a consensus to provide to policy makers was a logical strategy. Future reports, however, will need to take cognisance of the above criticisms and 'will need to be focused ... on making sure that all sources of uncertainty have been included and reflected in the numerical ranges [of probabilities of climate change] on which public discourse will inevitably focus' (Webster 2009: 39). And in the spirit of that hope, the report writers should recognize the various meanings of the concept and word 'vulnerability', and the suggestion of a framework to provide 'much-needed conceptual clarity and facilitate bridging the various approaches to researching vulnerability to climate change' (Füssel 2007: 155).

Switching to the specific case of drought, in the investigation of drought management strategies social scientists have tended to recognize motives which might have been overlooked by their natural science colleagues, and three cases provide some insights. The first, from the USA, provides an actual assessment of how two states in the USA faced their separate drought problems; the second is a review of reactions to the 1993–1994 drought in Bulgaria; while the third illustrates the agenda considered relevant to any general attempt to manage drought in the USA. In all cases there is recognition of the multi-faceted nature of the drought stress and the equally multi-faceted nature of the responses to the drought challenge.

Experience with Weather Modification in the USA

The USA has probably the longest history of attempts at weather modification, beginning in the nineteenth century and gathering momentum in the twentieth. In 1986 a report from the Science and Technology Centre of the Syracuse Research Corporation, in Syracuse, New York, provided evidence of projects to instigate weather modifications in two states attempting to cope with the 1976–1978 drought in Colorado and the 1982–1984 drought in Texas (Lambright 1986). The Colorado drought was a 'winter drought' in that a reduction in the winter snowfall led to problems for the ski industry in the winter and problems for state agriculture due to reduced snowmelt runoff in the following spring and summer. The Texas drought was a 'summer drought' in that the expected summer rainfall needed for agriculture was not received.

The report identified four stages in the apparent responses to drought and examined the responses in sequence:

> 1. Pre-drought ... the period before the drought occurs ... both the longer-term pre-drought period and the more immediate time of onset, when the drought is coming, but there is not full awareness.

> 2. Drought. This is the time when the drought's effects are fully felt and a sense of extremity prevails.

> 3. Aftermath ... period immediately after a drought. The drought is still fresh in the minds of the public and its representatives.

4. Normalcy. In this stage, weather is back to 'normal', in the sense of precipitation. The drought is history, and the drought problem is no longer on the policy agenda. (Ibid.: 20)

The Pre-Drought Period

For the pre-drought period the report showed that atmospheric scientists and water resource professionals were aware that both states were 'drought-prone', having experienced and remembered the 1950s droughts. However, both states had experienced a large influx of new settlers since the 1950s who were less likely to have the same level of awareness of drought threats. All the professionals were aware of the possibilities of weather modification and both states had a long history of private sector cloud seeding along with some federally funded Research and Development projects.

The Drought Period

The drought in Colorado was recognized in late December 1976 when snowfalls in the Rocky Mountains were much reduced; the ski industry felt the immediate effects and there were fears for the spring and summer snow melt runoff for agriculture down on the plains. The Governor of the State asked for expert advice in January 1977 and two strategies were suggested: the first was water conservation and the second was weather modification, i.e. cloud seeding. His decision to support the latter in particular was in spite of scientists' concern that 'there was uncertainty about whether the program would produce results'. But his view was that the program could serve a political purpose as well. In politics, he later declared, 'movement is action, and cloud seeding was a way to dramatize to the people that there was a drought'. The legislature took the hint and by February 'a first year state program of winter mountain [cloud] seeding was under way' (Ibid.: 3). The program was continued by private operators into a second year with further state and even federal funding. In the early winter months of 1978 the seeding had to be halted 'because there was too much snow falling and a resulting danger of avalanches. In April 1978, the program officially came to an end, and by fall [autumn] virtually all conservation measures also ended. The drought, which in the public eye had ended months before, was now officially over' (Ibid.: 4).

In Texas, the drought began in the west in the summer of 1982 and by 1984 had spread to the south central parts of the state. The state government asked for federal disaster assistance for some of the affected rural areas, but took no action itself on weather modification. In contrast, three large urban areas did consider weather modification in addition to general water conservation strategies. When approached, however, the state government was not sympathetic, taking the view of its scientists that 'a drought was not a useful time to conduct a [cloud seeding]

program, at least given the meteorological conditions with which Texas was faced'
i.e. there was a shortage of clouds (Ibid.: 4)!

The Aftermath

In Colorado, official cloud seeding efforts were terminated in 1978, with part of
the reason given 'that the sense of urgency stemming from the 1976–1978 drought
had by this time petered out' (Ibid.: 5). Yet this did not end attempts to continue
weather modification, but conflicting interests prevented any overall state policy
support for continued seeding. On the one hand, for example, was the state capital
Denver's Water Board support for continued seeding as part of general water
management, versus the cattlemen's fears over continued seeding leading to heavy
snowfalls blanketing their grazing lands, and the ski industry's anxiety to continue
seeding to guarantee its snow cover.

In Texas, however, there were parallel and relatively successful initiatives
from the three main urban areas (San Antonio, Corpus Christi and San Angelo)
for private seeding with city government funds continuing after the drought had
officially ended. There were local objections – from individual land owners not
wanting their air space to be targeted, and a hill community in the centre of the
target area fearing that their flash flood frequency would be increased merely for
the benefit of the distant cities. Compromises were reached, however, and private
seeding programs were established for a further four years with permission from
the state.

Normalcy

The report commented that 'Normalcy is a state of mind as well as weather.
Normalcy takes place when there is enough precipitation and enough time has
elapsed since the drought to remove mitigation from the policy agenda of a
government' (Ibid.: 7). The appearance of normalcy in fact may be accelerated
by the occurrence of a contrasting problem. Many a drought has been broken by a
major downpour and resulting flood, and this was the case in Colorado. In Texas,
however, at the time of the report, cloud seeding was still in progress although not
with equal vigour in all areas.

This report provided some fascinating insights into the ways society confronts
threats from drought. Colorado, with its Rocky Mountain western backbone (the
base for its established winter sports industry) and the dry but fertile edge of the
Great Plains in its eastern half (where ranching and farming had their individual
and different demands for moisture), had a brief common state government policy
on weather modification. Texas, in contrast, showed no state government support
for weather modification and the impetus came from urban communities situated
in both the drier west and more humid central and southern parts of the state. Why
were there such differences? The report suggested five possible factors:

1. The droughts came at different seasons in each case and as a result weather modification was not seen as equally valuable. In Colorado, increased precipitation was needed in the winter when there were still likely to be clouds over the mountains and when cloud seeding might be expected to produce results. Hence, in Colorado, weather modification was seen as a reasonable strategy by the experts, even if there were no guarantees on results. In Texas, in contrast, the summer skies over the plains in past droughts had offered no clouds to be seeded and officials were less sanguine about weather modification as a short term strategy.

2. The Colorado Governor personally saw weather modification as a good political strategy and could build popular support for it, but in Texas the officials saw water conservation as the way.

3. While the official Texan attitude did not favour weather modification as the immediate response to drought, there was considerable interest in it as a long term strategy to be used in the 'normal' seasons to build up ground water and surface water storage in anticipation of future droughts, particularly by cities in semi-arid areas facing an influx of population.

4. Support for extended state sponsored weather modification fell away rapidly in Colorado once the worst of the drought was past, as the various interests recognized its limitations for their future activities and the resulting opposition strengthened. Texas, in contrast, seems to have accepted at the local government level that weather modification was generally useful.

5. The continuation of weather modification programs in Texas, however, was not guaranteed and the report suggested that local opposition was increasing. Meanwhile in Colorado there had been no conclusive proof that seeding had been as successful as its proponents claimed.

The report concluded that the demand for weather modification 'came from urban settings that are growing rapidly and are short of water. As a formerly rural-based technology, weather modification's urban-orientation is new and of great policy significance' (Ibid.: 11). In essence the report provided carefully documented evidence of the presence of many stakeholders, each with their separate agendas for coping with the drought threat, and each with their varied opinions on the value of the physical experiments in rain-making which seemed to be the most popular scientific response to the threat.

Drought in Bulgaria: A Contemporary Analogy for Climate Change

The above title was of a comprehensive review of the impacts of, and official and community response to, drought in Bulgaria in 1993–1994, but set in the broader context of 1982–1994 events (Knight, Raev and Staneva 2004). Decreased precipitation produced runoff levels of less than a third of long term averages; the quality of drinking water declined; forests showed restricted growth; forest fires increased in frequency; agricultural harvests were reduced; hydro-electric power declined; water rationing was introduced in all major cities; and diseases such as viral hepatitis and dysentery increased. In other words, this was a fairly typical scenario for drought impacts. The responses were also typically crisis-oriented as there was no existing plan to cope with any drought threat. The fact that the nation was going through the transition from a centralized political system to a more democratic system did not help cooperation between central (urban) and peripheral (rural) officialdom, and resource planning was not helped by the fact Bulgaria's water was the lowest priced in Europe and as a result wastage had been endemic. For the future there was the need for a national plan to anticipate the next drought occurrence. This was to be achieved by 'the immediate creation of a [collaboration] of scholars, stakeholders, and policy – and decision – makers to design and execute a plan for monitoring the event and its impacts. Such a group might include people intimately involved in dealing with the drought, but as a body it should be independent of a management or regulatory role' (Ibid.: 313).

Concluding the study, the authors provided 50 recommendations for policy makers, ranging from those of the 'Highest Priority' (including 'a comprehensive, integrated national assessment of the potential consequences of climate change', and that 'impacts from climate change are incorporated in all aspects of environmental planning at all levels from local to national'), through 'Water Resources', 'Management of Forest Resources', 'Wild Fauna', 'Management of Agricultural Crops', 'The Economy', 'Health and Hygiene', 'Sociology and Ethics of Water' and 'The Politics of Water'. History will show whether the recommendations were followed, but at least the attempt to place drought mitigation in the broader context of resource management was to be applauded.

'Managing Drought and Water Scarcity in Vulnerable Environments: Creating a Road Map for Change in the US' [Geologists to the Front!]

The title above was that of a Conference of the Geological Society of America, held at Boulder, Colorado, in September 2006 (http://www.geosociety.org/meetings/06drought). A first question might be what interest had a geological society in drought? The aims of the conference provided some explanation. The goal was 'to create a forum for improving planning and the management of drought and water scarcity in the United States and to stimulate national debate through publication of a science – and policy – based discussion document'. Note the combination of 'science' and 'policy'. That the scope of the enquiry was to range far beyond the confines of geology was indicated by the suggested topics for investigation:

- Hydraulic aspects of drought (past, present and future).
- Biologic aspects of drought including quantitative ecosystem impacts.
- Economic aspects of drought (historical, contemporary, future).
- Risk-based approaches to drought, including probabilistic risk assessments.
- Qualitative and quantitative measures of confidence in drought analyses.

- Public policy approaches for managing and mitigating drought impacts.
- Facilitating collaboration of multiple stakeholders.
- Impact of global climate change on drought management and water scarcity.
- Enhanced drought prediction, monitoring, and impact assessment.

This broad agenda provided a stage for collaboration between the natural and social sciences on the management of droughts. History will tell us whether the actors were able to seize their opportunities.

Conclusion

It would appear then that there is some on-going collaboration between the natural and social sciences, along with involvement of the general public, in attempts to understand and advise upon the challenges of drought amid the other environmental problems facing societies. Certainly, such collaboration and involvement must improve our general understanding of the nature of the problems. Understanding, however, is only part of the process of resource management and, as we have seen above, the nature of that understanding can change quite rapidly. While those changes create their own problems for policy makers, the initiation and implementation of policies are much more complicated.

To illustrate some of those complications and at the same time to try to draw together some thoughts on the questions raised at the beginning of this chapter, I offer this précis of the history of the Bull Run Watershed, Oregon, USA. Concluding his study of the watershed, which is the main source of water for the City of Portland, Oregon, the investigator, Larson, summarized its management history. First came the US federal reservation for the city of the watershed and its forests in 1892; then the inauguration of logging of the forest in 1958 with the connivance of the US Forest Service; the cessation of logging as a result of a successful lawsuit in 1976 by a private citizen of Portland against the US Forest Service for violating the legal protection; the overturning of the ban and renewal of logging in 1977; a disastrous rainstorm in February 1996 which swept bared soil and logging debris into the reservoir and effectively made the water undrinkable, so that the city had to scramble to provide emergency ground water to cope with this 'man made drought'; to the final federal legislation of September 1996 which prohibited all further logging in the watershed (Larson 2009).

Summarizing his analysis, the writer suggested the following lessons, which in my experience have a wider relevance to the diverse societies which make up our global population:

- 'The systematic process of scientific research is not well-suited to resolving issues in which prevailing economic or political forces demand simple, prompt answers'.
- 'Scientists who seek nothing but truth in their investigations are often

ignored or, worse, defamed by those whose economic or political agendas are threatened'.

- 'The capacity of scientists to solve environmental issues fairly and expeditiously is usually overestimated'.
- 'Meanwhile, the public waits for these interminable conflicts to be resolved, confused by the barrage of technical information and disinformation, and thus unsure of whom to believe'.
- 'At stake is the region's economic prosperity on the one hand, and environmental quality and dwindling natural resources on the other – in other words, competing values'.
- 'In the end, resolution is often achieved not by scientific resolution and decision-making, but by people simply deciding what they value most' (Ibid.: 182).
- For the natural scientists it was not necessarily a defeat as there were natural scientists batting for both city and loggers; what was not certain was whether the public had been well served by them.

I think that the findings from Larson's study have borne out my own research into the scientific investigations into drought. The detailed subdivision of chemistry, mathematics and physics which is implicit in any investigations of the global climate allows the creation of detailed models of specific climate components. But the linking of those models and their findings into a holistic assessment of climate, its components and shifts through time, is incredibly complicated and hungry for computer time. When to that complex process is added the necessary collation and presentation of those findings in a form understandable to a non-specialist member of the public or policy maker, inevitably generalizations have to be made. Whether averages, modal values, probabilities or some other summary values are provided, the findings will be implicitly flawed unless they can demonstrate their scope and the limitations of their findings.

Such findings represent the inherent qualities of the data; their relevance to humanity is something else, and this is where social scientists and the general public must have their say. Management of resources is an exercise in power over nature aimed at producing benefits for humanity, and given humanity's diversity, ensuring benefits to all can be difficult. Not only the variety of legitimate stakeholders, but also the scarcity and value of the resource, and the political philosophy of society will all be requisite considerations. For in planning to meet drought's challenges, we need to remember that drought means so many different things to different people.

Chapter 11
The Coming Droughts

The Story so Far

There is a long history of the role of drought in human affairs around the world. Either by itself, or in association with other catastrophic events, with or without human mismanagement of the environment, drought seems to have been responsible for the collapse of established civilizations. Those historical collapses are a pertinent background to the concerns in the last two centuries over the socio-economic costs of droughts around the world. Defining droughts, however, has been difficult. Although scientists have suggested at least four different types of drought, the existence of many different stake-holders with interests and agendas relevant to drought impacts has produced a variety of definitions with no real hope of an acceptable universal one. In effect, the definitions have reflected different dimensions in the basic equation of the short-fall between moisture demand and supply.

Nonetheless, droughts have been recognized at various scales over time and space, with evidence provided from North America, Australia, Africa and China. The registration of those droughts in any official record, however, does not seem to have been automatic. In between the process of recognition and registration have been not only developments in the sciences observing the earth, but also the filters of human interest and even what has been suggested as a moral climatology, whereby the records have been filtered for possible political purposes. The causes of drought are varied, as Chapter 4 demonstrated. Causes of variations in precipitation events and associated soil moisture conditions can be related to the spatial patterns of air mass movements, ocean currents, and volcanic dust clouds, while surface water supplies can be affected by earthquakes or even human mismanagement. Crop harvests may reflect droughts but often many other factors as well, again including human mismanagement in the form of excessive demands upon the environment for moisture. The world can provide plenty of examples of apparent droughts brought about by human greed, criminal activities and warfare.

The broad scope of the impacts of drought is displayed in Chapter 5, which shows drought as a basic factor in the historical development of, and current characteristics of, the global ecosystems. And not all the impacts may be destructive. With the incidence of wildfires drought's role is explicit, and with the process of desertification and famine its role is traditionally strong but increasingly debated. The association with disease is less obvious but nonetheless clear, while the socio-economic impacts fuel the media headlines and energize political maneuvering at local, national and international levels.

Attempts to cope with the challenges of droughts began early and have continued throughout human history. These range from the fatalistic acceptance of events and their consequences, through to adaptation and innovation to modify those impacts, and finally to a belief in the ability to meet and overcome those challenges through the application of humanity's 'contriving brain and skilful hand'. There is a long history of common attempts to understand, to rationalize, and to predict the occurrence of drought, and to share the losses and thereby reduce their impacts. Traditionally regional and local knowledge of coping strategies focused upon reading the environmental clues as to the risk of drought, trialling plants which could survive, and breeding animals which could survive or at least be moved to less droughty locations. Irrigation systems ranging from the private well, through the communal dam or tank to 'hydraulic civilizations', were established early and have continued despite the incipient problems of losses through seepage, salinization of soils and the siltation of storages.

That long history has been extended into contemporary struggles to cope with drought, but characterized by the more sophisticated use of technology and mechanical power from fossil fuels. In part that process became necessary as the continental interiors of the Americas, Africa, Asia and Australia were invaded by Europeans from the sixteenth century onwards. These folk were less familiar with the extensive and drought-prone environments which confronted them. Yet contemporary strategies contain much from the past, and the intrusive technologies have had to recognize that the lessons of history and local knowledge still play a valuable role in drought mitigation. Also included from the past, but invigorated by the increased power of engineers and the globalization of trade and productive systems, are the political impacts of drought. As in the past, governments have been required to try to protect their citizens, but protection has become politicized as to whom it should be given and why; economists are still not united nor certain that it should be offered at all; and governments are unwilling to admit that they still appear to be powerless in the face of a catastrophic drought.

Part of the problem of coping with drought has been the fact that the recognition and registration of droughts has become politicized. Potential or actual victims claim questionable drought relief, distant observers use the occurrence of drought to berate neglectful governments, and speculative resource users on the margins of the good lands claim and use relief to cope with increasing drought risks as they press further into drought-prone country. As a result, justification for public funds for drought relief is more questioned now than ever, despite the fact that common memory seems to conveniently forget the mistakes of the past and greets each new drought with indignant surprise.

The common memory on drought, however, is non-the-less rich in the history of occurrences and of drought's imprint upon society and landscape. The scenarios cover the human versus nature conflicts with variable outcomes, and range from regional impacts on farms and pastoral stations up to science fiction accounts of doomsday global droughts. And yet some of those challenges seem to have been met, in part, by the application of humour.

For scientists, however, the investigation of droughts has been far from humorous. While the role of drought in the global ecology has been well documented and the means by which plants and animals have adapted to its stresses have been carefully researched and understood, the interface between drought and society has posed problems for investigators. Faced by a phenomenon which can be defined in many different ways, depending upon the moisture needed, the relevance of physical definitions alone has been questioned. As scientists have discovered, different societies have had differing views on the threats from drought. The original scientific classification of drought as a natural hazard has had to be rejected in the light of humanity's increasing demands for water. Indeed, drought's impacts have had to be recognized as beneficial in some instances, and the debates here have been intensified by suggestions of general climate change and a human role in that change, with scientists not being unanimous on either issue.

Recently the debates have become more acrimonious, especially when scientific data is seen to have relevance for official policies. As a result the data have been questioned and even the morals of some of the scientists have been seen as suspect. But what of the future? In the light of the issues raised above, can we claim to have successfully met the challenges from the Bull of Heaven or is it still snorting at the gate? Is drought merely a feature of nature?

Nature Pleads not Guilty

As part of an international scientific investigation into the role of drought in the environmental and social disaster of the Sahel in West Africa in the early 1970s, which as we have seen stimulated global concern for potential drought impacts, a series of three volumes appeared under the heading *Drought and Man: The 1972 Case History*. The first volume was entitled *Nature Pleads not Guilty* (Garcia 1981), and provided a Marxist view of why the drought alone could not be blamed for the disaster, arguing that capitalist exploitation of natural resources exacerbated the impacts of the droughts. One does not need to be a Marxist, however, to recognize that drought may not necessarily be solely a natural disaster; humanity plays a role. But whether we see drought as a purely natural disaster or the result of disastrous environmental management on our part, there seems to be little doubt that the future of drought is assured – events and impacts will continue to plague humanity. The certainty of such a prediction is based upon the evidence of:

1. Documented upward trends in world water use;
2. The strong and accumulating evidence that on-going trends in climate will affect the accessible water resources of the globe;
3. The technological bias in drought management strategies, which prevents a holistic approach to managing the world's water resources by concentrating more on increasing the supply rather than reducing the demand.

4. The fundamental technical and moral problems of juggling the end uses of
   water between multiple users and their conflicting demands.

Trends in World Water Use

> How is it that water, which is so very useful that life is impossible without it,
> has such a low price – while diamonds, which are quite unnecessary, have such
> a high price? (Opie 1994: 306)

> We have been god-like in our planned breeding of our domesticated plants and
> animals, but we have been rabbit-like in our unplanned breeding of ourselves.
> (Toynbee 1963: 29)

All the evidence over the last 30 years has indicated that demand for global water
is rising faster than supplies, and part of the reason lies in the explosion of the
human population. A report from the International Institute for Applied Systems
Analysis (IIASA) in 1994 noted: 'In 1990, 132 million people lived in countries
experiencing severe water stress. The projections suggest that within a generation
up to 15 times as many people could live in water-stressed lands, even without
shifts in climate' (International Institute for Applied Systems Analysis 1994:
9–10).

By the turn of the century UN concern for sustainable economic development
had recognized the role of water supplies and commissioned a 'Comprehensive
Assessment of the Freshwater Resources of the World'. The report noted that:

> Concern over the global implications of water problems was voiced as far back
> as the U.N. Conference on the Human Environment in Stockholm in 1972. It has
> been the focus of a number of meetings, including the U.N. Water Conference
> in Mar del Plata, Argentina, in 1977, the Global Consultation on Safe Water and
> Sanitation for the 1990s in New Delhi, India, 1990, the International Conference
> on Water and the Environment: Development Issues for the 21st Century, in
> Dublin, Ireland, and the U.N. Conference on the Environment and Development
> in Rio de Janeiro, Brazil, both in 1992 ... [and the] Interministerial Conference
> on Drinking Water Supply and Environments, Noorwijk, Netherlands, in 1994.
> (UN Economic and Social Council, Commission on Sustainable Development
> 1997: 8)

This report reminded readers of the concerns of the UN Conference on the
Environment and Development of 1992, that the holistic management of fresh
water as a finite and vulnerable resource and the integration of sectoral water plans
and programs within the framework of national economic and social policy were
to be of paramount importance for actions in the 1990s and beyond. Yet by 1997
little improvement in conditions had taken place, and as a result:

> There is clear and convincing evidence that the world faces a worsening series of local and regional water quantity and quality problems, largely as a result of poor water allocation, wasteful use of the resource, and lack of adequate management action. Water resources constraints and water degradation are weakening one of the resource bases on which human society is built. (UN Economic and Social Council, Commission on Sustainable Development 1997: 4)

In terms of demand, the link with projected global population growth (from 5.8 billion in 1999 to 8.3 billion by 2025) was obvious, with 'water use growing at more than twice the rate of population increase during this century … [while] By 2025 as much as two-thirds of the world population would be under stress conditions' (Ibid.). The link was re-emphasized in 2002 when the International Development Research Centre released a major review *In Focus: Water. Local Level Management* (Brooks 2002) and the Food and Agriculture Organization designated a 'World Food Day' entitled 'Water: source of food security'.

What seem to be the reasons for this increased demand and concern for future supplies? First is that the world's demand for water is growing with its population, and consequently water supplies per head in 2000 were estimated to be one third lower than they were in 1970 (UNU 1998). In fact, demand is growing faster as the growth of urban populations (expected to increase by 50–80 per cent in the next two generations) is putting greater pressure upon water supplies because they need more water, particularly for sanitation and for the expansion of industry. Providing those new metropolises with adequate water, sanitation, and clean air was forecast to be 'one of the most daunting and under-appreciated challenges of the first half of the 21st century' (BSDNRC 1999: 12). One commentator suggested that, 'as megalopolises of ten, twenty and thirty million inhabitants arise, water demand will bloat with them, feeding social unrest' (Cribb 2002: 28).

In the rural areas increased use of irrigation for food production also creates further demands. At the same time as these factors increase the demand for water available supplies are being reduced as a result of wastage and mismanagement in delivery systems, pollution and the consequent declining quality of potable supplies, possible climate change affecting rainfalls and the disruption of supply systems by earthquakes (UN Economic and Social Council, Commission on Sustainable Development 1997). Predictions around 2000 were that by 2050 'at least one in four people is likely to live in countries affected by chronic or recurring shortages of freshwater' (UN World Water Assessment Programme 2003: 10), and some estimates were even higher. As a result of these trends, 'Water could be the oil of the 21st century – a resource vital to all life, that is in increasingly short supply' (UNU 1998).

What Futures for Drought?

Given the debates and disagreements on the subject of climate change noted in the previous chapter, what can be extracted to help understand the future of droughts, given that some kind of climate change appears to be taking place?

Explanations of, and forecasts for, the future of droughts pose difficulties. Scientists as humans have their own beliefs and agendas and, as we have shown, can see the same facts differently. The editor of the journal *Climatic Change* commented in 2003 upon a controversial journal article, which he had published and which claimed to show that human actions began the 'Greenhouse Effect' (and thus human influence upon climate change) some thousands of years before the more popularly accepted date of the Industrial Revolution:

> Greenhouse sceptics will no doubt be aroused to ire but they will focus on an unflappable scientist who is at least as well read and certainly has less of an agenda than most of his will-be accusers. (Crowley, 2003: 260, commenting upon Ruddiman's paper 'The Anthropogenic Greenhouse Era began thousands of years ago')

John Zillman, then Director of the Australian Bureau of Meteorology, in 2005 reviewed the problems facing climatologists studying climate change. While the scientists were themselves divided on the facts and interpretations of climate change, governments nonetheless expected clear guidance as to the relevance of the phenomena and, with the rest of the community, could not understand why they did not seem to have all the answers. He noted that the scientists themselves were also struggling to recognize that the new theory of chaos in natural systems theory had to be built into their assessments of the nature of climate change. The theory implied that changes in natural systems might be relatively rapid and potentially catastrophic. Within the community also were specific interests or organizations which could be seen to be stakeholders in some of the arguments. Examples might be operators of thermal energy stations and motor vehicle manufacturers, who were accused of causing air pollution and thus possibly adding to climate warming. Zillman's own assessment was that future natural climate changes might well mask any human interference. Apart from a general warming of the continents faster than the oceans, and that there was a reasonably sure expectation of 'more hot days', he said that regional climate changes were very difficult to predict (Zillman, McKibbin and Kellow 2005: 23–4).

So what can be offered as conclusions? There seems no doubt that global temperatures are rising and will continue to rise and, bearing in mind the world's dependence upon food produced under irrigation (some 40 per cent), that the implied threat to global irrigated water supplies from increased evapotranspiration is real, particularly as the irrigated area is expanding and demand for water increasing. Further, increasing temperatures in the tropics are not expected to result in increased yields but rather reductions, although mid-latitude yields would benefit, provided precipitation kept pace (Easterling and Apps 2005).

Even the scientific definition of meteorological drought itself might need to be modified. In Australia Nicholls suggested in 2004 that a definition based upon rainfall deficiencies alone had so far 'showed a close association with droughts impacts as reported in the media', but that recent droughts (1982, 1994 and 2002) had come at a time of warmer temperatures despite similar rainfalls, so that 'It seems unlikely that such a close correspondence will continue into the future, unless the definition of drought is broadened to include the (changing) temperature aspect' (Nicholls 2004: 334). While not doubting this argument, we must be wary of accepting such a revised definition as recasting drought as a 'natural hazard' or Act of God.

What then seem to be the prognoses? In November 2006 the British Meteorological Office published *Effects of Climate Change in Developing Countries: Latest Science from the Hadley Centre* (Hadley Centre for Climate Prediction and Research 2006). The findings were quite specific:

- Global warming has continued and 2005 was the second warmest year globally on record.
- 'The fraction of the planet's land surface in drought has risen sharply since the start of the 1980s … this is likely to be due to human induced climate change'.
- 'We predict that by 2100, if significant mitigation does not take place, around half of the planet's land surface will be liable to drought' [particularly Africa, South America and southeast Asia].
- The future will show similar trends: 'More of the world is likely to be in drought. While increases in carbon dioxide concentration can actually enhance the productivity of plants, climate change will offset much of this enhanced growth. Increased incidence of fire may be made worse by more widespread drought, causing further damage to vegetation and increasing carbon emissions'.

In 2009, focusing upon the United Kingdom, the Climate Projections confirmed that the country is likely to see hotter, drier summers and warmer, wetter winters coupled with more frequent extreme weather such as flooding, heatwaves and droughts in the future.

The official view of the future then would seem to be bleak, and already reports of drought occurrence seem to be rising. However, this may be the result of improved global communications or because of increasing pressure upon land – not necessarily due to climate changes alone. The new reports also may be the result of economic necessity pushing farmers on to less suitable lands, or continued ignorance of the environmental impact of farming, or perhaps even greed on the part of the operators. Certainly in South Australia estimates of the wheat harvest needed to provide a profitable crop have risen over the years. When we recognize that in this winter rainfall country the correlation between growing season rainfall and wheat yield is strong (with a statistical correlation coefficient of about 0.7), more rain usually means more crop. In the 1880s 0.2 tonnes per hectare was sufficient to produce a profitable crop; by 1930 this had doubled to 0.4 tonnes; by 1960 this had risen to 0.6 tonnes and by the 1980s some 0.8 tonnes (Heathcote 1991). Currently this is close to 1 tonne per hectare. In effect farmers have been asking more of the rainfall each year to enable a profit to be made from the crop; the rain which would have produced a profitable crop in the 1880s or even 1960s will no longer pay the bills in the twenty-first century.

Almost certainly the future challenges of drought will pose some surprises. Speaking generally, Schneider forecast that 'global change science and policy making will have to deal with uncertainty and surprise for the foreseeable future' (Schneider 2003: 953). Surprises in this context result from the interaction between people's expectations and the behaviour of the environment and occur 'when perceived reality departs qualitatively from expectations' (Holling 1986: 294).

By way of explanation, focusing in on the North American Great Plains, researchers noted that:

> A review of the available paleoclimatic data indicates that the twentieth century droughts do not represent the full range of potential drought variability given a climate like that of today ... It is possible that the conditions that lead to severe droughts, such as those of the late sixteenth century, could recur in the future, leading to a natural disaster of a dimension unprecedented in the twentieth century ... The paleoclimatic data suggest a 1930s-magnitude Dust Bowl drought occurred once or twice a century over the past 300–400 years, and a decadal-length drought once every 500 years. (Woodhouse and Overpeck 1998: 2709–10)

Thus, there would seem to be scope for some future nasty surprises in the potential scope of climate changes. Will technology cope with the challenge of those changes?

The Technological Bias in Drought Management: Panacea or Problem?

The history of human attempts at drought management has always included a sizable technological component, whether or not the ultimate strategy was reversion to prayer or ritual. Traditional strategies, as we have seen, included water diversion and opportunistic crop plantings, and both of these are still in use. However, the last two centuries of human occupation of the globe have seen an increasing reliance upon technological solutions as the main defensive strategy. How successful has this strategy been? I argue that this reliance upon technology, to the virtual exclusion of other strategies, has rarely insulated society against the threat from droughts and in some cases has even increased society's vulnerability to drought's catastrophic impacts. There are several reasons for this assessment, some obvious, but others not so.

1. *Past Solutions Did Not Always Work.* One troubling characteristic of technological solutions to environmental problems generally, is that they are usually based upon past experiences and this has been the standard response to a new drought event. Yet there is no evidence to suggest that any future drought will have the same characteristics of intensity or duration as those on record. In fact, archaeologists and researchers into past climates keep reminding us that the geological record contains droughts with dimensions and impacts with orders of magnitude greater than any experienced within human history. If and when such an event occurs in the future, the technological response will fall back upon crisis management based upon limited experience, and society will bear the cost.

2. *The Dominance of Physics.* Drought mitigation in the technological context is concerned either with techniques to manipulate the collection and redistribution of water (especially as irrigation systems), techniques to reduce the demand for that water, or techniques to improve the efficiency of its distribution and use. These are basically mechanical problems involving physics and engineering. The questions as to who should receive the benefits and who might be disadvantaged by the process are rarely considered as part of the strategy. Such ethical questions, if considered at all, tend to be seen as the domain of politicians and power brokers, and are thus usually overlooked or ignored by technologists.

3. *Limited Scope for Future Water Diversions?* Quite apart from the ethical questions, however, is the question as to the future of the accessed water source. The success of deep drilling technology and associated pumping capacity has made available two major resources vital to the modern world, namely oil and water. Access to vast underground sources of water has made possible the creation of not only the traditional oases in the desert but the vast green fields of crops serviced by the centre-pivot irrigation schemes, which spread from the initial developments in mid-western USA in the 1960s to the fringes of the Saharan Desert in Libya in the 1980s and many other agriculturally thirsty areas of the world in between. Yet evidence is appearing that extraction in all cases is greater than natural replacement of ground water by regional rainfall and the aquifers, and as a result the irrigators are mining their water supplies. The International Water Management Institute has claimed that the annual global groundwater extraction of 250 sq.km, two thirds of which goes to irrigation, is 100 sq.km more than the annual replacement rate (Pearce 2006). This mining of global groundwater is a direct result of the seductive, but uncontrolled, application of technological expertise, which in one sense is hastening the advent of a global water shortage.

4. *Lack of Concern for the Impacts of Water Management?* Until relatively recently there was rarely any consideration of the effects of that manipulative process

upon either the source environment, or the intermediate (transit) environments, or the destination environments, of the diverted water supply. Such questions increasingly are seen as the domain of the environmentalists or 'greenies'. Thus it is not surprising that the political 'green' lobby against 'Big Dams' has had such success and storages seen once as marvels of modern engineering technology could be condemned as detrimental environmental engineering. In effect, until new water management strategies take into account the complex effects of water diversions, their benefits will continue to be questioned.

5. *Limited Life for the Projects?* There is abundant evidence that past technological strategies to alleviate drought threats have had limited life spans. The world is dotted with sediment-filled dams and abandoned irrigation channels and aqueducts. Less obvious but equally telling are the salted or waterlogged fields and the eroded croplands which dot the earth's surface. The failures may in part be blamed upon technical incompetence, for example the failure to appreciate sediment loads in the water supplies or the need for adequate flushing of salts from the fields, or lack of essential maintenance of the water conduits. However, the failures may just as importantly be the result of the collapse of institutions and the resulting loss of knowledge necessary to maintain the technology. Perhaps the best illustration is that of the Mongol invasion of the Tigris/Euphrates valley in the thirteenth century, where an invading nomadic pastoral society overran an elaborately organized irrigated agricultural society. But defunct dams and crumbling irrigation systems can be found also in the 'New World' of the Americas and Australia as well as in both Africa and Asia, and for similar reasons, being a combination of natural processes and human mismanagement.

6. *Seductive Short-Term Successes.* The seductive success in the short term of technological innovations to cope with water needs in drought, such as crop and livestock breeding to improve drought resistance, has led to exploitation of more climatically marginal and therefore more risky environments, with users assuming that drought risk has been removed. Such 'drought-proofing' claims have in the past been the argument for irrigation systems in the developed nations, which as we have seen are not proof against drought impacts. A second point is that the pressure of that increasing population will push desperate subsistence farmers, perhaps trialling newly developed (supposedly drought-resistant) crops, further into the arid lands, where sparse rains are more variable and less reliable. In such locations the very process of land clearance is likely to increase the vulnerability of the area to drought, as protective deep-rooted vegetation is removed. Commercial farmers, on the other hand, facing the cost-price squeeze of their markets, will be tempted to gamble more often on the expectations of good seasons and like the South Australian wheat farmers noted above, be wanting more rain than before to make a profit. As already mentioned, drought and not rain has 'followed the plough' as agriculture was extended into areas of more variable rainfall – a scenario already noted for Australia (de Kantzow and Sutton 1981) and given

an international context by Glantz (1994). Flexing their technological wonders, the budding farmers and pastoralists have moved further out into risky country, unconsciously dragging their future droughts behind them.

7. *Myopic Concentration upon Creating New Supplies.* Technological solutions have tended to focus initially upon increasing the supply of water rather than decreasing the demand. Increasing the supply was usually easier in the sense that existing water in another place was diverted to the needy area. Over the years the technology of canals, *qanats*, siphons and aqueducts enabled diversions from further and further afield. Such diversions rarely took note of their broader impacts, such as that upon either the source or the destination environments, and as a result both areas suffered significant environmental modifications. Given the increasing concern for the broader impacts of resource management, such impacts – whether of the big dams themselves or the surplus additional surface waters infiltrating into, and thus raising, ground water levels – have brought their own problems, which have tended to be seen once more as merely requiring further technological solutions. But technological solutions alone rarely recognize the ramifications of the proposed solutions. Ignored often are the details of who should receive the new water supplies; who will lose water access as a result of any diversions; and what will be the environmental consequences both at the source and destination of any diverted water? Management of water supplies rarely includes all aspects of any innovative water use. A more recent overview has noted the recognition of the importance of such flows at the turn of the century, but has documented the difficulties facing implementation of the new strategies since then (Pigram 2006). Interestingly, in January 2010, the South Australian government successfully negotiated with New South Wales to allow surplus flood waters from torrential downpours in northeastern NSW to be passed down the Murray River into South Australia, to restore an environmental flow in its parched lower reaches.

8. *Isolated Technological Solutions.* In the past then, it would seem that technology in isolation has not removed the threat or impact of drought and part of the reason seems to have been what might be labelled the 'technological isolation' of the basic thinking. The 1990s, however, saw the development of what was claimed to be a new scientific paradigm – that of 'complexity':

> Complex thought is aware that it is not possible to know everything; it does not aspire to absolute certainty but only to emphasize the connections between the various kinds of knowledge. The principles of reduction and abstraction that had allowed sciences to progress in the past have become dangerous due to the accelerated expansion of science in recent times. (Bianchi 1994: 12)

Implicit in this was the recognition of the many facetted nature of human experiences and the need to recognize 'a holistic vision or an ecological vision of the world … the idea of the world as an ensemble of integrated parts and in which men and societies participate in the cyclical processes of nature' (Ibid.: 13). From such a viewpoint technology's characteristics appear particularly inadequate. Improving the environment, in the sense of protecting it from drought, should be seen as 'not only a matter of technology, science, political economy and managerial capacity, it is also, and perhaps above all, an ethical question' (Ibid.: 13–14). To whom does the rain belong, who owns the streams and the lakes, and who owns the water in the ground?

9. Under-estimating the Real Costs of Technological Solutions? While it thus might be claimed that technology in the past has had a rather myopic view of the problem to be confronted, it can also be claimed with certainty to have underestimated the real costs, both initial and on-going, of the technological solutions adopted. Few irrigation schemes have ever paid for their construction costs, usually because governments absorbed those costs in the general budget, and the long term environmental costs have rarely been calculated, let alone compensated for. In the case of Australia, one economist claimed years ago that excessive investment in irrigation schemes (mainly by governments) was a misuse of national resources. Investment in improving the efficiency of dry-land farming would have paid more dividends with that same capital investment (Davidson 1969).

Yet, despite these problems, one of the major factors in the increasing global demand for water has been the technology behind the expansion and intensification of land areas receiving irrigation, which has been seen as the panacea for not only improving the productivity of the land but also as a means of protecting that land against drought. From 1900 to 1995 water withdrawals from surface and groundwater sources increased globally over six-fold, more than the two-fold increase in world population over the same period. Some 70 per cent of those withdrawals are for irrigation and the bulk (87 per cent) of such irrigated water goes for crop production. This demand for diverted water for agriculture has grown by 60 per cent since 1960 and has been seen as a major human achievement. However, perhaps it is also one of the most spectacular results of myopic technological water management, and a major factor in the spectacular desiccation of the Aral Sea, whose surface area has been halved since 1960 as normal supplies of fresh water in the Amur Darya and Syr Darya rivers were siphoned off by expanding irrigated areas, first by the Soviet Union and then by the successor countries of Kazakhstan, Turkmenistan, Uzbekistan and Kirgystan (UN Economic and Social Council, Commission on Sustainable Development 1997, Glantz ed. 1994). As I wrote this in 2008 a similar scenario was unfolding in the Murray-Darling Basin in southeastern Australia, where the major river system was being diverted for irrigation and its waters rarely reached the ocean. By 2009 continued droughts had so restricted river flows that some irrigated areas had been abandoned, and the optimal allocation of waters is still under debate.

10. Technology Has Been Too Successful? The success of technology's search for new sources of water has kept the price of water too low, so that there has been little incentive to save and every incentive to waste it. As two recent researchers suggested: 'Technology ... is a two edged sword, whose benefits can be substantially blunted by Jevon's paradox, the concept that increases in efficiency often lead to lower prices and hence to greater consumption of resources' (Hall and Day 2009: 236). A short term reduction in resource prices is not likely to lead to their more efficient management. To be fair, however, technology alone has not always been the sole culprit in the failure to cope with droughts. As we have seen in Chapter 4, Sicily's unsuccessful attempts to meet the threat from the drought of 2003 has been attributed to poor management, lack of infrastructure maintenance, water theft and Mafia corruption of the construction industry.

This would seem to be justification for casting a wide ranging and complex net to find answers to the threats from drought. A major question, perhaps, should be to ask ourselves whom, or what, are we trying to protect from this dreadful scourge?

Drought Protection for What/Whom? Water Supplies – End Use Juggling?

Before arguing the case for better drought mitigation policies, however, we need to remind ourselves of the varying contexts of past and present planning.

The Changing Goalposts?

In effect the goal posts for defining and mitigating drought have moved significantly over time, basically as the demand for all qualities of water grew rapidly. This growth reflected a variety of factors. First was the increased human knowledge of the earth's geology, which developed the skills to find new aquifers which provided increased supplies of quality water. Second were the increased demands for water, particularly the development of irrigation systems for agriculture associated with the transition of hunter-gathering to agricultural societies. When to these factors were added demands for water as a result of the Industrial Revolution, whether as the transport medium in the canals, or as raw material in food processing, or as an industrial coolant, the results were the creation of specialist engineers for its storage, transport, and distribution. Most recently, improved personal hygiene, urban sewage and drainage requirements, and a longer lived human population have seen an upsurge in demand for good quality water. All of these have combined over the centuries to increase the demands for global water supplies, thereby increasing effectively the vulnerability of global societies to drought.

The Changing Clients and Changing Criteria?

As a result, any policy for the mitigation of drought needs to recognize the fact that there are many and varied demands for water in the modern world. As a result the allocation of the periodically restricted supplies is becoming a major problem. The basic need of all life forms for water is an obvious first priority, but throughout

history humans have created pecking orders of priorities in their protective efforts
against droughts. Threats to human life, domestic livestock and crops, were the
basic driving forces and are still apparent in modern planning. But these have been
augmented by broader concerns for drought's impacts upon the survival of society,
upon the survival of specific plant and animal species, and upon the survival of
specific natural ecosystems and environments. In other words the stakeholders
in drought management have increased from farmers and pastoralists operating
locally, to lobby groups trying to protect specific and possibly non-economically
valuable plants or animals, to national governments concerned about economic
downturns in agricultural or industrial production and environmental degradation,
up to the United Nations' concerns at a global level. In such a context the
politicians or planners might have to weigh the survival of an endangered plant
or animal species against the economic survival of a local community, a scenario
increasingly common as the human population expands.

Changing Philosophical Contexts

The philosophical context of current planning has been identified as a third wave
of international concern for sustainable resource management, following on from
the 1940–1950s concern over non-renewable resources and the late 1960–1970s
concern over the environmental impacts of resource development (Chisholm
1992). Thus a commentator on the Australian drought policy of 1994 noted that:
'Emphasis is placed on how government policies, backed up by technologies to
aid drought prediction, monitoring, and management, are expected to lead to more
self-reliant management at the farm level, and the development of agricultural
systems that are physically, biologically, and financially sustainable' (White and
Bourdas 1997: 213).

The environmental context is the link between land cover/land use and climate,
where land clearance has the potential to increase evapo-transpiration and so reduce
effective precipitation, as has been shown for Australia (Xinmei et al. 1995). In
addition, the fact that climate change appears to be inevitable, already evident, and
possibly influenced by human activities, makes the environment doubly relevant,
both as instigator of stresses and victim of human mismanagement. In Australia,
changes in mean annual rainfall 'may be small, but concurrent changes in drought
frequency could be quite large, increasing in some areas and decreasing in others'.
As a result there may be large regional 'declines in stream runoff and related
increases in water salinity' (Hengeveld and Kertland 1998: 16).

For the Great Plains of the USA, one of the world's important 'bread baskets',
researchers have estimated that based upon past records future droughts may be
of a much greater severity and duration than so-far experienced (Woodhouse
and Overpeck 1998). If true, future planning to mitigate drought impacts will be
complex indeed.

The economic context is, of course, globalization. Given this context, how
might resource management be evaluated, in terms of a global market first and

national interests second, and how important are the interests of private capital? Comparing British and French attempts to cope with drought threats in the context of their contrasting water management policies, one commentator asked a basic question on the balance between private benefit and public good: 'How then do states achieve the shifting balance between, on the one hand, the free-market economic rationale that underlies the privatization process and, on the other hand, public demand, not only for clean and cheap water but also for transparent and accountable water services?' (Buller 1996: 462). Such a balance might well require a decision on the relative merits of the interests of the water company, its shareholders, and public water users.

As an example of the complex relationships involved, consider Australia, economically a small nation with little clout on the global market compared with say the USA. Thus a drought in Australia could lose local grain farmers their share of the global grain markets, but a drought in North America (affecting both US and Canadian grain production) might allow Australian farmers to regain their lost share. In the global environmental context, however, the United Nations might become seriously concerned if irrigation developments destroy the viability of a potential world heritage wetlands site in Australia's Channel Country or if the run off of excess fertilizers from Queensland sugar cane farms destroys corals on the Great Barrier Reef. Yet at the national scale, although Australia may be 'small' economically, it is large enough in area physically to be only partly affected by drought at any one time and this scale factor is a national bonus as we have seen.

But the political context in Australia poses similar problems to the British and French case above. Namely the current policy of the Commonwealth and State governments, as elsewhere in the Western World, is to retreat from responsibility for public welfare and to encourage the privatization of service provision. The push towards 'self-reliant management', and the withdrawal of drought relief from all but the most extreme cases (even if not yet achieved in practice), are signs for the future. If this trend is continued around the world, then the basic question will arise as to whether privately motivated managers will be as conversant with (or sensitive to) the national interest in resource management as a whole. Indeed the question might be can they privately meet the new challenge from drought?

Predicting the Coming Droughts: Where, When, and What Impact?

Given these contexts, what are my predictions? I believe that future harmful droughts are inevitable and the reasons for this view are relatively clear: human demand for water is rapidly increasing and even if the climate does not change, increasing demand alone will overwhelm available future water supplies.

Where will the future droughts appear? One good prediction is on the edge of the arid lands of the world as population pressures push desperate farmers ever further down the rainfall gradients. The current temperate zone grain production areas are also vulnerable now and will continue to be so into the future; even the

Great Plains of the USA have been given a gloomy forecast. We can also be certain that urban areas will experience water shortages as demand escalates, existing storages prove inadequate and gradually lose capacity as they silt-up, and popular opposition to further big dams grows. Even the Pacific Islands are not sacrosanct, as sea level rises will pose their own droughts by reducing the freshwater lens on which the coral atolls depend.

When will these droughts occur? Every year, somewhere in the world, drought will be stalking the landscape. At a continental level, drought-free periods may be longer but it is likely that somewhere on every continent, each year will provide evidence of drought impacts. At a regional level areas may be free of droughts for periods of a decade or longer, creating in the inhabitants a false sense of security and so, through neglect of protective strategies, an increased vulnerability to 'natural disasters' such as drought. Thus when drought does come, as it inevitably will, the collective memory of strategies to cope with it will be fuzzy or blank and society will be significantly damaged.

When those droughts appear what will be the consequences? One extreme result might be international conflict. The Vice-President of the World Bank (Ismail Serageldin) noted in 1995, while 'many of the wars of this century were about oil ... wars of the next century will be over water' (Selby 2005: 201). At least one author has forecast such international conflicts as a result (de Villiers 1999), and specific conflicts over water access between Australia and Indonesia have been anticipated by S. Wade *The Politics of Water*, 2002.

That such predictions have been made should not be surprising, since over 300 major river basins cross international boundaries and, given the propensity for technological manipulation of river flows by dams and diversionary structures, any alterations to what might be seen as the 'life-blood' of a nation are likely to be viewed with alarm by downstream users. In 1997, in fact, the United Nations adopted a Convention on the Law of Non-Navigational Uses of International Watercourses. This convention has yet to be ratified by sufficient states to give it any force, but in the meantime there are several international disputes which might escalate into warfare.

Examples include the history of conflicts in the Middle East (Gleick 1997), notably between Israel and her Arab neighbours, and between Turkey, Syria and Iraq. Others include the disputes between Namibia and Botswana over Namibia's proposal to take 17 million m^3 per year from the Okavango River – a diversion which Botswana claims will adversely affect the Okavango Delta and its tourist industry which earns vital foreign currency (Ashton 2000); between Spain and Portugal over the former's plans to take more water from the upper valleys of the Douro, Tejo and Guadiana rivers for diversion to the irrigation areas of southern Spain, in breach of a 1998 water sharing agreement with downstream Portugal (Theil 2004); and between Singapore and Malaysia over Malaysia's effective stranglehold on Singapore's drinking water supplies. Significantly, in 1998 *The Arid Lands Newsletter* featured a series of papers on 'Conflict Resolution and Transboundary Water Resources' in Volume 44.

Despite these examples of international tensions over water, there is disagreement over whether they are likely to escalate into armed conflict, with some arguing that water is a very different resource to oil, and that the gains that might be made though warfare are not worth the cost. One writer concludes: '… water is simply not important enough as a source of revenues, or as a source of security, for state elites to warrant going to war over it'. (Selby 2005: 391).

Future Drought Management – Some Thoughts

Hydromania

Water is far from a simple commodity,
Water's a sociological oddity,
Water's a pasture for science to play in.
Water's a mark of our dubious origin,
Water's a link with a distant futurity,
Water's a symbol of ritual purity,
Water is politics, water's religion,
Water is just about anyone's pigeon.
Water is frightening, water's endearing,
Water's a lot more than mere engineering.
Water is tragical, water is comical,
Water is far from the Pure Economical
So studies of water, though free from aridity,
Are apt to produce a good deal of turbidity.

(Boulding 1970)

With his customary insight and wit Kenneth Boulding spanned the spectrum of water's significance to human society. Something similar and equally challenging could be said for drought's relevance to mankind, past, present and future.

Can the impacts of future droughts be reduced? Can the challenge from the Bull of Heaven be met? Yes, I believe so, but it will take a considerable effort and better management than we have seen in the past. Some preliminary thoughts and suggestions are set out below.

As part of its program for the International Decade for Disaster Reduction which ended in December 1999, the International Union of Geodesy and Geophysics suggested that drought research needs included:

Improved forecasting models for early warning systems, a better infrastructure for data communication and for bringing the warnings to the people in the field and on the street, drought management plans based upon decent water economical analysis – including the protection of ecological values – and proper land development schemes as a basis for reducing the risk of a drought disaster. (Gottschalk 1999: 13)

How to achieve these goals? Several points are relevant.

1. We need to recognize that drought is not only a natural disaster; it is obvious by now that historically humanity has brought drought upon itself and future droughts are inevitable. In effect, drought has often followed the plough's excursions into the world's natural grasslands. That such will continue reflects the fact that pressures on current agricultural areas from increasing human populations will cause population movement into less productive, marginal areas. 'Consequently, we can expect to hear more in the future than we have in the past about droughts and their impacts on humankind' (Glantz 2000: 291). In other words, our management plans must recognize that to the increasing variability of the growing seasons on current agricultural lands as the result of possible climate changes, must be added the risks from increasing occupation of lands more vulnerable to meteorological droughts.

2. Our basic water management strategies need improvement to cope with the forecast increase in the frequency and severity of meteorological droughts resulting from imminent climate changes. Improved warning systems will need not only to consider precipitation forecasts but the ecological impacts of those forecasts. Strategies to improve water management will need incentives to conserve and protect supplies, e.g. by more realistic water pricing, and here Australia has been praised as being on the right track (Harris 2000: 13). Water loss from irrigation systems is still unacceptable; losses of 20 per cent through evaporation and seepage were still noted along the Murray River in 1999 (*The Australian*, 31 December 1999). We need also to recognize that irrigation itself is no guarantee of drought-proofing. By 2009, with the results of past mismanagement of water resources in the Murray-Darling Basin rapidly unfolding, irrigators with traditional as well as recently-acquired water entitlements were facing the possibility of either drastically reduced entitlements or strong pressure to sell all or part of those entitlements back to state and Commonwealth governments. This latter option was justified by governments as enabling either reallocation to other commercial uses, or to maintain environmental flows.

There is also considerable scope for improved efficiencies in water use and recycling in the urban context. Finally, we might learn from the American experience with 'Water Banks'. This is a system of water sales or credits by owners with surpluses, usually of ground water, to users in need. Both private and state run schemes are operating and are forecast to be an emerging tool for drought management in western USA (Miller 2000). Whether this continued commodification of water resources will accommodate non-commercial uses, say for environmental management, is problematic however.

3. Governments need to enforce the principle of self-reliant management on agricultural water users, to dissuade them from undertaking risky enterprises in the belief that they will be bailed out by official disaster relief if caught out by drought, i.e. the moral hazard problem, and give encouragement to those managers who have demonstrated that they can cope with droughts. If drought relief is provided, say in exceptional circumstances, however so defined, the relief must be both means- and needs-tested and, when required, made conditional on the adoption of responsible management practices. For too long we have listened to the cries of 'wolf' from, and provided relief support to, some managers who should not really be in the business.

In contrast, the basic question remains: why cannot we learn from, encourage and reward managers who don't need drought relief; why cannot we learn how they have survived the stresses and make their experiences more widely accessible? For these are the real heroes of the bullfight, the true matadors of the bullring! Whether they are farmers or pastoralists striving to support their families or agri-business managers trying to maintain a profitable business, as successful managers they should be acknowledged and supported.

4. We must realize that drought affects society as a whole and concerns every aspect of government, which requires collaboration between all government agencies. There are many interested parties/potential victims now. Whereas the droughts of the nineteenth century tended to be agricultural in their impacts and the relevant government agency needing to respond was the department of agriculture, now with drought impacts having moved into the suburbs, they require attention from many more government agencies and require more complex remedial actions. This shift of responsibilities, however, raises its own problems.

Contemporary droughts, in the ramifications of their relevance to officialdom, are now very similar in complexity to the processes of desertification, and both pose broad ecological problems. Randall Baker drew upon his experiences of international and national strategies to cope with the Sahel droughts and associated desertification of the 1960–1970s (Baker 1976). This research showed that the ramifications of the drought impacts became relevant over time to an increasing number of official agencies, whose activities were rarely coordinated and often overlapped, causing confusion and waste of resources. Subsequently he elaborated the idea to ecological and environmental problems generally, noting that 'there is a considerable dysfunction between the nature of ecological problems and the "problem-solving structures" within the public arenas. This dysfunction is … termed the "administrative trap"' (Baker 1989). In other words, there was no longer a single official agency or government department which could cope with, or effectively respond to, the responsibility for drought management. But with the increase in agencies involved came inefficiencies and even less effective management strategies! The secret of success in coping with drought is collaboration between government agencies, not the creation of another department.

5. Droughts are predictable surprises, but have multiple causes. As I pointed out many years ago, the onset of each drought, certainly in Australia and I suspect elsewhere in the world, has been greeted by 'indignant surprise'. A recent paper has elaborated this idea in the context of society's responses to climatic change in general and has provided interesting explanations. The argument is that the reason that societies are caught out/surprised by events is because those events are predictable surprises. These are 'an event or set of events that catch an organization off-guard, despite the leaders' prior awareness of all the information necessary to anticipate the events and their consequences' (Bazerman 2006: 180). Caught by surprise, societies tend to do nothing or, under pressure to do something, resort to crisis management with uncertain outcomes.

That societies are caught out is suggested to lie in a basic reason that events are usually seen as having only a single cause, and as we have seen for drought there are many causes. In addition, however, there are five further reasons offered as to why, in the case of the original context of climate change (but I would argue this for drought also) there tends to be no response to the predictable surprise of a possible change. First is the positive illusion that the problem does not exist or is not serious enough to merit action (possibly because technology will cope with it); second that events are interpreted in an egocentric, or self-serving manner (which tends to put responsibility for the event on someone else); third that society tends to 'overly discount the future, despite our contentions that we want to leave the world in good condition for future generations' (so the surprise is not considered to have any long term impacts); fourth that the status quo is preferred, even if change would be beneficial in the long run (traditional drought relief strategies echo this policy as we have seen); fifth that society is unwilling to react unless the problem has been previously experienced or 'witnessed through vivid data' (this seems to require a major disaster event to energize society). Surprises, such as droughts, it would seem, may not always provoke a response.

6. The scientific study of drought cannot be left to the natural scientists alone; social scientists have a similar and vitally necessary contribution to make. The wide ranging and complex nature of drought impacts upon the environment and societies, and the multiple interests in drought, require a holistic approach by the scientific communities. Specialise by all means, but ensure that findings and ideas are shared and discussed with other disciplines before decision makers are briefed.

To conclude let me revive the community's humourous response to drought and quote from two poems (again from Kenneth Boulding) presented at a world symposium on humanity's role in changing the face of the earth, as I think they have relevance to the problem of drought management. At the end of discussions on the 'Limits of the Earth', a participant claimed the floor and read the following poems by Boulding, composed during the session, into the records of the symposium (Thomas, 1956: 1087).

A Conservationist's Lament

> The world is finite, resources are scarce,
> Things are bad and will be worse.
> Coal is burned and gas exploded.
> Forests cut and soils eroded.
> Wells are dry and air's polluted,
> Dust is blowing, trees uprooted.
> Oil is going, ores depleted,
> Drains receive what is excreted.
> Land is sinking, seas are rising,
> Man is far too enterprising.
> Fire will rage with Man to fan it,
> Soon we'll have a plundered planet.
> People breed like fertile rabbits,
> People have disgusting habits.
> *Moral*:
> The evolutionary plan
> Went astray by evolving Man.

The Technologist's Reply

> Man's potential is quite terrific,
> You can't go back to the Neolithic.
> The cream is there for us to skim it,
> Knowledge is power, and the sky's the limit.
> Every mouth has hands to feed it,
> Food is found when people need it.
> All we need is found in granite
> Once we have the men to plan it.
> Yeast and algae give us meat,
> Soil is almost obsolete.
> Men can grow to pastures greener
> Till all the earth is Pasadena.
> *Moral*:
> Man's a nuisance, Man's a crackpot,
> But only Man can hit the jackpot.

Both messages have relevance for the management of drought, not least that followers of the two orthodoxies should get together, stop arguing and point scoring, and cooperate before it is too late.

The history of 'drought and the human story', as I have tried to demonstrate in this book, is the recognition of the many faces of drought, the many perceptions of what it is, and the extent of its role in and importance to societies. Over time, the

Bull of Heaven has brought significant and often catastrophic distress to humanity and global societies, but so far humanity has survived. That fact alone should encourage us to believe that because in the past humanity by and large has had the ability and means to protect itself, we should therefore be able to adapt to current and future droughts. But there are many more of us now, with many more varied needs for water's vital life support, and for us to survive in the future will need not only the traditional human abilities and skills, but also collaboration between scientists, governments and the general public.

That collaboration should have three basic aims: the first being the curbing of our increasing demands for water by reducing wastage through losses in our distribution systems, encouraging improved efficiencies in the use of water, and encouraging the recycling of water for multiple rather than single functions – these are challenges for the technologists and natural scientists. The second aim is to emphasize that because water is not equally available around the world and has been grossly undervalued in the past, its vital social and economic value must be recognized and built into local and national economies – these are challenges for political economists and social scientists. The third aim accepts the awakening recognition of the equally vital role of water in the global environment independent of any human economic needs, and focuses upon monitoring trends in natural supply to see how well that supply meets environmental as well as human demands. These are challenges for both natural and social scientists working together to recognize and evaluate the positives and negatives of the data.

In fact, the somewhat arbitrary allocation of challenges suggested above plays down the need for an holistic approach to the basic challenge of drought. We can all play a role and participate in that challenge, for we are all potential matadors in the global bullring! From the households curbing their use of domestic water and recycling grey water onto the garden plot to the city engineers transferring urban rainwater from roofs and gutters to recycling storages; from the laboratory scientists trialing plants and animals with drought survival characteristics to the field scientists taking the pulse of the global ecology; from the climatologists trying to make sense of ever changing atmospheric patterns to the lawyers drawing up instruments for equitable legal access to water and property rights; from the officials sorting out the merits of drought relief applications to the governments encouraging public awareness and developing policies to promote water saving economic developments and rewards for successful individual or community successes in mitigating drought impacts; from the environmental lobby groups highlighting private misuse of resources to the international alliances promoting the protection of endangered species and their, as yet unknown, benefits for humanity; here are great challenges and opportunities for us all to meet the threat from drought.

There is much to be done and we can and must all contribute before we can really claim that the Bull of Heaven has been tamed.

Bibliography

ABARE 1997. Effect of interest changes on farm sector incomes. *ABARE Current Issues*, 6.

ABARE 2006. *Australian Stock and Livestock report*, 27 October. Canberra: ABARE.

ABS 1988. *Year Book Australia 1988*. Canberra: Australian Bureau of Statistics.

Adams, F. 1893. *The Australians: A Social Sketch*. London: T. Fisher Unwin.

ADB 1983. *Australian Dictionary of Biography*. Vol. 9. Carlton: Melbourne University Press.

Adger, W.N., Dessai, S., Goulden, M., Hulme, M., Lorenzoni, I., Nelson, D.R., Naess, L.O., Wolf, J. and Wreford, A. 2009. Are there social limits to adaptation to climate change? *Climatic Change*, 93(3–4), 335–354.

Agnew, C. and Anderson, E. 1992. *Water Resources in the Arid Realm*. London: Routledge.

Agnew, C.T. 2000. Using the SPI to identify drought. *Drought Network News*, 12(1), 6–12.

Allan, R.J. 2000. ENSO and climatic variability in the last 150 years, in *El Niño and the Southern Oscillation*, edited by H. Diaz, and V. Markgraf. Cambridge: Cambridge University Press, 3–55.

Allan, R.J. and D'Arrigo, R.D. 1999. Persistent ENSO sequences: how unusual was the 1990–1995 El Niño? *The Holocene*, 9(1), 101–118.

Allan, R.J. and Heathcote, R.L. 1987. The 1982–83 drought in Australia, in *The Societal Impacts Associated with the 1982–83 Worldwide Climate Anomalies*, edited by M. Glantz, R. Katz, and M. Krenz. Boulder: National Center for Atmospheric Research, 19–23.

Allan, W. 1965. *The African Husbandman*. Edinburgh: Oliver and Boyd.

Alvarez, J.A., Rossi, G., Vagliasindi, F. and Vela, A. (eds) 2005. *Drought Management and Planning for Water Resources*. Boca Raton [Florida]: CRC Press.

Amiran, D.H.K. and Ben-Arieh, Y. 1963. Sedentarization of Beduin in Israel. *Israel Exploration Journal*, 13(3), 161–183.

Anderson, J.R. 1979. Impacts of climatic variability on Australian agriculture: a review. *Review of Marketing and Agricultural Economics*, 47(3), 147–177.

Andrzejewski, B.W. 1970. The roobd'oon of Sheikh Aqib Abdullah Jama: a Somali prayer for rain. *African Language Studies*, 11, 21 34.

Anon. 1966. Drought. *Current Affairs Bulletin*, 38(4), 51–64.

Anon. 1978. *An Australian Prayer Book*. Sydney: Synod of Church of England.

Anyamba, A., Tucker, C.J., Huerte, A.R. and Boken, V.K. 2005. Monitoring drought using coarse-resolution polar-orbiting satellite data, in *Monitoring and Predicting Agricultural Drought*, edited by V.K. Boken, A.P. Cracknell and R.L. Heathcote. New York: Oxford University Press, 57–78.

Archer, M., Burnley, I., Dodson, J., Harding, R., Head, L. and Murphy, P. 1998. *From Plesiosaurs to People: 100 Million Years of Environmental History.* Canberra: Department of the Environment.

Armstrong, K. 2005. *A Short History of Myth.* Edinburgh: Canongate.

Arons, N.G. 2004. *Waiting for Rain. The Politics and Poetry of Drought in Northeast Brazil.* Tucson: University Arizona Press.

Ashton, P. 2000. Southern African water conflicts: are they inevitable or are they preventable?, in *Water for Peace in the Middle East and Southern Africa.* Geneva: Green Cross International. Available at www.hidropolitik.hacettepe. edu.tr/middleeast.pdf.

Aubreville, A. 1949. *Climats, Forêts et Désertification de l'Afrique Tropicale.* Paris: Société d'Editions Géographiques, Maritimes et Coloniales.

Australia 1918. *Final Report of the Dominions Royal Commission.* London: HMSO.

Australia 1937. Report of the Board of Inquiry Appointed to Inquire into the Land and Land Industries of the Northern Territory of Australia. *Commonwealth of Australia Parliamentary Papers*, 3(4), 813–925.

Australia 1945. Bradfield Scheme for 'Watering the Inland': meteorological aspects. *Commonwealth Bureau of Meteorology Bulletin*, 34.

Australia 1996. *Commonwealth Disaster Relief: Review of the Natural Disaster Relief Arrangement.* Canberra: Department of Finance, Primary Industry and Environment Branch.

Babugara, A. 2009. Vulnerability of children in drought. *Natural Hazards Observer*, 33(5), 9–13.

BAE 1986. *Crop and Rainfall Insurance: A BAE submission to the IAC.* Canberra: Commonwealth Bureau of Agricultural Economics.

Bagla, P. 2002. Climate forecasting: drought exposes cracks in India's monsoon model. *Science*, 297(5585), 1265–1267.

Baker, A.R.H. 1966. The Kentish iugum: its relationship to soils at Gillingham. *English Historical Review*, 81, 74–79.

Baker, R. 1966. *Sahara Conquest.* London: Lutterworth.

Baker, R. 1976. Innovation technology transfer and nomadic pastoral societies, in *The Politics of Natural Disaster: The Case of the Sahel Drought*, edited by M.H. Glantz. New York: Praeger, 176–85.

Baker, R. 1984. Protecting the environment against the poor: the historical roots of soil erosion. *The Ecologist*, 14(2), 53–60.

Baker, R., 1989. Institutional innovation, development and environmental management: an administrative trap. *Public Administration and Development*, 9, 29–47.

Balfour, P.G. 2006. *Bagpipes in Babylon. A Lifetime in the Arab World and Beyond.* London: I.B. Tauris and Co. Ltd.

Ballard, J.G. 2001. *The Drought.* 2nd Edition. London: Flamingo.

Barlow, M. 2006. A ware-tight defence. *The Advertiser* [Adelaide], 30 December, 3.

Barlow, M., Cullen, H. and Lyon, B. 2002. Drought in central and southwest Asia: La Niña, the warm pool, and Indian Ocean precipitation. *Journal of Climate*, 15(7), 697–700.

Barnett, T., Manone, R., Pennell, W., Stammer, D., Semter, B. and Washington, W. 2004. The effects of climate change on water resources in the West: introduction and overview. *Climatic Change*, 62, 1–11.

Barr, J. 1999. Drought assessment: the 1997–98 El Niño drought in PNG and Solomon Islands. *Australian Journal of Emergency Management*, 14(2), 31–37.

Bazerman, M.H. 2006. Climate change as a predictable surprise. *Climatic Change*, 77(1–2), 179–193.

Beatley, T. 1988. Ethical dilemmas in hazard management. *Natural Hazards Observer*, 12(5), 1–3.

Beavan, B. 1978. *A Sunburnt Country.* Adelaide: Rigby.

Beinart, W. 1996. Environmental destruction in Southern Africa, in *Time-Scales and Environmental Change*, edited by T. Driver and G. Chapman. London: Routledge, 149–168.

Benson, L., Petersen, K. and Stein, J. 2007. Anasazi (Pre-Columbian Native-American) migrations during the middle 12th and late 13th centuries – were they drought induced? *Climatic Change*, 83(1–2), 187–213.

Bentley, P. 1941. *The English Regional Novel.* London: Allen and Unwin.

Berger, A. (ed.) 1981. *Climate Variations and Variability.* Dordrecht: Reidel.

Bianchi, E. 1994. The environment and complexity: the role of ethics and the HDGECP, in *Global Change Perception*, edited by E. Bianchi. Milan: Angelo Guerini e Associati, 11–20.

Bita, N. 2010. Water's quick fix a long term drain. *The Weekend Australian*, 23–24 January, 5.

Blaikie, P., Cannon, T., Davis, I. and Wisner, B. 1994. *At Risk. Natural Hazards, People's Vulnerability, and Disasters.* London: Routledge.

Blakeburn, P. 1993. *The American Experience in Combating Desertification and its Impacts.* United Nations INCD Secretariat: Paper to the Information Sharing Segment of the INCD, Nairobi 24 May–28 May.

Boer, R. and Subbiah, A.R. 2005. Agricultural drought in Indonesia, in *Monitoring and Predicting Agricultural Drought*, edited by V.K. Boken, A.P. Cracknell and R.L. Heathcote. New York: Oxford University Press, 330–344.

Boken, V.K., Cracknell A.P. and Heathcote R.L. (eds) 2005. *Monitoring and Predicting Agricultural Drought.* New York: Oxford University Press.

Boken, V.K. 2005. Agricultural drought and its monitoring and prediction: some concepts, in *Monitoring and Predicting Agricultural Drought*, edited by V.K.

Boken, A.P. Cracknell and R.L. Heathcote. New York: Oxford University Press, 3–10.

BOM 1965. *General Remarks on Drought by the Director of Meteorology.* Melbourne: Bureau of Meteorology.

BOM 2001. *Living with Drought.* Melbourne: Bureau of Meteorology.

Bonyhady, T. 2000. *The Colonial Earth.* Melbourne: Miegunyah Press.

Boorstin, D.J. 1985. *The Discoverers. A History of Man's Search to Know his World and Himself.* 2nd Edition. New York: Vintage Books.

Boorstin, D.J. 1998. *The Seekers: The Story of Man's Continuing Quest to Understand the World.* New York: Random House.

Borrell, A.K. 1994. *Lessons and Opportunities from the Drought: Proceedings of a Symposium, University of Southern Queensland, Toowoomba, 23 November.* Parkville: Australian Institute of Agricultural Science.

Boulding, K. 1970. Water (poem), in *Australian Journal of Agricultural Economics,* 14(2), 104–105.

Bowden, M.J., Kates R.W. et al. 1981. The effects of climate fluctuations on human populations: two hypotheses, in *Climate and History,* edited by T.M.L. Wigley, M.J. Ingram and G. Farmer. Cambridge: Cambridge University Press, 479–513.

Bowen, E.G. 1964. Linear and planetary trails in the solar wind. *Journal of Geophysical Research,* 69, 49–69.

Box, J. 2002. Survey of Greenland instrumental temperature records: 1873–2002. *International Journal of Climatology,* 22, 1829–1847.

Bradley, R.S. and Jones, P.D. 1992. *Climate since AD 1500.* New York: Routledge.

Braudel, F. 1992. *The Mediterranean and the Mediterranean World in the Age of Philip II.* London: University of California Press.

Brezhnev, L. 1978. *The Virgin Lands.* Moscow: Progress Publishers.

Broecker, W.S. 1995. Chaotic climate. *Scientific American,* 273, 44–50.

Brook, K.D. and Carter, J.O. 1996. A prototype national drought alert strategic information criteria information system for Australia. *Drought Network News,* 8(2), 13–16.

Brooks, D.B. 2002. *Water. Local level Management.* Ottawa: IDRC.

Brooks, R.H. 1975. Drought and public policy in northeastern Brazil. *Ekistics,* 230, 30–35.

Brown, L.R. and Finsterbusch, G.W. 1972. *Man and his Environment: Food.* New York: Harper and Row.

Brown, R.H. 1948. *Historical Geography of the United States.* New York: Harcourt Brace.

Bryson, R.A. and Murray, T.J. 1977. *Climates of Hunger: Mankind and the World's Changing Climate.* Madison: University Wisconsin Press.

BSDNRC 1999. *Our Common Journey: A Transition Toward Sustainability.* Washington: National Academy Press. Available at: http://www.nap.edu/books/0309067839/html.

Buller, H. 1996. Privatization and Europeanization: the changing context of water supply in Britain and France. *Journal of Environmental Planning and Management*, 39(4), 461–482.

Bullock, P. and Le Houe'rou, H. 1996. Land degradation and desertification, in *Climate Change 1995. Impacts, Adaptations and Mitigation of Climate Change: Scientific-Technical Analyses*, edited by R. Watson, M. Zinyowera, R. Moss. New York: Cambridge University Press, 173–189.

Bunbury, A. 2002. *Arid Arcadia. Art of the Flinders Ranges*. Adelaide: Art Gallery of South Australia.

Bureau of Rural Sciences 2000. *Exceptional Circumstances*. [Online: Department of Agriculture, Fisheries and Forestry]. Available at: http:// www.brs.gov.au/ docs/rural_science/agrifood/exceptcirc.html.

Burnley, I. 1986. What we can learn from drought and depression. *Inside Australia*, 2(3), 7–10.

Burstyn, V. 2005. *Water Inc*. London: Verso.

Burton, I. and Kates, R.W. 1964. The perception of natural hazards in resource management. *Natural Resources Journal*, 3, 412–441.

Burton, I., Kates, R.W. and White, G.F. 1978. *The Environment as Hazard*. New York: Oxford University Press.

Butler, J.R.G. and Doessel, D.P. 1988. *The Economic Role of the Commonwealth Government in Natural Disasters*. Brisbane: School of Management, Queensland Institute of Technology, 17.

Butler, J.R.G. and Doessel, D.P. 1989. The economic role of the Commonwealth Government in natural disasters, in *Flood Insurance and Relief in Australia*, edited by D.I. Smith and J.W. Handmer. Canberra: Centre for Resource and Environmental Studies, Australian National University, 133–144.

Butlin, N.G. 1962. *Australian Domestic Product and Foreign Borrowing 1861–1938/39*. London: Cambridge University Press.

Butt, T.A., McCarl, B.A., Angerer, J., Dyke, P.T. and Stuth, J.W. 2005. The economic and food security implications of climate change in Mali. *Climatic Change*, 68(3), 355–378.

Butterill, L.C. 2003. Government responses to drought in Australia, in *Beyond Drought. People, Policy and Perspectives*, edited by L. Fisher and M. Fisher. Collingwood: CSIRO, 49–65.

Buxton, R. and Stafford Smith, M. 1996. Managing drought in Australia's rangelands: four weddings and a funeral. *The Rangeland Journal*, 18(2), 292–308.

Byrnes, J. (compiled by) 1987. *Rural Australia Symposium 1987: Contributed Papers to the National Symposium Albury*. Armidale: Commonwealth Department of Primary Industry and the University of New England.

Calder, R. 1951. *Men Against the Desert*. London: Allen and Unwin.

Caldwell, J.C. and Caldwell, P. 1990. High fertility in sub-Saharan Africa. *Scientific American*, 262(5), 82–89.

California [State] 1989. *Drought Assistance – A Report to the Legislature in Response to Senate Bill 32.* State of California: The Resources Agency, Department of Water Resources.

Campbell, A. 1991. *Planning for Sustainable Farming.* Melbourne: Lothian Books.

Carr, J.T. 1966. *Texas Droughts, Causes, Classification and Prediction.* Texas Water Development Board Report, 30, 6–9.

Cary, J. 1992. Lessons from the past and present attempts to develop sustainable land use. *Review of Marketing and Agricultural Economics,* 60(2), 277–284.

Castro, J. de. 1953. *The Geography of Hunger.* London: Victor Gollancz.

Chambers, L.E. and Griffiths, G.M. 2008. The changing nature of temperature extremes in Australia and New Zealand. *Australian Meteorological Magazine,* 57, 13–35.

Changnon, S.A. 2002. Impacts of the mid-western drought forecasts of 2000. *Journal Applied Meteorology,* 41, 1042–1052.

Changnon, S.A. 2005. Economic impacts of climate conditions on the US: Past, present and future. *Climatic Change,* 68(1–2), 1–9.

Chatterton, L. and Chatterton, B. 1996. *Sustainable Dryland Farming. Combining Farmer Innovation and Medic Pasture in a Mediterranean Climate.* Cambridge: Cambridge University Press.

Chisholm, A.H. 1992. Australian agriculture: a sustainability story. *Australian Journal of Agricultural Economics,* 36(1), 1–29.

Clark, W.C. 1985. Scales of climate impacts. *Climatic Change,* 7, 5–27.

Cloudsley-Thompson, J.L. 1965. *Desert Life.* Oxford: Pergamon Press.

Cocks, D. 1992. *Use with Care. Managing Australia's Natural Resources in the Twenty-First Century.* Kensington: University of NSW Press.

Corlett, D. 2009. Survival of the flattest. *Australasian Science,* 30(4), 21–24.

Costanza, R., Steffen, W., Graumlich, L., Hibbard, K. and Schimel, D. 2005. Sustainability or collapse? Society in the 21st century. *Global Change Newsletter,* 64, 19–22.

Cotton, W.R. 1997. *Weather Modification by Cloud Seeding – A Status Report 1989–1997.* Available at http://rams.atmos.colostate.edu/gkss.html.

Coughlan, M. 1985. Drought in Australia, in Australian Academy of Technical Sciences, *Natural Disasters in Australia.* Parkville: Australian Academy of Technical Sciences.

Cousins, N. 1979. *An Anatomy of an Illness as Perceived by the Patient.* New York: Norton.

Cox, E. 1978. *The Great Drought of 1976.* London: Hutchinson.

Cribb, J. 2002. Underground movement for water banks. *Australasian Science,* 23(10), 28–30.

Crowley, T.J. 2003. When did global warming start? *Climatic Change,* 61(3), 259–260.

Crutzen, P.J. 2006. Albedo enhancement by stratospheric sulfur injections: a contribution to resolve a policy dilemma? *Climatic Change,* 77, 211–220.

Cupper, M. 2007. Ice Age cold case. *Australasian Science*, 28(2), 16–17.

Cutter, S.L. 1993. *Living with Risk: The Geography of Technological Hazards.* London: Edward Arnold.

D'Arrigo, R. and Wilson, R. 2008. El Niño and Indian Ocean influences on Indonesian drought: implications for forecasting rainfall and crop productivity. *International Journal of Climatology*, 28, 611–616.

da Cunha, E. 1901. *Os Sertões (Campanha de Canudos).* Rio de Janiero: Franscisco Alves.

Dahlin, B.H. 1983. Climate and prehistory in the Yucatan Peninsula. *Climatic Change*, 5, 245–263.

Dai, A., Trenberth, K.E.and Qian, T. 2004. A global dataset of Palmer Drought Severity Index for 1870–2002: relationship with soil moisture and effects of surface warming. *Journal of Hydrometeorology*, 5(6), 1117–1130.

Dalby, D. and Harrison-Church, R.J. (eds) 1973. *Drought in Africa. Report of the 1973 Symposium.* London: School of Oriental and African Studies, University London.

Daly, D. 1994. *Wet as a Shag, Dry as a Bone: Drought in a Variable Climate.* Brisbane: Department of Primary Industries Queensland.

Davidson, B.R. 1969. *Australia Wet or Dry? The Physical and Economic Limits to Irrigation.* Carlton: Melbourne University Press.

Davies, P. 2000. *The Devil's Music: In the Eye of the Hurricane.* London: Michael Joseph.

Davis, M. 2001. *Late Victorian Holocausts. El Niño Famines and the Making of the Third World.* London: Verso.

Davison, G., McCarty, J.W. and McLeary, A. (eds) 1987. *Australians 1888.* Broadway: Fairfax, Syme and Weldon Associates.

de Freitas C.R. 2002. Perceived change in risk of natural disasters caused by global warming. *Australian Journal of Emergency Management*, 17(3), 34–38.

de Jager, J.M., Howard, M.D. and Fouche, H.J. 2000. Computing drought severity and forecasting its future impact on grazing in a GIS, in *Drought: A Global Assessment*, edited by D. Wilhite. London: Routledge, 1, 269–278.

de Kantzow, D.R. and Sutton, B.G. (eds)1981. *Cropping at the Margins: Potential for Overuse of Semi-Arid Lands.* Sydney: Australian Institute of Agricultural Science and Water Research Foundation.

de Villiers, M. 1999. *Water Wars: Is the World's Water Running Out?* New York: Weidenfeld.

Degaetano, A.T. 1999. A temporal comparison of drought impacts and responses in the New York City metropolitan area. *Climatic Change*, 42, 539–560.

deMenocal, P. 1997. *New African Climate Study Key to Global Cycles.* San Francisco: The EnviroNews Service.

Dessai, S. O'Brien, K. and Hulme, M. 2007. Editorial: on uncertainty and climate change. *Global Environmental Change*, 17(1), 1–3.

Diamond, J. 2005. *Collapse: How Societies Choose to Fail or Succeed.* New York: Viking.

Diaz, L.N. 2005. Monitoring agricultural drought using El Niño and Southern Oscillation data, in *Monitoring and Predicting Agricultural Drought*, edited by V.K. Boken, A.P. Cracknell and R.L. Heathcote. New York: Oxford University Press, 28–39.

Dick, E. 1975. *Conquering the Great American Desert: Nebraska*. Lincoln, Nebraska: Nebraska State History Society.

Dillon, J.L. 1985. Policy and technology in drought mitigation, in Australian Academy of Technical Sciences, *Natural Disasters in Australia*. Parkville: Australian Academy of Technical Sciences, 299–321.

Doornkamp, J.C. Gregory, K.J. and Burn, A.S. 1980. *Atlas of Drought in Britain 1975–76*. London: Institute of British Geographers.

Doulton, H. and Brown, K. 2009. Ten years to prevent catastrophe? Discourses of climate change and International development in the UK press? *Global Environmental Change*, 19, 191–202.

Downing, T.E. and Stowell, Y. 2003. Household food security and coping with climatic variability in developing countries, in *Handbook of Weather, Climate, and Water: Atmospheric Chemistry, Hydrology, and Societal Impacts*, edited by T. Potter and B. Coleman. New York: John Wiley and Sons, 719–741.

Drabek, T.E. 1986. *Human System Responses to Disaster. An Inventory of Sociological Findings*. New York: Springer-Verlag.

Dregne, H.E., Tucker, C.J. and Newcomb, W.W. 1991. Expansion and contraction of the Sahara desert from 1980 to 1990. *Science*, 253, 299–301.

Driver, T.S. and Chapman, G.P. (eds) 1996. *Time-Scales and Environmental Change*. London: Routledge.

Drought Policy Review Task Force 1989. *Managing for Drought: Interim Report July 1989*. Australian Government Publishing Service, Canberra.

Drought Policy Review Task Force 1990. *Final Report: Drought Policy Review Task Force*. Canberra: Australian Government Publishing Service.

Dupree H. and Roder W. 1974. Coping with drought in a preindustrial, preliterate farming society, in *Natural Hazards Research: Concepts, Methods, and Policy Implications*, edited by G.F. White. New York: Oxford University Press, 115–119.

Durkheim, E. 1965. *The Elementary Forms of Religious Life*. New York: Free Press.

Dvorak, J. 2007. Volcano myths and rituals. *American Scientist*, 95, 8–9.

Dyer, J.A. 2000. Drought monitoring for famine relief in Africa, in *Drought: A Global Assessment*, edited by D. Wilhite. London: Routledge, 1, 223–233.

Eamus, D. 2006. Groundwater isn't a get-out-of-jail-free card. *Australasian Science*, 27(5), 39–41.

Easterling, W. 1987. Drought as hazard: can we plan for it? *Natural Hazards Observer*, 11(4).

Easterling, W. and Apps, M. 2005. Assessing the consequences of climate change for food and forest resources: a view from the IPCC. *Climatic Change*, 70, 165–189.

Eddy, J.A. 1977. Climate and the changing sun. *Climatic Change*, 1, 173–190.

Edmonds, J. and Rosenberg, N.J. 2005. Climate change impacts for the conterminous USA: an integrated assessment summary. *Climatic Change*, 69, 151–162.

Enfield, G.H. and Nash, D.J. 2002. Drought, desiccation and discourse: missionary correspondence and nineteenth-century climate change in central Southern Africa. *Geographical Journal*, 168(1), 33–47.

Fagan, B. 2000. *The Little Ice Age.* London: Perseus Books.

Fagan, B. 2004. *The Long Summer: How Climate Changed Civilization.* London: Granta Books.

Fairhead, J. and Leach, M. 1996. Reframing forest history, in *Time-Scales and Environmental Change*, edited by T. Driver and G. Chapman. London: Routledge, 169–195.

Fei, J. and Zhou, J. 2006. The possible climatic impact in China of Iceland's Eldgja eruption inferred from historical sources. *Climatic Change*, 76, 443–457.

Fei, J., Zhou, J. and Hou, Y. 2007. Circa AD626 volcanic eruption, climatic cooling, and collapse of Eastern Turkic Empire. *Climatic Change*, 81, 469–475.

Ffolliott, P.F. et al. 2002. Dryland environments. *Arid Lands Newsletter*, 52, 1–14.

Flannery, K.V. 1971. Origins and ecological effects of early domestication in Iran and the Near East, in *The Domestication and Exploitation of Plants and Animals*, edited by P.J. Ucko and G.W. Dimbleby. London: Duckworth, 73–100.

Flannery, T.F. 1994. *The Future Eaters. An Ecological History of the Australasian Lands and People.* Port Melbourne: Reed Books.

Flannery, T.F. 2006. *We are the Weather Makers.* London: Penguin.

Fleuret, A. 1986. Indigenous responses to drought in sub-Saharan Africa. *Disasters*, 10(3), 224–229.

Flohn, H. 1969. Ein geophysikalisches Eiszeit-Modell. *Eiszeitaller und Gegenwart*, 20, 204–231.

Foley, J.C. 1957. *Droughts in Australia: Review of Records from the Earliest Years of Settlement to 1955.* Melbourne: Commonwealth Bureau of Meteorology Bulletin, 43.

Franke, R. and Chasin, B. 1980. *Seeds of Famine: Ecological Destruction and the Development Dilemma in the West African Sahel.* Montclair, NJ: Allanheld, Osmun and Co.

Franke, R. and Chasin, B. 1981. Peasants, peanuts, profits and pastoralists. *The Ecologist*, 11(4), 156–168.

Franklin, M. 2009. Double benefit in drought policy shift. *The Australian*, 2 November, 8.

Frazer, J.G. 1957. *The Golden Bough.* London: Macmillan.

Freeberne, M. 1962. Natural calamities in China, 1949–61. *Pacific Viewpoint*, 3(2), 33–72.

Friedrich, W.L. 2000. Fire in the Sea. *The Santorini Volcano: Natural History and Legend of Atlantis* (translated by A.R. McBirney). Cambridge: Cambridge University Press.

Frith, H.J. 1971. Management for wildlife conservation. *Australian Conservation Newsletter*, 3(4), 1–2.

Froude, J.A. 1886. *Oceania or England and Her Colonies*. London: Longmans.

Fu, G., Chen, S., Liu, C. and Shepard, D. 2004. Hydro-climatic trends of the Yellow River Basin for the last 50 years. *Climatic Change*, 65, 149–178.

Füssel, M.-H. 2007. Vulnerability: a generally applicable conceptual framework for climate change research. *Global Environmental Change*, 17(2), 155−167.

Gaind, R. 1995. Curtains open on a silent classic. *The Canberra Times*, 1 December, 14.

Garcia, R.V. 1981. *Nature Pleads Not Guilty*. New York: Pergamon.

Garcia, R.V. and Escudero, J.C. 1982. *Drought and Man, Vol. 2, The Constant Catastrophe: Malnutrition, Famines and Drought*. New York: Pergamon.

Gardner, B.L. and Kramer, R.A. 1986. Experience with crop insurance programs in the United States, in *Crop Insurance for Agricultural Development: Issues and Experience*, edited by P. Hazell, C. Pomaredo and A. Valdes. Baltimore: Published for the International Food Policy Research Institute by the Johns Hopkins University Press, 195–222.

Gergis, J.L. and Fowler, A.M. 2009. A history of ENSO events since A.D.1525: implications for future climate change. *Climatic Change*, 92, 343–387.

Ghosh, S. and Head, L. 2009. Retrofitting the Suburban Garden: morphologies and some elements of sustainability potential of two Australian residential suburbs compared. *Australian Geographer*, 40(3), 319–346.

Gibbs, W.J. 1979. *The Impact of Climate on Australian Society and Economy*. Report of Conference held November 1978, Phillip Island, Victoria: Department of Science, CSIRO.

Gibbs, W.J. and Maher, J.V. 1967. *Rainfall Deciles as Drought Indicators*. Melbourne: Commonwealth Bureau of Meteorology, Bulletin 48.

Giglioli, I. and Swyngedouw, E. 2008. Let's drink to the great thirst! Water and the politics of fractured techno-natures in Sicily. *International Journal of Urban and Regional Research*, 32(2), 392–414.

Gill, M. 1999. Cycles of fire; cycles of life, in National Academies Forum, *Fire! The Australian Experience: Proceedings from the National Academies Forum Seminar held at the University of Adelaide, SA, 30 September–1 October*. Canberra: National Academies Forum, 51–57.

Glacken, C.J. 1967. *Traces on the Rhodian Shore. Nature and Culture in Western Thought*. Berkeley: University of California Press.

Glantz, M.H. 1976. *The Politics of Natural Disaster: The Case of the Sahel Drought*. New York: Praeger.

Glantz, M.H. 1977. The U.N. and desertification: dealing with a global problem, in *Desertification: Environmental Degradation in and around Arid Lands*, edited by M.H. Glantz. Boulder: Westview Press, 1–15.

Glantz, M.H. 1982. Consequences and responsibilities in drought forecasting: the case of Yakima, 1977. *Water Resources Research*, 18, 3–13.

Glantz, M.H. 1987. Drought and economic development in sub-saharan Africa, in *Planning for Drought: Toward a Reduction of Societal Vulnerability*, edited by D.A. Wilhite and W.E. Easterling. Boulder: Westview Press, 297–316.

Glantz, M.H. (ed.) 1994. *Drought Follows the Plow.* Cambridge: Cambridge University Press.

Glantz, M.H. 2000. Drought follows the plough: a cautionary tale, in *Drought: A Global Assessment*, edited by D.A. Wilhite. London: Routledge, 2, 285–291.

Glantz, M.H. 2001. *Currents of Change: Impacts of El Niño and La Niña on Climate and Society.* Cambridge: Cambridge University Press.

Glantz, M.H. 2003. *Climate Affairs: A Primer.* Washington: Island Press.

Glantz, M.H. and Katz, R.W. 1987. African drought and its impacts: revived interest in a recurrent phenomenon. *Desertification Control Bulletin*, 14, 22–30.

Glantz, M., Katz, R. and Kranz, M. (eds) 1987. *The Societal Impacts associated with the 1982–83 Worldwide Climate Anomalies.* Boulder: National Center for Atmospheric Research.

Glantz, M.H. and Orlovsky N. 1984. Desertification: a review of the concept, in *Encyclopaedia of Climatology*, edited by J.E. Oliver and R. Fairbridge. New York: Hutchins.

Glantz, M.H. and Orlovsky N.S. 1986. Desertification: anatomy of a complex environmental process, in *Natural Resources and People*, edited by K.A. Dahlberg and J.W. Bennet. Boulder: Westview Press, 213–229.

Glantz, M.H., Rubinstein, A.Z. and Zonn, I. 1994. Tragedy in the Aral Sea Basin. *Global Environmental Change*, June, 174–198.

Gleick, P.H. 1997. Water and conflict in the twenty-first century: the Middle East and California, in *Decentralization and Coordination of Water Resource Management*, edited by D.D. Parker and Y. Tsur. New York: Kluwer Academic, 411–428.

Goldsberg, I.A. (ed.) 1972. *Agroclimatic Atlas of the World.* Moscow-Leningrad: Hydrometizdat.

Gornitz, V. 1985. A survey of anthropogenic vegetation changes in West Africa during last century – climate implications. *Climatic Change*, 7, 285–325.

Gottschalk, L. (ed.) 1999. *Hydrological Extremes: Understanding, Predicting, Mitigating: Proceedings of an International Symposium held during IUGG 99.* Birmingham: the XXII General Assembly of the International Union of Geodesy and Geophysics, 18–30 July.

Gould, S.J. 1999. Commentary: tragic optimism for a millennium dawning, in *99 Britannica Book of the Year*, Chicago: Encyclopaedia Britannica Incorporated, 6–9.

Graumlich, L.J. and Ingram, M. 2000. Drought in the context of the last 1,000+ years: some surprising implications, in *Drought: A Global Assessment,* edited by D. Wilhite. London: Routledge, 1, 234–242.

Gregory, G. 1984. *Country Towns and the Drought.* Armidale: Australian Rural Adjustment Unit, University of New England.

Griffin, G.F. and Friedel, M.H. 1985. Discontinuous change in Central Australia: some implications of major ecological events for land management. *Journal of Arid Environments*, 9(1), 63–80.

Griffin, D.W., Kellogg, C.A., Garrison, V.H. and Shinn, E.A. 2002. The global transport of dust. *American Scientist*, 90, 228–235.

Hadley Centre for Climate Prediction and Research 2005. *Climate Change, Rivers and Rainfall. Recent Research on Climate Change Science from the Hadley Centre.* Exeter: Meteorological Office.

Hadley Centre for Climate Prediction and Research 2006. *Effects of Climate Change on Developing Countries. Latest Science from the Hadley Centre.* Exeter, Meteorological Office.

Ha-Duong, M., Swart, R. and Bernstein, L. 2007. Editorial. Uncertainty management in the IPCC: agreeing to disagree. *Global Environmental Change*, 17(1), 8–11.

Hall, C.A.S. and Day, J.W. 2009. Revisiting the limits to growth after peak oil. *American Scientist*, 97, 230–237.

Hamblyn, R. 2001. *The Invention of Clouds. How an Amateur Meteorologist Forged the Language of the Skies.* London: Picador.

Hansen, T. 1964. *Arabia Felix: The Danish Expedition of 1761–1767.* London: Collins.

Hare, F.K. 1982. Climate: yesterday, today and tomorrow, in *The Great Ideas Today.* Chicago: Britannica Great Books, 51–103.

Hare, F.K. 1983. *The Experiment of Life: Science and Religion.* Toronto: University College Toronto.

Hare, F.K., Kates, R.W. and Warren, A. 1977. The making of deserts: climate, ecology, and society. *Economic Geography*, 53(4), 332–346.

Hartigan, P.J. 1965. Said Hanrahan, in *Australian Bush Ballads*, edited by D. Stewart and N. Keesing. Sydney: Angus and Robertson, 141–143.

Haug, G.H., Günter, D., Peterson, L.C., Sigman, D.M., Hughen, K.A. and Aeschlimann, B. 2003. Climate and the collapse of the Mayan civilisation. *Science* 299, 1731–1735.

Haylock, H.J.K. and Ericksen, N.J. 2000. From State dependency to self-reliance: agricultural drought policies and practices in New Zealand, in *Drought: A Global Assessment*, edited by D. Wilhite. London: Routledge, 2, 105–114.

Hazell, P., Pomareda, C. and Valdes, A. (eds) 1986. *Crop Insurance for Agricultural Development: Issues and Experience.* Baltimore: Published for the International Food Policy Research Institute by the Johns Hopkins University.

Heathcote, R.L. 1965. *Back of Bourke. A Study of Land Appraisal and Settlement in Eastern Australia.* Carlton: Melbourne University Press.

Heathcote, R.L. 1967. The effects of the past droughts on the national economy, in *Report of the ANZAAS Symposium on Drought.* Melbourne: Bureau of Meteorology, 27–45.

Heathcote, R.L. 1969. Drought in Australia: a problem of perception. *Geographical Review*, 59(2), 175–194.

Heathcote, R.L. 1979. The threat from natural hazards in Australia, in *Natural Hazards in Australia*, edited by R.L. Heathcote and B.G. Thom. Canberra: Australian Academy of Science, 3–10.

Heathcote, R.L. 1980. An administrative trap? Natural hazards in Australia: a personal view. *Australian Geographical Studies*, 18(2), 194–200.

Heathcote, R.L. 1981. Goyder's Line – a line for all seasons?, in *People and Plants in Australia*, edited by D.J. and S.G.M. Carr. Academic Press, 295–321.

Heathcote, R.L. 1983. *Arid Lands: Their Use and Abuse.* Australia: Longmans.

Heathcote, R.L. 1986a. Climate and famine: differing interpretations of the linkages. *Australian Overseas Disaster Relief Organisation Newsletter*, 4(4), 6–8.

Heathcote, R.L. 1986b. Drought mitigation in Australia: Reducing the losses but not removing the hazard. *Great Plains Quarterly*, 6, 225–237.

Heathcote, R.L. 1987. Images of a desert? Perceptions of arid Australia. *Australian Geographical Studies*, 25(1), 3–25.

Heathcote, R.L. 1988. Drought in Australia: still a problem of perception? *GeoJournal*, 16(4), 387–397.

Heathcote, R.L. 1990. Historical experience of climate change in Australia: Lessons from analogy, in *Global Change, the Human Dimensions*, edited by H.C. Brookfield and L. Doube. Canberra: Academy of the Social Sciences in Australia, 53–67.

Heathcote, R.L. 1991. Managing the droughts? Perception of resource management – drought hazard. *Vegetatio*, 91, 219–230.

Heathcote, R.L. 1994. Land cover and land use on the Eyre Peninsula 1880s to 1990s, in *Land Use and Land Cover in Australia: Living with Global Change*, edited by B.G. Thom. Chichester: Wiley.

Heathcote, R.L. 1994. Manifest destiny, mirage and Mabo: contemporary images of the rangelands. *Rangeland Journal*, 16, 155–166.

Heathcote, R.L. 2002. *Braving the Bull of Heaven: Drought Management Strategies, Past, Present and Future.* Brisbane; Royal Geographical Society of Queensland.

Hengeveld, H. and Edwards, P. 2000. 1998 in review – an assessment of new research developments relevant to the science of climate change. *Climate Change Newsletter*, 12(3), 1–13.

Hengeveld, H. and Kertland, P. 1998. An assessment of new research developments relevant to the science of climate change. *Climate Change Newsletter*, 10(3), 3–36.

Herbst, P.H., Bredenkamp, D.B. and Barker, H.M.G. 1966. A technique for the evaluation of drought from rainfall data. *Journal of Hydrology*, 4(3), 264–272.

Hewitt, K. 1997. *Regions of Risk. A Geographical Introduction to Disasters.* Harlow: Addison Wesley Longman.

Hinchley, M.T. (ed.) 1979. *Proceedings of the Symposium on Drought in Botswana.* Gabarone: Botswana Society and Clark University.

Hiscock, P. and Wallis, L.A. 2005. Pleistocene settlement of deserts from an Australian perspective, in *Desert Peoples: Archaeological Perspectives*, edited by P. Veth, M. Smith and P. Hiscock. Oxford: Blackwell Publishing, 34–57.

Hitchcock, R.K. 1979. The traditional response to drought in Botswana, in *Proceedings of the Symposium on Drought in Botswana*, edited by M.T. Hinchley. Gabarone: Botswana Society and Clark University, 91–97.

Hobson, S. and Short, R. 1993. A perspective on the 1991–92 drought in South Africa. *Drought Network News*, 5(1), 3–6.

Holden, W.C. 1928. West Texas drouths. *Southwestern Historical Quarterly*, 32(2), 103–123.

Hollhuber, D. 1974. *Zur Perzeption des Trockenheitrisikos: Die Durre in Sahel*. Karlsruhe: Geographisches Institut bei Universitat Karlsruhe No.6.

Holling, C.S. 1986. The resilience of terrestrial ecosystems: local surprise and global change, in *Sustainable Development of the Biosphere*, edited by W.C. Clark and R.E. Munn. Cambridge: Cambridge University Press, 292–317.

Holmes, J.M. 1960. *The Geographical Basis of Keyline*. Sydney: Angus and Robertson.

Hooghiemstra, H. 2004. Fossil pollen in marine sediments: an African key for distribution patterns. *Global Change Newsletter*, 60, 4–7.

Houghton, J.T., Ding, Y., Griggs, D.J., Noguer, M., van der Linden, P.J., Dai, X. Maskell, K. and Johnson, C.A. (eds) 2001. *Climate Change 2001: The Scientific Basis. Contribution of Working Group I to the Third Assessment Report of the Intergovernmental Panel on Climate Change*. Cambridge and New York: Cambridge University Press.

Huang, Z. and Zhang,W. 2004. Climatic fluctuation and disasters during the recent 100 years in China's tropics. *Journal of Geographical Sciences*, 14, 12–20.

Hudson, N. 2002. Frogs muscle in on astronaut problem. *Australasian Science*, 23(9), 32–33.

Huggett, R.J. 1997. *Environmental Change. The Evolving Ecosphere*. Routledge, London.

Hulme, M. and Kelly M. 1997. Exploring the links between desertification and climate change, in *Environmental Management: Readings and Case Studies*, edited by L.A. Owen and T. Unwin. Blackwell, 213–230.

Humphreys, W.J. 1913. Volcanic dust and other factors in the production of climatic changes and their possible relation to ice ages. *Journal of the Franklin Institute*, (Aug), 131– 172.

Hunt, B.G. and Elliott, T.I. 2005. A simulation of the climatic conditions associated with the collapse of Mayan civilization. *Climatic Change*, 69, 393–407.

Huntington, E. 1907. *The Pulse of Asia*. New York: Houghton Mifflin.

IAC 1986. *Crop and Rainfall Insurance*. Canberra: Australian Government Publishing Service.

International Institute for Applied Systems Analysis. 1994. *International Institute for Applied Systems Analysis Report*. International Institute for Applied Systems Analysis.

IPCC 2001. Summary for Policymakers, in *Climate Change 2001: The Scientific Basis. Contribution of Working Group I to the Third Assessment Report of the Intergovernmental Panel on Climate Change*, edited by J.T. Houghton, Y. Ding, D.J. Griggs, M. Noguer, P.J. van der Linden, X. Dai, K. Maskell, and C.A. Johnson. Cambridge and New York: Cambridge University Press.

IPCC 2007a. Summary for Policymakers, in *Climate Change 2007: The Physical Science Basis. Contribution of Working Group I to the Fourth Assessment Report of the Intergovernmental Panel on Climate Change*, edited by S. Solomon, D. Qin, M. Manning, Z. Chen, M. Marquis, K.B. Averyt, M.Tignor and H.L. Miller. Cambridge and New York: Cambridge University Press.

IPCC 2007b. *Climate Change 2007: Synthesis Report. Contribution of Working Groups I, II and III to the Fourth Assessment Report of the Intergovernmental Panel on Climate Change* [Core Writing Team, Pachauri, R.K and Reisinger, A. (eds)]. Geneva: IPCC.

Isdale, P.J. et al. 1998. Paleohydrological variation in a tropical river catchment using fluorescent bands. *The Holocene*, 8, 1–8.

IUGG nd. *Water Benefits and Burden*. Washington: International Union of Geodesy and Geophysics, International Decade for Disaster Relief Program.

Jacks, G.V. and Whyte R.O. 1938. *The Rape of the Earth: A World Survey of Soil Erosion*. London: Faber and Faber.

Jacobsen, T. and Adams R.M. 1958. Salt and silt in ancient Mesopotamian agriculture. *Science*, 128(3334), 1251–1258.

Johnson, R.W. 1993. Volcanic eruptions and atmospheric change. *Climate Change Newsletter*, 5(4), 3.

Johnson, T.C., Talbot M.R. and Odada E.O. 2005. A successful decade of East African lake research. *Global Change Newsletter*, 64, 7–9.

Joy, C.S. 1991. *The Cost of Natural Disasters in Australia*. Climate Change Impacts and Adaption Workshop, Macquarie University, 13–15 May 1991.

Junger, S. 2001. *Fire*. London: Fourth Estate.

Jungk, R. 1958. *Brighter Than a Thousand Suns. A Personal History of the Atomic Scientists*. New York: Harcourt, Brace.

Kates, R.W. 1977. Drought: can we learn from experience? *Natural Hazard Observer*, 1(4), 1–2.

Kates, R.W. 1979. *The Australian Experience: Summary and Prospect, Natural Hazards in Australia*. Canberra: Australian Academy of Sciences, 511–520.

Kates, R.W. 1981. *Drought Impact in the Sahelian-Sudanic Zone of West Africa: A Comparative Analysis of 1910–15 and 1968–74*. Worcester: Center for Technology, Environment and Development, Clark University.

Kates, R.W. 1995. Lab-notes from the Jeremiah experiment: hope for a sustainable transition. *Annals of the Association of American Geographers*, 85(4), 623–640.

Kates, R.W. and Weinberg A.M. (eds) 1986. *Hazards: Technology and Fairness*. Washington, DC: National Academies Press.

Keast, A. 1959. Australian birds: their zoogeography and adaptation to an arid continent, in *Biogeography and Ecology in Australia*, edited by A. Keast, R.L. Crocker and C.S. Christian. Den Haag: Junk, 89–114.

Keys, D. 2000. *Catastrophe. An Investigation into the Origins of the Modern World.* London: Arrow Books.

Kiesecker, J.M., Belden L., Shea K. and Rubbo M.J. 2004. Amphibian decline and emerging disease. *American Scientist*, 92(2), 138–147.

Kirk, W. 1951. Historical geography and the concept of the behavioural environment. *Indian Geographical Journal*, Silver Jubilee Vol., 152–160.

Kirkpatrick, E.M. and Schwartz, C.M. 1982. *Chambers Idioms.* Edinburgh: Chambers.

Kirkpatrick, J. 1994. *A Continent Transformed. Human Impact on the Natural Vegetation of Australia.* Melbourne: Oxford University Press.

Kluckhorn F.R. and Strodtbeck, F.L. 1961. *Variations in Value Orientations.* Oxford: Row, Peterson.

Knight, C.G., Raev, I. and Staneva, M.P. 2004. *Drought in Bulgaria: A Contemporary Analog for Climate Change.* Aldershot: Ashgate.

Knutson, C., Svoboda M. and Hayes M. 2006. Analysing tribal drought management: a case study of the Hualapai Tribe, in *Quick Response Report No. 183.* Lincoln: Natural Hazards Center, University of Nebraska.

Kogan, F. 2000. Global drought detection and impact assessment from space, in *Drought: A Global Assessment*, edited by D. Wilhite. London: Routledge, 1, 196–209.

Kogan, F.N. 2005, NOAA/AVHRR Satellite Data-based Indices for monitoring agricultural droughts, in *Monitoring and Predicting Agricultural Drought*, edited by V.K. Boken, A.P. Cracknell and R.L. Heathcote. New York: Oxford University Press, 79–88.

Kohler, T.A., Varien, M.D., Wright A.M., and Kuckelman, A.M.W.K.A. 2008. Mesa Verde migrations. *American Scientist*, 96(2), 146–153.

Kokot, D.F. 1955. Desert encroachment in South Africa. *African Soils*, 3(3), 404–409.

Köppen, W. 1931. *Die Klima der Erde.* Berlin: Deutscher Verlag.

Kutzleb, C.R. 1968. *Rain Follows the Plow: The History of an Idea.* University of Colorado, PhD thesis in History.

Lambright, H. 1986. *The Prediction and Modification of Drought: A State Decision Making Study.* Syracuse NY: Syracuse Research Corporation.

Larson, D.W. 2009. The battle of Bull Run. *American Scientist*, 97, 182–184.

Lavorel, S. 2003. Global change, fire, society and the planet. *Global Change Newsletter*, 53, 2–6.

Lawrence, G., Vanclay, F. and Furze, B. 1992. *Agriculture, Environment and Society. Contemporary Issues for Australia.* South Melbourne: Macmillan.

Le Comte, D. 2009. Meteorology and climate, in *Encyclopaedia Britannica Book of the Year 2009*, 227–228.

Lee, D.M. 1979. *On Monitoring Rainfall Deficiencies in Semi Desert Regions.* Australia: Bureau of Meteorology.

Lehner, B., Doll, P., Alcamo, J., Heinrichas, T. and Kaspar, F. 2006. Estimating the impact of global change in flood and drought risks in Europe: a continental, integrated analysis. *Climatic Change*, 75, 273–299.

Levy, M.A., Thorkelson, C., Vörösmarty, C., Douglas, E. and Humphreys, M. 2005. Freshwater availability anomalies and outbreak of internal war: results from a global spatial time series analysis, in *Human Security and Climate Change: An International Workshop,* Oslo, 21–23 June.

Lewis, N.B. 1975. *Hundred Years of State Forestry.* South Australia: 1875–1975. Adelaide: Government Printer.

Li, H., Wang, X. and Gao, Y. 2004. Analysis and assessment of land desertification in Xinjiang based on RS and GIS. *Journal of Geographical Sciences*, 14(2), 159–166.

Linnerooth-Bayer, J., Hochrainer, S., Mechler, R. and Suarez, P. 2007. A favourable climate for insurance options. *Winter*, 16–17.

Linnerooth-Bayer, J., Mechler, R. and Bals, C. 2008–09. Insuring against climate change options. *Winter*, 2008–09, 11.

Linnerooth-Bayer, J., Suarez, P., Victor, M. and Mechler, R. 2009. Drought insurance for subsistence farmers in Malawi. *Natural Hazards Observer*, 33(5), 6–8.

Liu, H., Li, X., Fischer, G. and Sun, L. 2004. Study of the impacts of climate change on China's agriculture. *Climatic Change*, 65, 125–148.

Liverman, D.M. 2000. Adaptation to drought in Mexico, in *Drought: A Global Assessment*, edited by D. Wilhite. London: Routledge, 2, 35–45.

Livingstone, D.N. 2002. Race, space and moral climatology: Notes towards a genealogy. *Journal of Historical Geography*, 28(2), 159–180.

Lloyd, A.G. and Mauldon, R.G. 1986. Agricultural instability and government policies: the Australian experience, in *Crop Insurance for Agricultural Development: Issues and Experience*, edited by P. Hazell, C. Pomaredo and A. Valdes. Baltimore: Published for the International Food Policy Research Institute by the Johns Hopkins University Press, 156–177.

Lockeretz, W. 1978. The lessons of the Dust Bowl. *American Scientist*, 66, 560–569.

Lohmann, U. and Wild, M. 2005. Solar dimming. *Global Change Newsletter*, 63, 21–23.

Long, J. 2007. Megafauna theory faces extinction. *Australasian Science*, 28(2), 18–19.

Lowdermilk, W.C. 1953. Floods in Deserts, in *Desert Research*. Jerusalem: Research Council of Israel, 365–374.

Ludden, L.P. 1895. *Report of the Nebraska State Relief Commission to the Governor of the State of Nebraska.* Omaha: Nebraska State Legislature.

Ludwig, F., Milroy, S.P. and Asseng, S. 2009. Impacts of recent climate change on wheat production in Western Australia. *Climatic Change*, 92, 495–517.

Ludwig, J., Tongway, D., Freundenberger, D., Noble, J. and Hodgkinson, K. (eds). 1997. *Landscape Ecology Function and Management: Principles from Australia's Rangelands*. Collingwood: CSIRO.

Luke, R.H. and McArthur, A.G. 1978. *Bushfires in Australia*. Canberra: Australian Government Publishing Service.

Luntz, S. 2004. The rain in WA falls mainly beyond the plain. *Australasian Science*, 25(7), 8.

Luntz, S. 2007a. Ancient megafauna killed by drought. *Australasian Science*, 28(1), 6.

Luntz, S. 2007b. Cloud seeding revisited. *Australasian Science*, 28(5), 11.

Luntz S. 2009. Cloud seeding works – in Tasmania. *Australasian Science*, 30(2), 5.

Mabbutt, J.A. 1987. Implementation of the plan of action to combat desertification: Progress since UNCOD. *Land Use Policy*, 4(4), 371–388.

Macdonald, L.H. 1986. *Natural Resources Development in the Sahel: The Role of the United Nations System*. Tokyo: United Nations University Press.

Maddison, A. 2001. *The World Economy: A Millennium Perspective*. Paris: OECD.

Magalhaes, A. and Magee, P. 1994. The Brazilian Nordeste (Northeast), in *Drought Follows the Plow*, edited by M. Glantz. Cambridge: Cambridge University Press, 59–76.

Maine, C.E. 1977. *Thirst! The Searing Novel of the Ultimate Drought*. London: Sphere Books.

Mainguet, M.M. 1999. Desertification, in *Encyclopedia of Environmental Science*, edited by D.E. Alexander and R.W. Fairbridge. Boston: Kluwer, 125–129.

Males, W., Poulter, D. and Murtough, G. 1987. Off farm income and rural adjustment, in *Rural Australia Symposium 1987: Contributed Papers to the National Symposium Albury N.S.W. 6–8 July 1987*, compiled by J. Byrnes. Armidale: Rural Development Centre, University of New England, 1–15.

Malin, J.C. 1955. *The Contriving Brain and the Skilful Hand in the United States*. Lawrence: J.C. Malin.

Malin, J.C. 1956. *The Grassland of North America*. Lawrence: J.C. Malin.

Malingreau, J.P. 1984. *Remote Sensing and Forest Fire Monitoring in Indonesia*. New York: Ford Foundation.

Manins, P., Allan, R., Beer, T., Fraser, P., Holper, P., Suppiah, R. and Walsh, K. 2001. *Atmosphere, Australia State of the Environment Report 2001 (Theme Report)*. Canberra: CSIRO on behalf of the Department of the Environment and Heritage.

Maracchi, G., Sirotenko, O. and Bindi, M. 2005. Impacts of present and future climate variability on agriculture and forestry in the temperate regions: Europe. *Climatic Change*, 70(1–2), 117–135.

Marsh, G.P. 1864. *Man and Nature*. Reprinted in 1965 as *The Earth as Modified by Human Action*. Cambridge, Mass.: Belknap Press.

Martin, R. 1994. *Australia in Crisis. The Drought*. Ringwood: Corella.

Mathur, K. and Jayal, N.G. 1993. *Drought, Policy and Politics in India. The Need for a Long-term Perspective.* New Delhi: Sage.

Mayar, J. 1977. Panel on food, nutrition and population interactions: An introduction to the significance of the conference. *Annals of the New York Academy of Sciences*, 300, 5–16.

McBryde, F.W. 1982."Drought" as a Seasonal Phenomenon. *The Professional Geographer*, 34(3), 347.

McCabe, J.T. 2004. *Cattle Bring Us to Our Enemies: Turkana Ecology, Politics, and Raiding in a Disequilibrium System.* Ann Arbor: University of Michigan Press.

McCabe, J.T. 2009. Extreme events and social institutions: lessons from east Africa. *Natural Hazards Observer*, 34(1), 7–8.

McCarthy, T. 2001. High Noon in the West. *Time Magazine*, 158(2), 24–31.

McKeon, G.M. 1997. Development of a national drought alert strategic information service, in *Indicators of Drought Exceptional Circumstances: Proceedings of a Workshop held in Canberra on 1 October 1996*, edited by D.H. White and V.M. Bordas. Canberra: Bureau of Resource Sciences, 63–66.

McKibben, B. 1988. Is the world getting hotter? *The New York Review*, 35(19), 7–11.

McLennan, W. (compiled by) 1996. *Australian Agriculture and the Environment.* Canberra: Australian Bureau of Statistics.

McMahon, B. 2008. Climate change report like a disaster novel, says Australian minister. *The Guardian*, 7 July, 15.

Meehl, G.A. and Hu, A. 2006. Megadroughts in the Indian Monsoon Region and Southwest North America and a mechanism for associated multidecadal Pacific sea surface temperature anomalies. *Journal of Climate*, 19(9), 1605–1623.

Menzhulin, G.V., Savvateyev, S.P., Cracknell, A.P. and Boken, V.K. 2005. Climate change, global warming, and agricultural droughts, in *Monitoring and Predicting Agricultural Drought*, edited by V.K. Boken, A.P. Cracknell and R.L. Heathcote. New York: Oxford University Press, 429–449.

Mercer, D. 2009. Coastal water factories and millennium 'water dreaming'. *Dialogue* [Academy of the Social Sciences in Australia], 28(1), 57–58.

Mercer, D. and Marden, P. 2006. Ecologically sustainable development in a 'quarry' economy: one step forward, two steps back. *Geographical Research*, 44(2), 183–203.

Mersha, E. and Boken, V.K. 2005. Agricultural drought in Ethiopia, in *Monitoring and Predicting Agricultural Drought*, edited by V.K. Boken, A.P. Cracknell and R.L. Heathcote. New York: Oxford University Press, 227–237.

Michaels, P.J. 2004. *Meltdown. The Predictable Distortion of Global Warming by Scientists, Politicians, and the Media.* Washington, DC: Cato Institute.

Miller, K.A. 2000. Managing supply variability: the use of water banks in the western United States, in *Drought: A Global Assessment*, edited by D.A. Wilhite. London: Routledge, 2, 70–86.

Moctezuma, E.M. and Olguin, F.S. 2002. *Aztecs*. London: Royal Academy of Arts.

Moggridge, B. 2007. Groundwater dreaming. *Australasian Science*, 28(3), 32–34.

Monteil, V. 1959. Nomads and nomadian in the Arid Zone. *Australian Social Science Journal*, 4, 594.

Moran, E.F., Adams, R., Bakoyema, B., Iorini, S. and Boucek, B. 2006. Human strategies for coping with El Niño related drought in Amazônia. *Climatic Change*, 77(3–4), 343–361.

Moreton, C. 2005. Is this the end of the world? *The Independent*, 16 October, 32–34.

Morgan, G. and McCrystal, J. 2008. *Poles Apart: Beyond the Shouting, Who's Right about Climate Change?* Melbourne: Scribe.

Morgan, W.B. and Solarz, J.A. 1994. Agricultural crisis in Sub-Saharan Africa: development constraints and policy problems. *The Geographical Journal*, 160(1), 57–73.

Mortimore, M. 1989. *Adapting to Drought; Farmers, Famines and Desertification in West Africa*. Cambridge: Cambridge University Press.

Mortimore, M. 1998. *Roots in the African Dust. Sustaining the Sub-Saharan Drylands*. Cambridge: Cambridge University Press.

Morton, S.R. and Price, P.C. (eds) 1994. *R&D for the Sustainable Use and Management of Australia's Rangelands: Proceedings of a National Workshop and Associated Papers*. Canberra: The Land and Water Resources Research and Development Corporation.

Mumford, L. 1963. *The Condition of Man*. 2nd Edition. London: Mercury.

Murton, B. 1984. Spatial and temporal paths of famine in Southern India before the Famine Codes, in *Famine as a Geographical Phenomenon*, edited by B. Curry and G. Hugo. Lancaster: Reidel Publishers, 71–90.

Nakagawa, M. et al. 2000. Impact of severe drought associated with the 1997–1998 El Niño in a tropical forest in Sarawak. *Journal of Tropical Ecology*, 16, 355–367.

Namias, J. 1983. Some causes of United States drought. *Journal of Climate and Applied Meteorology*, 22, 30–39.

Napper, N. 1967. Poems of the Mallee. *Pinnaroo Border Times*.

Nash, R. 1990. *The Rights of Nature: A History of Environmental Ethics*. Madison: University Wisconsin Press.

Newitt, M.D.D. 1988. Drought in Mozambique 1823–1831. *Journal of Southern African Studies*, 15(1), 15–35.

Neylon, J. 2009. Dry Brush. Aridity is a key theme in Australian art. *The Adelaide Review*, January, 28–29.

Nicholls, N. 1983. Predictability of the 1982 Australian drought. *Search*, 14(5–6), 154–155.

Nicholls, N. 1985. Impact of the Southern Oscillation on Australian crops. *International Journal of Climatology*, 5, 553–560.

Nicholls, N. 1986. A method for predicting Murray Valley Encephalitis in South Eastern Australia using the South Oscillation. *Australian Journal of Experimental Biology and Medical Science*, 64, 587.

Nicholls, N. 1999. Cognitive illusions, heuristics, and climate prediction. *Bulletin of the American Meteorological Society*, 80, 1385–1396.

Nicholls, N. 2004. The changing nature of Australian droughts. *Climatic Change*, 63(3), 323–336.

Nicholson, S.E. and Farrar, T.J. 1994. The influence of soil type on the relationships between NDVI, rainfall, and soil-moisture in semiarid Botswana. *Remote Sensing of Environment*, 50(2), 107–120.

Nix, H. 1994. The Brigalow, in *Australian Environmental History: Essays and Cases*, edited by S. Dovers. Melbourne: Oxford University Press, 198–234.

Nobre, P. and Cavalcanti, I.F.A. 2000. The prediction of drought in the Brazilian Nordeste: progress and prospects for the future, in *Drought: A Global Assessment*, edited by D. Wilhite. London: Routledge Publishers, 1, 68–82.

O'Brien, B.J. 1988. *Draft Environmental Guidelines for Tourist Developments*. Perth: Western Australian Tourist Commission.

O'Connor, P. 1995. Moral meteorology. China and the powerful mythology of weather. *ANU Reporter*, 28 June, 12.

Odingo, R.S. 1992. Implementation of the Plan of Action to Combat Desertification (PACD) 1978–1991. *Desertification Control Bulletin*, 21, 6–14.

Oguntoyinbo, J. 1986. Drought prediction. *Climatic Change*, 9, 79–90.

O'Keefe, P. 1979. Comments on W. Torry, anthropological studies in hazardous environments: past trends and new horizons. *Current Anthropology*, 20(3), 531–532.

Oliver, M. and Tobin, J. 1989. Drought assistance – some comments. *Range Management Newsletter*, November, 16–19.

O'Meagher, B., Stafford-Smith, M. and White, D. 2000. Approaches to integrated drought risk management, in *Drought: A Global Assessment*, edited by D. Wilhite. London: Routledge Publishers, 2, 115–128.

Opie, J. 1994. The drought of 1988, the global warming experiment, its challenge to irrigation in the Old Dust Bowl Region, in *The History of Agriculture and the Environment*, edited by D. Helms and D. Bowers. Iowa: The Agricultural History Society, 279–306.

Oram, P.A. 1985. Sensitivity of agricultural production to climatic change. *Climatic Change*, 7, 129–152.

Orlove, B.S., Chiang, J.C.H. and Cane, M.A. 2002. Ethno climatology in the Andes. *American Scientist*, 90, 428–435.

Palmer, W.C. 1965. Meteorological drought. *US Weather Bureau Research Paper*, 45.

Parry, M. 1990. *Climate Change and World Agriculture*. London: Earthscan.

Passmore J. 1980. *Man's Responsibility for Nature: Ecological Problems and Western Traditions*. 2nd Edition. London: Duckworth.

Paterson, T. 2006. Climate change blamed for shrinking Elbe. *The Independent*, 21 January, 24.

Pearce, F. 2008. Climate of suspicion. *The Guardian*, 7 June, 33.

Perry, M. 1963. *Australia's First Frontier: The Spread of Settlement in New South Wales 1788–1829*. Parkville: Melbourne University Press.

Peterson, L.C. and Haug, G.H. 2005. Climate and the collapse of Maya Civilisation. *American Scientist*, 93(4), 322–329.

Pick, J.H. and Alldis, V.R. 1944. *Australia's Dying Heart: Soil Erosion and Station Management in the Inland.* Parkville: Melbourne University Press.

Piervitali, E. and Colacino, M. 2001. Evidence of drought in Western Sicily during the period 1565–1915 from Liturgical Offices. *Climatic Change*, 49, 225–238.

Pigram, J.J. 2006. *Australian Water Resources from Use to Management.* Collingwood: CSIRO.

Pitman, M.G. 1986. Plant growth, drought and salinity, in *Plant Growth, Drought and Salinity*, edited by N.C. Turner and J.B. Passioura. Melbourne: CSIRO.

Pittock, A.B. 1986. Climate predictions and social responsibility – guest editorial. *Climatic Change*, 8, 203–207.

Pockley, P. 2009. The evaporation paradox. *Australasian Science*, 30(10), 12–13.

Pohl, O. 2003. Disease dustup: Dust clouds may carry infectious organisms across oceans. *Scientific American*, Features, June.

Poli, C. 1994. Tolerance towards diversity: Ways of life and risk perception, in *Global Change Perception*, edited by E. Bianchi. Milan: Angelo Guerini e Associati, 83–88.

Porter, P.W. 1978. Geography as human ecology: a decade of progress in a quarter century. *American Behavioural Scientist*, 22(1), 15–39.

Potts, M.D. 2003. Drought in a Borneo ever wettrainforest. *Journal Ecology*, 91(3), 467–474.

Powell, A.A. 1963. *A National Fodder Reserve for the Wool Industry: An Economic and Statistical Analysis*. Sydney: Department of Agricultural Economy.

Powell, J.M. 1986. 'Abideth forever?' Global use of semiarid lands in the interwar years. *Great Plains Quarterly*, 6, 151–170.

Powell, J.M. 1989. *Watering the Garden State: Water, Land and Community in Victoria 1834–1988.* Sydney: Allen and Unwin.

Pratt, M., Cerda, M.S., Boulayha, M. and Sponberg, K. 2005. Harnessing radio and internet systems to monitor and mitigate agricultural droughts in rural African communities, in *Monitoring and Predicting Agricultural Drought*, edited by V.K. Boken, A.P. Cracknell and R.L. Heathcote. New York: Oxford University Press, 276–291.

Pulwarty, R.S., Wilhite, D.A., Diodato, D.M. and Nelson, D.I. 2007. Drought in changing environments: creating a roadmap, vehicles and drivers. *Natural Hazards Observer*, 31(5), 10–11.

Pyne, S.J. 1991. *Burning Bush: A Fire History of Australia*. New York: Henry Holt and Co.

Qian, W. and Zhu, Y. 2001. Climate change in China from 1880 to 1998 and its impact on the environmental condition. *Climatic Change*, 50, 419–444.

Queensland Government. 1984. *Queensland Year Book 1984*. Brisbane: Australian Bureau of Statistics.

Rao, V., Giarolla, E., Kayano, M.T. and Franchito, S.H. 2006. Is the recent increasing trend of rainfall over northeast Brazil related to sub-Saharan drought? *Journal of Climate*, 19(17), 4448–4453.

Rapp, A. 1987. Reflections on desertification 1977–1987: Problems and prospects. *Desertification Control Bulletin*, 15, 27–33.

Ratcliffe, F.N. 1963. *Flying Fox and Drifting Sand: The Adventures of a Biologist in Australia*. Sydney: Angus and Robertson.

Reifenberg, A. 1955. *The Struggle Between the Desert and the Sown: The Rise and Fall of Agriculture in the Levant*. Jerusalem: Publishing Department of the Jewish Agency.

Renault, M. 1972. *The Persian Boy*. London: Pantheon.

Reserve Bank Australia. 1967. *Proceedings of Bankers' Conference on Drought: Contributed Papers and Plenary Discussions*. Bankers' Conference on Drought, Martin Place, Sydney, 17–19 October.

Reynolds, J.F., Stafford-Smith, D.M. and Lambin, E. 2003. ARIDnet: seeking novel approaches to desertification and land degradation. *Global Change Newsletter*, 54, 5–9.

Rhodes, S.L. 1991. Rethinking desertification: what do we know and what have we learned? *World Development*, 19(9), 1137–1143.

Richmond, A.J. and Baron, W.R. 1989. Precipitation range carrying capacity and Navaho livestock raising, 1870–1975. *Agricultural History*, 63(2), 217–230.

Riebsame, W.E., Changnon S.A. and Karl, T.R. 1991. *Drought and Natural Resources Management in the United States: Impacts and Implications of the 1987–89 Drought*. Boulder: Westview Press.

Righarts, M. 2009. Climate in conflict. *Natural Hazards Observer*, 33(4), 11–13.

Riney-Kehrberg, P. 1993. From the horse's mouth: Dust Bowl farmers and their solutions to the problem of aridity, in *The History of Agriculture and the Environment*, edited by D. Helms and D. Bowers. Washington, D.C.: The Agricultural History Society, 137–150.

Roberts, W.O. and Lansford, H. 1979. *The Climate Mandate*. San Francisco: W.H. Freeman.

Rolls, E. 1994. More a new planet than a new continent, in *Australian Environmental History: Essays and Cases*, edited by S. Dovers. Melbourne: Oxford University Press, 22–36.

Ropelewski, C. and Folland, C.K. 2000. Prospects for the prediction of meteorological drought, in *Drought: A Global Assessment*, edited by D. Wilhite. London: Routledge, 1, 21–40.

Rosenberg, J.A. and Edmonds, N.J. 2005. Climate change impacts for the conterminous USA: an integrated assessment: From mink to the 'Lower 48'. *Climatic Change*, 69(1), 1–6.

Rosenberg, N.J. and Easterling, W.E. 1987. *Climate Resources Program; Resources for the Future*. Washington, D.C.: Resources for the Future.

Rossby, C.G. 1941. The scientific basis of modern meteorology, in *Yearbook of Agriculture*. Washington, D.C.: US Department of Agriculture, 599–655.

Rossini, E. 1981. Climatic impacts on agriculture, water resources and economy, in *Climatic Variations and Variability: Facts and Theories*, edited by A. Berger. Dordrecht: Reidel, 723–736.

Roy, M.G., Hough, M.H., and Starr, J.R. 1978. Some agricultural effects of the drought of 1975–76 in the United Kingdom. *Weather*, 33, 64–74.

Ruch, W. (ed.) 1998. *The Sense of Humor. Explanations of a Personality Characteristic*. London: Routledge.

Rucker, R.A. 1990. The effects of farm relief legislation on private lenders and borrowers: the experience of the 1930s. *American Journal of Agricultural Economics*, 72(1), 24–35.

Ruddiman, W.F. 2005. *Plows, Plagues and Petroleum: How Humans took Control of Climate*. Princeton: Princeton University Press.

Russell, C.S., Arey, D.G. and Kates, R.W. 1970. *Drought and Water Supply: Implications of the Massachusetts Experience for Municipal Planning*. Baltimore: Published for Resources for the Future by the Johns Hopkins Press.

Sainath, P. 1996. *Everybody Loves a Good Drought: Stories from India's Poorest Villages*. New Delhi: Penguin Books India.

Salick, J. and Ross, N. 2009. Traditional peoples and climate change. *Global Environmental Change*, 19, 137–139.

Salvati, L., Zitti, M., Ceccarelli, T. and Perini, L. 2009. Developing a synthetic index of land vulnerability to drought and desertification. *Geographical Research*, 47(3), 280–291.

Sandford, S. 1979. Towards a definition of drought, in *Proceedings of the Symposium on Drought in Botswana*, edited by M.T. Hinchley. Gabarone: Botswana Society and Clark University.

Santo, F.E., Guerreiro, R., Pires, V.C., Pessanha, L.E.V. and Gomes, I.M. 2005. Monitoring agricultural drought in mainland Portugal, in *Monitoring and Predicting Agricultural Drought*, edited by V.K. Boken, A.P. Cracknell and R.L. Heathcote. New York: Oxford University Press, 181–195.

Saskatchewan 1982. Historical perspective and assessment of drought programs, in *Study Element 7*. Saskatchewan Drought Studies. Regina, Saskatchewan: Prairie Farm Rehabilitation Administration and Environment Saskatchewan.

Savino, J. and Jones, M.D. 2007. *Supervolcano: The Catastrophic Event that Changed the Course of Human History*. Franklin Lakes: Career Press.

Schama, S. 1995. *Landscape and Memory*. Australia: Harper Collins.

Schmidt-Nielsen, K. 1964. *Desert Animals: Physiological Problems of Heat and Water*. New York: Oxford University Press.

Schneider, S.H. 2003. Imaginable surprise, in *Handbook of Weather, Climate, and Water*. Hoboken: Wiley, 947–954.

Schultz, C.B. and Stout, T.M. 1977. Drought and the model of a quaternary terrace-cycle. *Transactions of the Nebraska Academy of Science*, 4, 191–201.

Sears, P.B. 1949. *Deserts on the March*. London: Routledge and Kegan Paul.

Seddon, J.A. and Briggs, S.V. 1998. Lake and lakebed cropping in the Western Division of NSW. *Rangeland Journal*, 20(2), 237–254.

Selby, J. 2005. Oil and water: the contrasting anatomies of resource conflicts. *Government and Opposition*, 40, 200–24.

Sen, A.K. 1981. *Poverty and Famines: An Essay on Entitlement and Deprivation*. Oxford: Clarendon Press.

Shannon, G. 1994. Is a 'cold spot' in the Indian Ocean intensifying Australia's drought? *Search*, 25(10), 295–297.

Shantz, H.L. 1927. Drought resistance and soil moisture. *Ecology*, 8, 145–157.

Shan-yu, Y. 1943. The geographical distribution of floods and droughts in Chinese history, 206 BC–AD 1911. *Far Eastern Quarterly*, 11(4), 357–378.

Shaw, J. and Sutcliffe, J. 2003. Ancient dams, settlement archaeology and Buddhist propagation in central India: the hydrological background. *Hydrological Sciences Journal*, 48(2), 277–291.

Shields, A.J. 1979. The Brisbane floods of January 1974, in *Natural Hazards in Australia*, edited by R.L. Heathcote and B.G. Thom. Canberra: Australian Academy of Sciences, 439–447.

Simons, P. 2005. How the weather won it for Wellington. *The Times*, 18 June, 27.

Sinclair, A. and Fryxell, J.M. 1985. The Sahel of Africa: Ecology of a disaster. *Canadian Journal of Zoology*, 63, 987–994.

Smith, D.I. 1993. Drought policy and sustainability: lessons from South Africa. *Search*, 24(10), 292–295.

Smith, D.I. and Callaghan, S.D. 1988. *Climatic and Agricultural Drought, Payments and Policy: A Study of New South Wales*. CRES Working Paper 1988/16. Canberra: Centre for Resource and Environmental Studies, Australian National University.

Smith, D.I., Hutchinson, M.F. and McArthur, R.J. 1992. *Climatic and Agricultural Drought, Payments and Policy*. Canberra: Centre for Resource and Environmental Studies, Australian National University.

Smith, G. 2003. *Sidney Nolan. Desert and Drought*. Melbourne: National Gallery of Victoria.

Smith, M. 2005. Desert archaeology, linguistic stratigraphy and spread of Western Desert language, in *Desert Peoples: Archaeological Perspectives*, edited by P. Veth, M. Smith and P. Hiscock. Oxford: Blackwell, 222–242.

Snow, C.P. 1959. *The New Men*. New York: Penguin Books.

South Australian Parliamentary Papers 1917. *Report of the Agricultural Settlement Committee*. Adelaide: Government Printer.

South Australian Parliamentary Papers 1931. *Report of the Agricultural Settlement Committee*. Adelaide: Government Printer.

Speers, A. 1999. The true cost of water. *Australasian Science*, 20(3), 40–41.

Spitz, P. 1980. Drought and self-provisioning, in *Climatic Constraints and Human Activities*, edited by J.H. Ausubel and A.K. Biswas. Oxford: Pergamon, 125–147.

Spooner, B. 1987. The paradoxes of desertification. *Desertification Control Bulletin*, 15, 40–45.

Stebbing, E.P. 1935. The encroaching Sahara: the threat to the West African colonies. *Geographical Journal*, 85(6), 506–519.

Stebbing, E.P. 1937. The forests of West Africa and the Sahara: a study of modern conditions. *Geographical Journal*, 90, 550–552.

Steila, D. 1981. A note on climatic terminology. *The Professional Geographer*, 33, 373.

Steila, D. 1983. Quantitive vs qualitative drought assessment. *The Professional Geographer*, 35, 192–194.

Steinbeck, J. 1989. *The Grapes of Wrath*. New York: Viking.

Sternberg, H.O. 1952. *Land Use and the 1951 Drought in Ceara*. Paper to the 17th International Congress IGU, Washington.

Stewart, C. 2009. 2000–09: The roaring noughties. *The Weekend Australian*, 26–27 December, 3.

Stewart, T.R. and Glantz, M.H. 1985. Expert judgment and climate forecasting: a methodological critique of 'Climate Change to the year 2000'. *Climatic Change*, 7, 150–183.

Stigter, C.J., Dawei, Z., Onyewotu, L.O.Z. and Xurong, M. 2005. Using traditional methods and indigenous technologies for coping with climate variability. *Climatic Change*, 70, 255–271.

Stommel, H. and Stommel, E. 1979. The year without a summer. *Scientific American*, 240(6), 134–140.

Sturm, M., Perovich, D.K., and Serreze, M.C. 2003. Meltdown in the North. *Scientific American*, 289(4), 42–49.

Subbiah, A.R. 2000. Response strategies of local farmers in India, in *Drought: A Global Assessment*, edited by D. Wilhite. London: Routledge, 2, 29–34.

Susman, P., O'Keefe, P. and Wisner, B. 1983. Global disasters, a radical interpretation, in *Interpretation of Calamity: From the Viewpoint of Human Ecology*, edited by K. Hewitt. Boston: Unwin Hyman.

Swart, R., Bernstein, L., Ha-Duong, M. and Peterson, A. 2009. Agreeing to disagree: uncertainty management in assessing climate change, impacts and responses by the IPCC. *Climatic Change*, 92, 1–29.

Swearingen, W. and Bencherifa, A. 2003. Drought in northwest Africa, in *Handbook of Weather, Climate, and Water: Atmospheric Chemistry, Hydrology, and Societal Impacts*, edited by T. Potter and B. Coleman. New York: Wiley, 777–787.

Swift, J. 1973. Disaster and a Sahelian nomad economy, in *Drought in Africa 2 – Secheresse en Afrique,* edited by D. Dalby, R.J. Harrison-Church, F. Bezzaz. London: International African Institute in association with the Environment Training Programme.

Tadesse, T. 2008. Ethiopia accuses aid agencies of exaggerating drought. *The Guardian*, 21 June, 28.

Tannehill, I.R. 1947. *Drought. Its Causes and Effects*. Princeton: Princeton University Press.

Tarhule, A. and Woo, M.K. 1997. Towards an interpretation of historical droughts in northern Nigeria. *Climatic Change*, 37, 601–616.

Tennakoon, M.U.A. 1980. Desertification in the Dry Zone of Sri Lanka, in *Perception of Desertification*, edited by R.L. Heathcote. Tokyo: United Nations University.

Thackeray, A.I. 2005. Perspectives on Later Stone Age Hunter-Gatherer archaeology in arid southern Africa, in *Desert Peoples: Archaeological Perspectives*, edited by P. Veth, M. Smith and P. Hiscock. Oxford: Blackwell, 161–176.

Thesiger, W. 1964. *The March Arabs*. London: Butler and Tanner.

Thiel, A. 2004. Transboundary resource management in the EU: transnational welfare maximization and transboundary water sharing on the Iberian peninsula? *Journal of Environmental Planning and Management*, 47(3), 331–350.

Thomas, D.S.G. 1993. Sandstorm in a teacup? Understanding desertification. *Geographical Journal*, 159(3), 318–331.

Thomas, W.L. (ed.) 1956. *Man's Role in Changing the Face of the Earth*. Chicago: University of Chicago Press.

Thomas, W.L. 1957. *Land, Man and Culture in Mainland Southeast Asia*. Geography PhD Thesis, Yale University.

Thompson, S. 1958. *Motif of Folk Literature*. Bloomington: Indiana University Press.

Thornthwaite, C.W. 1973. Drought. *Encyclopaedia Britannica*, 7, 699–701.

Thunell, L.H. 2008. Clean up our act on water. *The Advertiser Review*, 15 November, 3.

Tiffany, C. 2005. *Everyman's Rules Scientific Living*. Sydney: Picador.

Tindale, N.B. 1974. *Aboriginal Tribes of Australia, Their Terrain, Environmental Controls, Distribution, Limits, and Proper Names*. Canberra: ANU Press.

Tolba, M.K. 1984. *Harvest of Dust UNITERRA: Special Report on Desertification*. Nairobi: United Nations Environment Program.

Torry, W.I. 1979. Hazards, hazes and holes: a critique of the environment as hazard and general reflections on disaster research. *Canadian Geographer*, 23, 368–383.

Toynbee, A. 1934. *A Study of History*. London: Oxford University Press.

Toynbee, A. 1963. *Introduction: The Genesis of Civilisations, A Study of History*. New York: Oxford University Press, 1.

Truby, J. and Boulas, L. 2001. The need for sustained drought preparedness on a national basis. *Natural Hazard Observer*, 26(2), 14.

Tucker, C.J., Dregne, H.E. and Newcomb, W.W. 1991. Expansion and contraction of the Sahara Desert from 1980 to 1990. *Science*, 253(5017), 299–300.

Tudge, C. 1977. *The Famine Business*. Harmondsworth: Penguin Books.

Ummenhofer, C. and Gupta, A.S. 2009. Why the Big Dry. *Australasian Science*, 30(5), 28–29.

UN 1977. *World Distribution of Arid Regions*. Paris: UNESCO.

UNCOD 1977. *World Map of Desertification at a Scale of 1:25,000,000*. Paris: FAO and UNESCO.

UN Economic and Social Council, Commission on Sustainable Development 1997. *Comprehensive Assessment of the Freshwater Resources of the World. Report of the Secretary-General*. Report E/CN.17/1997/9. Available at www.un.org/esa/dsd/resources/res_docucsd_05.shtml.

UNEP 1992. *World Atlas of Desertification*. London: Edward Arnold.

Unganai, L.S. and Bandason, T. 2005. Monitoring agricultural drought in southern Africa, in *Monitoring and Predicting Agricultural Drought*, edited by V.K. Boken, A.P. Cracknell and R.L. Heathcote. New York: Oxford University Press, 266–275.

UNU 1998. Water for sustainable growth: 'nor any drop to drink. *Work in Progress*, 15(2). Available at http://www.unu.edu/hq/ginfo/wip/wip-win98.html#water.

UN World Water Assessment Programme 2003. *The 1st UN World Water Development Report: Water for People, Water for Life*. United Nations Educational, Scientific and Cultural Organization (UNESCO). Available at http://www.unesco.org/water/wwap/wwdr/wwdr1/table_contents/index.shtml.

Updike, J. 1978. *The Coup*. New York: Knopf.

USACE 1993. *Lessons Learned from the California Drought (1987–1992)*. Institute for Water Resources, Water Resources Support Center, US Army Corps of Engineers.

US Congress 1994. *Report of the Bipartisan Task Force on Disasters*. US House of Representatives, 14 December.

US NDPC 1999. *Proposed Needs and Options from the National Drought Policy Commission's Monitoring and Prediction Working Group in Reference to The National Drought Policy Act of 1998*. Washington, D.C.: National Drought Policy Commission.

USA NDPC 2000. *Preparing for Drought in the 21st Century*. Washington, D.C.: National Drought Policy Commission.

Vita-Finzi, C. 1971. Geological opportunism, in *The Domestication and Exploitation of Plants and Animals*, edited by P.J. Ucko and G.W. Dimbleby. London: Duckworth, 31–34.

Vita-Finzi, C. 1973. *Recent Earth History*. London: Macmillan.

VMT 2007. *Our Water Mark: Australians Making a Difference in Water Reform* Melbourne: Victorian Women's Trust.

Vogel, C. 2003. Drought in South Africa, in *Handbook of Weather, Climate, and Water: Atmospheric Chemistry, Hydrology, and Societal Impacts*, edited by T. Potter and B. Coleman. New York: Wiley, 833–849.

Waddell, E. 1983. Coping with frosts, governments and disaster experts. Some reflections based on a New Guinea experience and a perusal of the relevant literature, in *Interpretation of Calamity: From the Viewpoint of Human Ecology*, edited by K. Hewitt. Boston: Unwin Hyman, 33–43.

Wahlquist, A. 2003. Media representations and public perceptions of drought, in *Beyond Drought*, edited by L.C.B. Fisher and M. Fisher. Collingwood: CSIRO, 67–86.

Wannan, B. 1957. *Tales from Back O'Bourke*. Melbourne: Bronzewing Books.

Ward, D.R. 2002. *Water Wars. Drought, Flood, Folly, and the Politics of Thirst*. New York: Riverhead Books.

Warner, R. 2009. Secular regime shifts, global warming and Sydney's water supply. *Geographical Research*, 47(3), 227–241.

Warrick, R.A., Trainer, P.B., Baker, E.J. and Brinkmann, W.A.R. 1975. *Drought Hazard in the United States: A Research Assessment*. Colorado: Institute of Behavioural Science, University of Colorado.

Waser, K. 2001. Literatures both written and beyond words. *Arid Lands Newsletter*, 50, 3–4.

Waterstone, M. 1991. Review of Mortimore: adapting to drought; famines and desertification in West Africa. *The Professional Geographer*, 43(2), 252.

Watson, A. 1999. A global survey of CO_2. *IGBP Newsletter*, 37, 6–7.

Watson, L. 1985. *Heaven's Breath: A Natural History of the Wind*. Sevenoaks: Coronet.

Watts, M. 1983. On the poverty of theory: natural hazards research in context, in *Interpretation of Calamity: From the Viewpoint of Human Ecology*, edited by K. Hewitt. Boston: Unwin Hyman, Chapter 13.

Watts, M. 1984. The demise of the moral economy: food and famine in a Sudano-Sahelian region in historical perspective, in *Life Before the Drought*, edited by E. Scott. London: Allen and Unwin, 124–148.

Webber, M. 1994. Politics, science and the control of nature, in *Restoring the Land: Environmental Values, Knowledge and Action*, edited by L. Cosgrove, D. Evans and D. Yencken. Carlton: Melbourne University Press, 116–131.

Webster, J.B. 1979. Drought and migration: the Lake Malawi littoral as a region of refuge, in *Proceedings of the Symposium on Drought in Botswana*, edited by M.T. Hinchley. Gabarone: Botswana Society and Clark University, 148–157.

Webster, M. 2009. Uncertainty and the IPCC. An editorial comment. *Climatic Change*, 92, 37–40.

Weiss, B. 1982. The decline of late Bronze Age civilization as a possible response to climate change. *Climatic Change*, 4(2), 173–198.

Wells, R. 2007. Extinction: who's at fault?, in *The Advertiser: Change: The Planet's Future*, Part 1, August, 10–11.

Welsch, R.L. 1978. *The Summer it Rained: Water and Plains Pioneer Humour*. Lincoln: Nebraska Water Resources Center, University Nebraska.

West, B. and Smith, P. 1996. Drought, discourse, and Durkheim: a research note. *Australian New Zealand Journal of Sociology*, 32(1), 93–102.

West, B. and Smith, P. 1997. Natural disasters and national identity: time, space and mythology. *Australian New Zealand Journal of Sociology*, 33(2), 205–215.

Westoby, M., Walker, B. and Noy-Meir, I. 1989. Opportunistic management for rangelands not at equilibrium. *Journal of Range Management*, 42(4), 266–274.

Whetton, P. 1989. SST anomalies. *Drought Network News*, 1(2), 4–5.

White, D.H. 1996. *Objective scientific and economic criteria for estimating the extent and severity of drought*, in Proceedings 2nd Australian Conference on Agricultural Meteorology, Brisbane, 1–4 October, 78–82.

White, D.H. 1997. *Objective Criteria for Exceptional Circumstances Declarations*: *Improving Scientific and Economic Inputs to Decision Making*. Available at: http://www.brs.gov.au/brs/apurb/drought/droutnet.html.

White, D.H. and Bordas, V.M. 1997. *Indicators of Drought Exceptional Circumstances*: *Proceedings of a Workshop held in Canberra on 1 October 1996*. Canberra: Bureau of Resource Sciences.

White, G.F. 1974. *Natural Hazards: Local, National, Global*. New York: Oxford University Press.

White, K. and Mattingly, D.J. 2006. Ancient lakes of the Sahara. *American Scientist*, 94, 58–65.

White, L. Jr. 1967. The historical roots of our ecologic crisis. *Science*, 155(3767), 1203–1207.

White, W.B., Gershunov, A, Annis, J.L., McKeon, G. and Syktus, J. 2004. Forecasting Australian drought using Southern Hemisphere modes of sea-surface temperature variability. *International Journal of Climatology*, 24, 1911–1927.

Whitwell, G. and Sydenham, D. 1991. *A Shared Harvest. The Australian Wheat Industry, 1939–1989*. South Melbourne: Macmillan Education Australia.

Wilhite, D.A. 1989. Planning for drought. What states are doing, what the nation must do. *Natural Hazards Observer*, 13(5), 1–2.

Wilhite, D.A. (ed.) 1993a. *Drought Assessment, Management, and Planning: Theory and Case Studies*. Dordrecht: Kluwer Academic.

Wilhite, D.A. 1993b. Planning for drought: a methodology, in *Drought Assessment, Management, and Planning: Theory and Case Studies*, edited by D.A Wilhite. Dordrecht: Kluwer Academic, 87–108.

Wilhite, D.A. 2000a. Drought as a natural hazard: Concepts and definitions, in *Drought: A Global Assessment*, edited by D. Wilhite. London: Routledge, 1:3–18.

Wilhite, D. (ed.) 2000b. *Drought: A Global Assessment*. London: Routledge.

Wilhite, D.A. 2003. Drought in the U.S. Great Plains, in *Handbook of Weather, Climate, and Water*, edited by T. Potter and B. Coleman. Hoboken: Wiley, 743–758.

Wilhite, D.A. and Easterling, W.E. 1987. *Planning for Drought: Toward a Reduction of Societal Vulnerability*. Boulder: Westview Press.

Wilhite, D. and Glantz, M.H. 1985. Understanding the drought phenomena: The role of definitions. *Water International*, 10, 111–120.

Wilhite, D.A., Svoboda, M.D., and Hayes, M.J. 1983. Government response to drought in the United States: with particular reference to the Great Plains. *Journal of Climate and Applied Meteorology*, 22, 40–50.

Wilhite, D.A., Svoboda, M.D., and Hayes, M.J. 2005. Monitoring drought in the United States: Status and trends, in *Monitoring and Predicting Agricultural Drought*, edited by V.K. Boken, A.P. Cracknell and R.L. Heathcote. New York: Oxford University Press, 121–131.

Wilhite, D.A. and Vanyarkho, O. 2000. Drought: Pervasive impacts of a creeping phenomena, in *Drought: A Global Assessment*, edited by D. Wilhite. London: Routledge, 1, 245–255.

Williams, M. 1974. *The Making of the South Australian Landscape*. Sydney: Academic Press.

Williams, M. 1976. Planned and unplanned changes in the marginal lands of South Australia. *Australian Geographer*, 13(4), 271–281.

Williams, M. 1990. Forests, in *The Earth as Transformed by Human Action: Global and Regional Changes in the Biosphere Over the Past 300 Years*, edited by B.L. Turner II, W.C. Clark, R.W. Kates, J.F. Richards, J.T. Matthews and W.B. Meyer. New York: Cambridge University Press, 179–201.

Williams, M. 2003. *Deforesting the Earth: From Prehistory to Global Crisis*. Chicago: University of Chicago Press.

Williams, M.A.J. et al. 1993. *Quaternary Environments*. London: Arnold.

Windust, A. 1995. *Drought Garden. Management and Design*. Mandurang: Allscape.

Winstanley, D. 1983. Desertification: a climatological perspective, in *Origin and Evolution of Deserts*, edited by S.G. Wells and D.R. Haragan. Albuquerque: University of New Mexico Press, 185–212.

Winterhalder, B. 1980. Environmental analysis in human evolution and adaptation research. *Human Ecology*, 8(2), 135–170.

Wittfogel, K.A. 1957. *Oriental Despotism: A Comparative Study of Total Power*. New Haven: Yale University Press.

Wittfogel, K.A. 1970. *Agriculture: A Key to the Understanding of Chinese Society, Past and Present*. Canberra: Australian National University Press.

WMO 1979. World Climate Programme. *World Meteorological Organization Bulletin*, 28(1), 42–44.

WMO 1986. *Report on Drought and Countries Affected by Drought During 1974–1985, WCP-118*. New York: World Meteorological Organisation, World Climate Data Program.

Wood, M. 1985. *In Search of the Trojan War*. New York: Facts on File.

Woodhouse, C.A. and Overpeck, J.T. 1998. 2000 years of drought variability in the central United States. *Bulletin of the American Meteorological Society*, 79(12), 2693–2714.

Worster, D. 1979. *Dust Bowl: The Southern Plains in the 1930s*. New York: Oxford University Press.

Wright, W.J. 2005. Agricultural drought policy and practices in Australia. In *Natural Disasters and Extreme Events in Agriculture: Impacts and Mitigation*, edited by M.V.K. Sivakumar, R.P. Motha and H.P. Das. Berlin and New York: Springer, 195–217.

Xinmei, J., Lyons, T.J. and Smith, R.C.G. 1995. Meteorological impact of replacing native perennial vegetation with annual agricultural species. *Hydrological Processes*, 9, 645–658.

Ye, D., Wenjie, D. and Yundi, J. 2003. The northward shift of climatic belts in China during the last 50 years. *Global Change Newsletter*, 53, 7–9.

Zeigler, D.J., Johnson, J.H. and Brunn, S.D. 1983. *Technological Hazards*. Washington, D.C.: Association of American Geographers.

Ziegler, P. 1969. *The Black Death*. New York: Harper and Row.

Zemp, M., Hoelzle, M. and Haeberli, W. 2009. Six decades of glacier mass-balance observations: a review of the worldwide monitoring network. *Annals of Glaciology*, 50(50), 101–111.

Zhu, Z. and Liu, S. 1981. *Desertification Process in North China and Its Regional Control*. Beijing: Chinese Forestry Publishing House.

Zillman, J.W., McKibbin, W.J. and Kellow, A. 2005. *Uncertainty and Climate Change: The Challenge for Policy*. Occasional Paper Series 2/2005. Canberra: Academy of the Social Sciences in Australia.

Index